Contents

KU-737-617

Introduction to Robotics

Mechanics and Control

Third Edition

Swindon C...

John J. Craig

PEARSON
Prentice
Hall

Pearson Education International

Vice President and Editorial Director, ECS: *Marcia J. Horton*
Associate Editor: *Alice Dworkin*
Editorial Assistant: *Carole Snyder*
Vice President and Director of Production and Manufacturing, ESM: *David W. Riccardi*
Executive Managing Editor: *Vince O'Brien*
Managing Editor: *David A. George*
Production Editor: *James Buckley*
Director of Creative Services: *Paul Belfanti*
Art Director: *Jayne Conte*
Cover Designer: *Bruce Kenselaar*
Art Editor: *Greg Dulles*
Manufacturing Manager: *Trudy Pisciotti*
Manufacturing Buyer: *Lisa McDowell*
Senior Marketing Manager: *Holly Stark*

© 2005 Pearson Education, Inc.
Pearson Prentice Hall
Pearson Education, Inc.
Upper Saddle River, NJ 07458

Printed in the United States of America

10 9 8 7 6 5 4 3 2 1

ISBN 0-13-123629-6

Pearson Education Ltd., *London*
Pearson Education Australia Pty. Ltd., *Sydney*
Pearson Education Singapore, Pte. Ltd.
Pearson Education North Asia Ltd., *Hong Kong*
Pearson Education Canada, Ltd., *Toronto*
Pearson Educación de Mexico, S.A. de C.V.
Pearson Education—Japan, *Tokyo*
Pearson Education Malaysia, Pte. Ltd.
Pearson Education, Inc., *Upper Saddle River, New Jersey*

Preface

Scientists often have the feeling that, through their work, they are learning about some aspect of themselves. Physicists see this connection in their work; so do, for example, psychologists and chemists. In the study of robotics, the connection between the field of study and ourselves is unusually obvious. And, unlike a science that seeks only to analyze, robotics as currently pursued takes the engineering bent toward synthesis. Perhaps it is for these reasons that the field fascinates so many of us.

The study of robotics concerns itself with the desire to synthesize some aspects of human function by the use of mechanisms, sensors, actuators, and computers. Obviously, this is a huge undertaking, which seems certain to require a multitude of ideas from various "classical" fields.

Currently, different aspects of robotics research are carried out by experts in various fields. It is usually not the case that any single individual has the entire area of robotics in his or her grasp. A partitioning of the field is natural to expect. At a relatively high level of abstraction, splitting robotics into four major areas seems reasonable: mechanical manipulation, locomotion, computer vision, and artificial intelligence.

This book introduces the science and engineering of mechanical manipulation. This subdiscipline of robotics has its foundations in several classical fields. The major relevant fields are mechanics, control theory, and computer science. In this book, Chapters 1 through 8 cover topics from mechanical engineering and mathematics, Chapters 9 through 11 cover control-theoretical material, and Chapters 12 and 13 might be classed as computer-science material. Additionally, the book emphasizes computational aspects of the problems throughout; for example, each chapter that is concerned predominantly with mechanics has a brief section devoted to computational considerations.

This book evolved from class notes used to teach "Introduction to Robotics" at Stanford University during the autumns of 1983 through 1985. The first and second editions have been used at many institutions from 1986 through 2002. The third edition has benefited from this use and incorporates corrections and improvements due to feedback from many sources. Thanks to all those who sent corrections to the author.

This book is appropriate for a senior undergraduate- or first-year graduate-level course. It is helpful if the student has had one basic course in statics and dynamics and a course in linear algebra and can program in a high-level language. Additionally, it is helpful, though not absolutely necessary, that the student have completed an introductory course in control theory. One aim of the book is to present material in a simple, intuitive way. Specifically, the audience need not be strictly mechanical engineers, though much of the material is taken from that field. At Stanford, many electrical engineers, computer scientists, and mathematicians found the book quite readable.

Directly, this book is of use to those engineers developing robotic systems, but the material should be viewed as important background material for anyone who will be involved with robotics. In much the same way that software developers have usually studied at least some hardware, people not directly involved with the mechanics and control of robots should have some such background as that offered by this text.

Like the second edition, the third edition is organized into 13 chapters. The material will fit comfortably into an academic semester; teaching the material within an academic quarter will probably require the instructor to choose a couple of chapters to omit. Even at that pace, all of the topics cannot be covered in great depth. In some ways, the book is organized with this in mind; for example, most chapters present only one approach to solving the problem at hand. One of the challenges of writing this book has been in trying to do justice to the topics covered within the time constraints of usual teaching situations. One method employed to this end was to consider only material that directly affects the study of mechanical manipulation.

At the end of each chapter is a set of exercises. Each exercise has been assigned a difficulty factor, indicated in square brackets following the exercise's number. Difficulties vary between [00] and [50], where [00] is trivial and [50] is an unsolved research problem.[1] Of course, what one person finds difficult, another might find easy, so some readers will find the factors misleading in some cases. Nevertheless, an effort has been made to appraise the difficulty of the exercises.

At the end of each chapter there is a programming assignment in which the student applies the subject matter of the corresponding chapter to a simple three-jointed planar manipulator. This simple manipulator is complex enough to demonstrate nearly all the principles of general manipulators without bogging the student down in too much complexity. Each programming assignment builds upon the previous ones, until, at the end of the course, the student has an entire library of manipulator software.

Additionally, with the third edition we have added MATLAB exercises to the book. There are a total of 12 MATLAB exercises associated with Chapters 1 through 9. These exercises were developed by Prof. Robert L. Williams II of Ohio University, and we are greatly indebted to him for this contribution. These exercises can be used with the MATLAB Robotics Toolbox[2] created by Peter Corke, Principal Research Scientist with CSIRO in Australia.

Chapter 1 is an introduction to the field of robotics. It introduces some background material, a few fundamental ideas, and the adopted notation of the book, and it previews the material in the later chapters.

Chapter 2 covers the mathematics used to describe positions and orientations in 3-space. This is extremely important material: By definition, mechanical manipulation concerns itself with moving objects (parts, tools, the robot itself) around in space. We need ways to describe these actions in a way that is easily understood and is as intuitive as possible.

---

[1] I have adopted the same scale as in *The Art of Computer Programming* by D. Knuth (Addison-Wesley).

[2] For the MATLAB Robotics Toolbox, go to http:/www.ict.csiro.au/robotics/ToolBox7.htm.

Chapters 3 and 4 deal with the geometry of mechanical manipulators. They introduce the branch of mechanical engineering known as kinematics, the study of motion without regard to the forces that cause it. In these chapters, we deal with the kinematics of manipulators, but restrict ourselves to static positioning problems.

Chapter 5 expands our investigation of kinematics to velocities and static forces.

In Chapter 6, we deal for the first time with the forces and moments required to cause motion of a manipulator. This is the problem of manipulator dynamics.

Chapter 7 is concerned with describing motions of the manipulator in terms of trajectories through space.

Chapter 8 many topics related to the mechanical design of a manipulator. For example, how many joints are appropriate, of what type should they be, and how should they be arranged?

In Chapters 9 and 10, we study methods of controlling a manipulator (usually with a digital computer) so that it will faithfully track a desired position trajectory through space. Chapter 9 restricts attention to linear control methods; Chapter 10 extends these considerations to the nonlinear realm.

Chapter 11 covers the field of active force control with a manipulator. That is, we discuss how to control the application of forces by the manipulator. This mode of control is important when the manipulator comes into contact with the environment around it, such as during the washing of a window with a sponge.

Chapter 12 overviews methods of programming robots, specifically the elements needed in a robot programming system, and the particular problems associated with programming industrial robots.

Chapter 13 introduces off-line simulation and programming systems, which represent the latest extension to the man–robot interface.

I would like to thank the many people who have contributed their time to helping me with this book. First, my thanks to the students of Stanford's ME219 in the autumn of 1983 through 1985, who suffered through the first drafts, found many errors, and provided many suggestions. Professor Bernard Roth has contributed in many ways, both through constructive criticism of the manuscript and by providing me with an environment in which to complete the first edition. At SILMA Inc., I enjoyed a stimulating environment, plus resources that aided in completing the second edition. Dr. Jeff Kerr wrote the first draft of Chapter 8. Prof. Robert L. Williams II contributed the MATLAB exercises found at the end of each chapter, and Peter Corke expanded his Robotics Toolbox to support this book's style of the Denavit–Hartenberg notation. I owe a debt to my previous mentors in robotics: Marc Raibert, Carl Ruoff, Tom Binford, and Bernard Roth.

Many others around Stanford, SILMA, Adept, and elsewhere have helped in various ways—my thanks to John Mark Agosta, Mike Ali, Lynn Balling, Al Barr, Stephen Boyd, Chuck Buckley, Joel Burdick, Jim Callan, Brian Carlisle, Monique Craig, Subas Desa, Tri Dai Do, Karl Garcia, Ashitava Ghosal, Chris Goad, Ron Goldman, Bill Hamilton, Steve Holland, Peter Jackson, Eric Jacobs, Johann Jäger, Paul James, Jeff Kerr, Oussama Khatib, Jim Kramer, Dave Lowe, Jim Maples, Dave Marimont, Dave Meer, Kent Ohlund, Madhusudan Raghavan, Richard Roy, Ken Salisbury, Bruce Shimano, Donalda Speight, Bob Tilove, Sandy Wells, and Dave Williams.

The students of Prof. Roth's Robotics Class of 2002 at Stanford used the second edition and forwarded many reminders of the mistakes that needed to get fixed for the third edition.

Finally I wish to thank Tom Robbins at Prentice Hall for his guidance with the first edition and now again with the present edition.

J.J.C.

CHAPTER 1

Introduction

1.1 BACKGROUND
1.2 THE MECHANICS AND CONTROL OF MECHANICAL MANIPULATORS
1.3 NOTATION

1.1 BACKGROUND

The history of industrial automation is characterized by periods of rapid change in popular methods. Either as a cause or, perhaps, an effect, such periods of change in automation techniques seem closely tied to world economics. Use of the **industrial robot**, which became identifiable as a unique device in the 1960s [1], along with computer-aided design (CAD) systems and computer-aided manufacturing (CAM) systems, characterizes the latest trends in the automation of the manufacturing process. These technologies are leading industrial automation through another transition, the scope of which is still unknown [2].

In North America, there was much adoption of robotic equipment in the early 1980s, followed by a brief pull-back in the late 1980s. Since that time, the market has been growing (Fig. 1.1), although it is subject to economic swings, as are all markets.

Figure 1.2 shows the number of robots being installed per year in the major industrial regions of the world. Note that Japan reports numbers somewhat differently from the way that other regions do: they count some machines as robots that in other parts of the world are not considered robots (rather, they would be simply considered "factory machines"). Hence, the numbers reported for Japan are somewhat inflated.

A major reason for the growth in the use of industrial robots is their declining cost. Figure 1.3 indicates that, through the decade of the 1990s, robot prices dropped while human labor costs increased. Also, robots are not just getting cheaper, they are becoming more effective—faster, more accurate, more flexible. If we factor these *quality adjustments* into the numbers, the cost of using robots is dropping even faster than their price tag is. As robots become more cost effective at their jobs, and as human labor continues to become more expensive, more and more industrial jobs become candidates for robotic automation. This is the single most important trend propelling growth of the industrial robot market. A secondary trend is that, economics aside, as robots become more capable they become *able* to do more and more tasks that might be dangerous or impossible for human workers to perform.

The applications that industrial robots perform are gradually getting more sophisticated, but it is still the case that, in the year 2000, approximately 78% of the robots installed in the US were welding or material-handling robots [3].

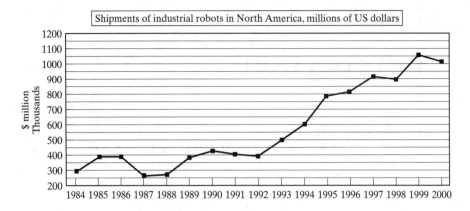

FIGURE 1.1: Shipments of industrial robots in North America in millions of US dollars [3].

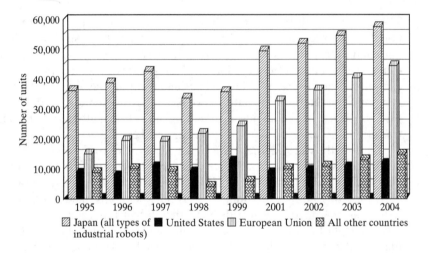

FIGURE 1.2: Yearly installations of multipurpose industrial robots for 1995–2000 and forecasts for 2001–2004 [3].

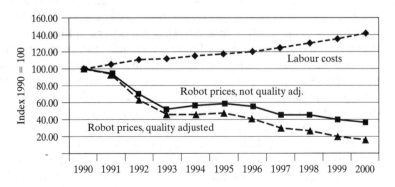

FIGURE 1.3: Robot prices compared with human labor costs in the 1990s [3].

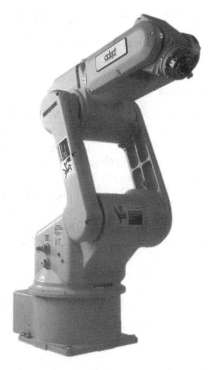

FIGURE 1.4: The Adept 6 manipulator has six rotational joints and is popular in many applications. Courtesy of Adept Technology, Inc.

A more challenging domain, **assembly** by industrial robot, accounted for 10% of installations.

This book focuses on the mechanics and control of the most important form of the industrial robot, the **mechanical manipulator**. Exactly what constitutes an industrial robot is sometimes debated. Devices such as that shown in Fig. 1.4 are always included, while numerically controlled (NC) milling machines are usually not. The distinction lies somewhere in the sophistication of the programmability of the device—if a mechanical device can be programmed to perform a wide variety of applications, it is probably an industrial robot. Machines which are for the most part limited to one class of task are considered **fixed automation**. For the purposes of this text, the distinctions need not be debated; most material is of a basic nature that applies to a wide variety of programmable machines.

By and large, the study of the mechanics and control of manipulators is not a new science, but merely a collection of topics taken from "classical" fields. Mechanical engineering contributes methodologies for the study of machines in static and dynamic situations. Mathematics supplies tools for describing spatial motions and other attributes of manipulators. Control theory provides tools for designing and evaluating algorithms to realize desired motions or force applications. Electrical-engineering techniques are brought to bear in the design of sensors and interfaces for industrial robots, and computer science contributes a basis for programming these devices to perform a desired task.

1.2 THE MECHANICS AND CONTROL OF MECHANICAL MANIPULATORS

The following sections introduce some terminology and briefly preview each of the topics that will be covered in the text.

Description of position and orientation

In the study of robotics, we are constantly concerned with the location of objects in three-dimensional space. These objects are the links of the manipulator, the parts and tools with which it deals, and other objects in the manipulator's environment. At a crude but important level, these objects are described by just two attributes: position and orientation. Naturally, one topic of immediate interest is the manner in which we represent these quantities and manipulate them mathematically.

In order to describe the position and orientation of a body in space, we will always attach a coordinate system, or **frame**, rigidly to the object. We then proceed to describe the position and orientation of this frame with respect to some reference coordinate system. (See Fig. 1.5.)

Any frame can serve as a reference system within which to express the position and orientation of a body, so we often think of *transforming* or *changing the description of* these attributes of a body from one frame to another. Chapter 2 discusses conventions and methodologies for dealing with the description of position and orientation and the mathematics of manipulating these quantities with respect to various coordinate systems.

Developing good skills concerning the description of position and rotation of rigid bodies is highly useful even in fields outside of robotics.

Forward kinematics of manipulators

Kinematics is the science of motion that treats motion without regard to the forces which cause it. Within the science of kinematics, one studies position, velocity,

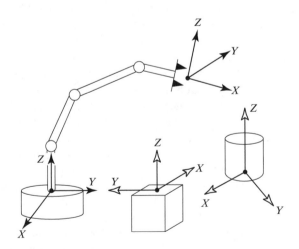

FIGURE 1.5: Coordinate systems or "frames" are attached to the manipulator and to objects in the environment.

acceleration, and all higher order derivatives of the position variables (with respect to time or any other variable(s)). Hence, the study of the kinematics of manipulators refers to all the geometrical and time-based properties of the motion.

Manipulators consist of nearly rigid **links**, which are connected by **joints** that allow relative motion of neighboring links. These joints are usually instrumented with position sensors, which allow the relative position of neighboring links to be measured. In the case of rotary or **revolute** joints, these displacements are called **joint angles**. Some manipulators contain sliding (or **prismatic**) joints, in which the relative displacement between links is a translation, sometimes called the **joint offset**.

The number of **degrees of freedom** that a manipulator possesses is the number of independent position variables that would have to be specified in order to locate all parts of the mechanism. This is a general term used for any mechanism. For example, a four-bar linkage has only one degree of freedom (even though there are three moving members). In the case of typical industrial robots, because a manipulator is usually an open kinematic chain, and because each joint position is usually defined with a single variable, the number of joints equals the number of degrees of freedom.

At the free end of the chain of links that make up the manipulator is the **end-effector**. Depending on the intended application of the robot, the end-effector could be a gripper, a welding torch, an electromagnet, or another device. We generally describe the position of the manipulator by giving a description of the **tool frame**, which is attached to the end-effector, relative to the **base frame**, which is attached to the nonmoving base of the manipulator. (See Fig. 1.6.)

A very basic problem in the study of mechanical manipulation is called **forward kinematics**. This is the static geometrical problem of computing the position and orientation of the end-effector of the manipulator. Specifically, given a set of joint

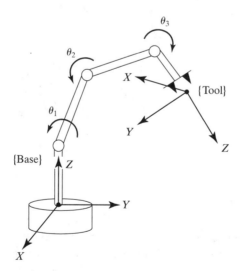

FIGURE 1.6: Kinematic equations describe the tool frame relative to the base frame as a function of the joint variables.

angles, the forward kinematic problem is to compute the position and orientation of the tool frame relative to the base frame. Sometimes, we think of this as changing the representation of manipulator position from a **joint space** description into a **Cartesian space** description.[1] This problem will be explored in Chapter 3.

Inverse kinematics of manipulators

In Chapter 4, we will consider the problem of **inverse kinematics.** This problem is posed as follows: Given the position and orientation of the end-effector of the manipulator, calculate all possible sets of joint angles that could be used to attain this given position and orientation. (See Fig. 1.7.) This is a fundamental problem in the practical use of manipulators.

This is a rather complicated geometrical problem that is routinely solved thousands of times daily in human and other biological systems. In the case of an artificial system like a robot, we will need to create an algorithm in the control computer that can make this calculation. In some ways, solution of this problem is the most important element in a manipulator system.

We can think of this problem as a *mapping* of "locations" in 3-D Cartesian space to "locations" in the robot's internal joint space. This need naturally arises anytime a goal is specified in external 3-D space coordinates. Some early robots lacked this algorithm—they were simply moved (sometimes by hand) to desired locations, which were then recorded as a set of joint values (i.e., as a location in joint space) for later playback. Obviously, if the robot is used purely in the mode of recording and playback of joint locations and motions, no algorithm relating

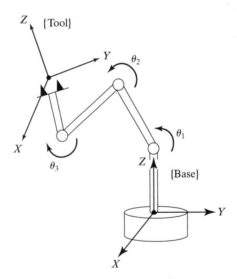

FIGURE 1.7: For a given position and orientation of the tool frame, values for the joint variables can be calculated via the inverse kinematics.

---

[1]By *Cartesian space*, we mean the space in which the position of a point is given with three numbers, and in which the orientation of a body is given with three numbers. It is sometimes called *task space* or *operational space*.

joint space to Cartesian space is needed. These days, however, it is rare to find an industrial robot that lacks this basic inverse kinematic algorithm.

The inverse kinematics problem is not as simple as the forward kinematics one. Because the kinematic equations are nonlinear, their solution is not always easy (or even possible) in a closed form. Also, questions about the existence of a solution and about multiple solutions arise.

Study of these issues gives one an appreciation for what the human mind and nervous system are accomplishing when we, seemingly without conscious thought, move and manipulate objects with our arms and hands.

The existence or nonexistence of a kinematic solution defines the **workspace** of a given manipulator. The lack of a solution means that the manipulator cannot attain the desired position and orientation because it lies outside of the manipulator's workspace.

Velocities, static forces, singularities

In addition to dealing with static positioning problems, we may wish to analyze manipulators in motion. Often, in performing velocity analysis of a mechanism, it is convenient to define a matrix quantity called the **Jacobian** of the manipulator. The Jacobian specifies a **mapping** from velocities in joint space to velocities in Cartesian space. (See Fig. 1.8.) The nature of this mapping changes as the configuration of the manipulator varies. At certain points, called **singularities**, this mapping is not invertible. An understanding of the phenomenon is important to designers and users of manipulators.

Consider the rear gunner in a World War I–vintage biplane fighter plane (illustrated in Fig. 1.9). While the pilot flies the plane from the front cockpit, the rear gunner's job is to shoot at enemy aircraft. To perform this task, his gun is mounted in a mechanism that rotates about two axes, the motions being called azimuth and elevation. Using these two motions (two degrees of freedom), the gunner can direct his stream of bullets in any direction he desires in the upper hemisphere.

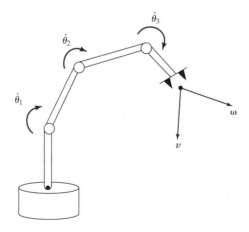

FIGURE 1.8: The geometrical relationship between joint rates and velocity of the end-effector can be described in a matrix called the Jacobian.

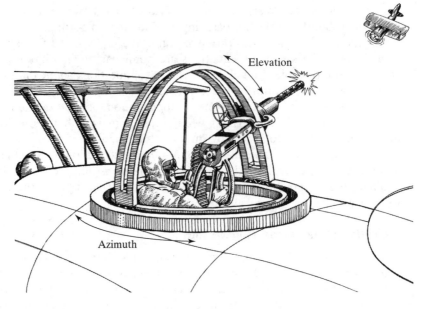

FIGURE 1.9: A World War I biplane with a pilot and a rear gunner. The rear-gunner mechanism is subject to the problem of singular positions.

An enemy plane is spotted at azimuth one o'clock and elevation 25 degrees! The gunner trains his stream of bullets on the enemy plane and tracks its motion so as to hit it with a continuous stream of bullets for as long as possible. He succeeds and thereby downs the enemy aircraft.

A second enemy plane is seen at azimuth one o'clock and elevation 70 degrees! The gunner orients his gun and begins firing. The enemy plane is moving so as to obtain a higher and higher elevation relative to the gunner's plane. Soon the enemy plane is passing nearly overhead. What's this? The gunner is no longer able to keep his stream of bullets trained on the enemy plane! He found that, as the enemy plane flew overhead, he was required to change his azimuth at a very high rate. He was not able to swing his gun in azimuth quickly enough, and the enemy plane escaped!

In the latter scenario, the lucky enemy pilot was saved by a *singularity*! The gun's orienting mechanism, while working well over most of its operating range, becomes less than ideal when the gun is directed straight upwards or nearly so. To track targets that pass through the position directly overhead, a very fast motion around the azimuth axis is required. The closer the target passes to the point directly overhead, the faster the gunner must turn the azimuth axis to track the target. If the target flies directly over the gunner's head, he would have to spin the gun on its azimuth axis at infinite speed!

Should the gunner complain to the mechanism designer about this problem? Could a better mechanism be designed to avoid this problem? It turns out that you really can't avoid the problem very easily. In fact, any two-degree-of-freedom orienting mechanism that has exactly two rotational joints cannot avoid having this problem. In the case of this mechanism, with the stream of bullets directed straight up, their direction aligns with the axis of rotation of the azimuth rotation. This means that, at exactly this point, the azimuth rotation does not cause a change in the direction of the stream of bullets. We know we need two degrees of freedom to orient the stream of bullets, but, at this point, we have lost the effective use of one of the joints. Our mechanism has become **locally degenerate** at this location and behaves as if it only has one degree of freedom (the elevation direction).

This kind of phenomenon is caused by what is called a **singularity of the mechanism**. All mechanisms are prone to these difficulties, including robots. Just as with the rear gunner's mechanism, these singularity conditions do not prevent a robot arm from positioning anywhere within its workspace. However, they can cause problems with *motions* of the arm in their neighborhood.

Manipulators do not always move through space; sometimes they are also required to touch a workpiece or work surface and apply a static force. In this case the problem arises: Given a desired contact force and moment, what set of **joint torques** is required to generate them? Once again, the Jacobian matrix of the manipulator arises quite naturally in the solution of this problem.

Dynamics

Dynamics is a huge field of study devoted to studying the forces required to cause motion. In order to accelerate a manipulator from rest, glide at a constant end-effector velocity, and finally decelerate to a stop, a complex set of torque functions must be applied by the joint actuators.[2] The exact form of the required functions of actuator torque depend on the spatial and temporal attributes of the path taken by the end-effector and on the mass properties of the links and payload, friction in the joints, and so on. One method of controlling a manipulator to follow a desired path involves calculating these actuator torque functions by using the dynamic equations of motion of the manipulator.

Many of us have experienced lifting an object that is actually much lighter than we expected (e.g., getting a container of milk from the refrigerator which we thought was full, but was nearly empty). Such a misjudgment of payload can cause an unusual lifting motion. This kind of observation indicates that the human control system is more sophisticated than a purely kinematic scheme. Rather, our manipulation control system makes use of knowledge of mass and other dynamic effects. Likewise, algorithms that we construct to control the motions of a robot manipulator should take dynamics into account.

A second use of the dynamic equations of motion is in **simulation**. By reformulating the dynamic equations so that acceleration is computed as a function of actuator torque, it is possible to simulate how a manipulator would move under application of a set of actuator torques. (See Fig. 1.10.) As computing power becomes more and more cost effective, the use of simulations is growing in use and importance in many fields.

In Chapter 6, we develop dynamic equations of motion, which may be used to control or simulate the motion of manipulators.

---

[2]We use *joint actuators* as the generic term for devices that power a manipulator—for example, electric motors, hydraulic and pneumatic actuators, and muscles.

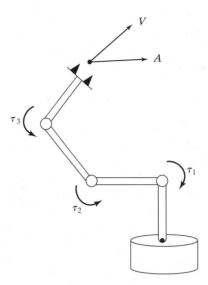

FIGURE 1.10: The relationship between the torques applied by the actuators and the resulting motion of the manipulator is embodied in the dynamic equations of motion.

Trajectory generation

A common way of causing a manipulator to move from here to there in a smooth, controlled fashion is to cause each joint to move as specified by a smooth function of time. Commonly, each joint starts and ends its motion at the same time, so that the manipulator motion appears coordinated. Exactly how to compute these motion functions is the problem of **trajectory generation**. (See Fig. 1.11.)

Often, a path is described not only by a desired destination but also by some intermediate locations, or **via points**, through which the manipulator must pass en route to the destination. In such instances the term **spline** is sometimes used to refer to a smooth function that passes through a set of via points.

In order to force the end-effector to follow a straight line (or other geometric shape) through space, the desired motion must be converted to an equivalent set of joint motions. This **Cartesian trajectory generation** will also be considered in Chapter 7.

Manipulator design and sensors

Although manipulators are, in theory, universal devices applicable to many situations, economics generally dictates that the intended task domain influence the mechanical design of the manipulator. Along with issues such as size, speed, and load capability, the designer must also consider the number of joints and their geometric arrangement. These considerations affect the manipulator's workspace size and quality, the stiffness of the manipulator structure, and other attributes.

The more joints a robot arm contains, the more dextrous and capable it will be. Of course, it will also be harder to build and more expensive. In order to build

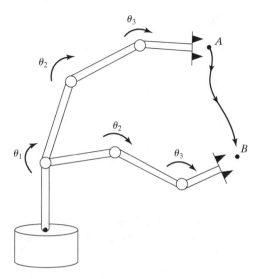

FIGURE 1.11: In order to move the end-effector through space from point A to point B, we must compute a trajectory for each joint to follow.

a useful robot, that can take two approaches: build a **specialized robot** for a specific task, or build a **universal robot** that would able to perform a wide variety of tasks. In the case of a specialized robot, some careful thinking will yield a solution for how many joints are needed. For example, a specialized robot designed solely to place electronic components on a flat circuit board does not need to have more than four joints. Three joints allow the position of the hand to attain any position in three-dimensional space, with a fourth joint added to allow the hand to rotate the grasped component about a vertical axis. In the case of a universal robot, it is interesting that fundamental properties of the physical world we live in dictate the "correct" minimum number of joints—that minimum number is six.

Integral to the design of the manipulator are issues involving the choice and location of actuators, transmission systems, and internal-position (and sometimes force) sensors. (See Fig. 1.12.) These and other design issues will be discussed in Chapter 8.

Linear position control

Some manipulators are equipped with stepper motors or other actuators that can execute a desired trajectory directly. However, the vast majority of manipulators are driven by actuators that supply a force or a torque to cause motion of the links. In this case, an algorithm is needed to compute torques that will cause the desired motion. The problem of dynamics is central to the design of such algorithms, but does not in itself constitute a solution. A primary concern of a **position control system** is to compensate automatically for errors in knowledge of the parameters of a system and to suppress disturbances that tend to perturb the system from the desired trajectory. To accomplish this, position and velocity **sensors** are monitored by the **control algorithm**, which computes torque commands for the actuators. (See

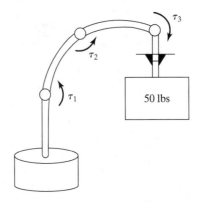

FIGURE 1.12: The design of a mechanical manipulator must address issues of actuator choice, location, transmission system, structural stiffness, sensor location, and more.

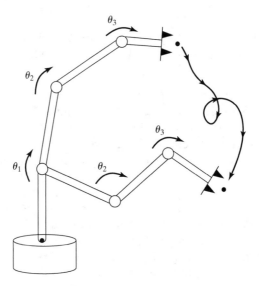

FIGURE 1.13: In order to cause the manipulator to follow the desired trajectory, a position-control system must be implemented. Such a system uses feedback from joint sensors to keep the manipulator on course.

Fig. 1.13.) In Chapter 9, we will consider control algorithms whose synthesis is based on linear approximations to the dynamics of a manipulator. These linear methods are prevalent in current industrial practice.

Nonlinear position control

Although control systems based on approximate linear models are popular in current industrial robots, it is important to consider the complete nonlinear dynamics of the manipulator when synthesizing control algorithms. Some industrial robots are now being introduced which make use of **nonlinear control** algorithms in their controllers. These nonlinear techniques of controlling a manipulator promise better performance than do simpler linear schemes. Chapter 10 will introduce nonlinear control systems for mechanical manipulators.

Force control

The ability of a manipulator to control forces of contact when it touches parts, tools, or work surfaces seems to be of great importance in applying manipulators to many real-world tasks. **Force control** is complementary to position control, in that we usually think of only one or the other as applicable in a certain situation. When a manipulator is moving in free space, only position control makes sense, because there is no surface to react against. When a manipulator is touching a rigid surface, however, position-control schemes can cause excessive forces to build up at the contact or cause contact to be lost with the surface when it was desired for some application. Manipulators are rarely constrained by reaction surfaces in all directions simultaneously, so a mixed or **hybrid** control is required, with some directions controlled by a **position-control law** and remaining directions controlled by a **force-control law**. (See Fig. 1.14.) Chapter 11 introduces a methodology for implementing such a force-control scheme.

A robot should be instructed to wash a window by maintaining a certain force in the direction perpendicular to the plane of the glass, while following a motion trajectory in directions tangent to the plane. Such split or **hybrid** control specifications are natural for such tasks.

Programming robots

A robot programming language serves as the interface between the human user and the industrial robot. Central questions arise: How are motions through space described easily by the programmer? How are multiple manipulators programmed

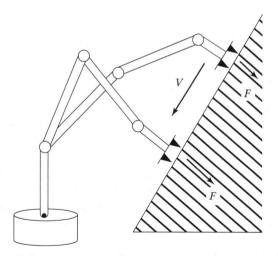

FIGURE 1.14: In order for a manipulator to slide across a surface while applying a constant force, a hybrid position–force control system must be used.

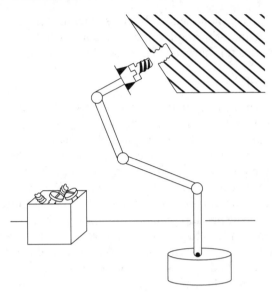

FIGURE 1.15: Desired motions of the manipulator and end-effector, desired contact forces, and complex manipulation strategies can be described in a *robot programming language.*

so that they can work in parallel? How are sensor-based actions described in a language?

Robot manipulators differentiate themselves from **fixed automation** by being "flexible," which means programmable. Not only are the movements of manipulators programmable, but, through the use of sensors and communications with other factory automation, manipulators can *adapt* to variations as the task proceeds. (See Fig. 1.15.)

In typical robot systems, there is a shorthand way for a human user to instruct the robot which path it is to follow. First of all, a special point on the hand (or perhaps on a grasped tool) is specified by the user as the **operational point**, sometimes also called the **TCP** (for Tool Center Point). Motions of the robot will be described by the user in terms of desired locations of the operational point relative to a user-specified coordinate system. Generally, the user will define this reference coordinate system relative to the robot's base coordinate system in some task-relevant location.

Most often, paths are constructed by specifying a sequence of **via points**. Via points are specified relative to the reference coordinate system and denote locations along the path through which the TCP should pass. Along with specifying the via points, the user may also indicate that certain speeds of the TCP be used over various portions of the path. Sometimes, other modifiers can also be specified to affect the motion of the robot (e.g., different smoothness criteria, etc.). From these inputs, the trajectory-generation algorithm must plan all the details of the motion: velocity profiles for the joints, time duration of the move, and so on. Hence, input to the trajectory-generation problem is generally given by constructs in the robot programming language.

The sophistication of the user interface is becoming extremely important as manipulators and other programmable automation are applied to more and more demanding industrial applications. The problem of programming manipulators encompasses all the issues of "traditional" computer programming and so is an extensive subject in itself. Additionally, some particular attributes of the manipulator-programming problem cause additional issues to arise. Some of these topics will be discussed in Chapter 12.

Off-line programming and simulation

An **off-line programming system** is a robot programming environment that has been sufficiently extended, generally by means of computer graphics, that the development of robot programs can take place without access to the robot itself. A common argument raised in their favor is that an off-line programming system will not cause production equipment (i.e., the robot) to be tied up when it needs to be reprogrammed; hence, automated factories can stay in production mode a greater percentage of the time. (See Fig. 1.16.)

They also serve as a natural vehicle to tie computer-aided design (CAD) data bases used in the design phase of a product to the actual manufacturing of the product. In some cases, this direct use of CAD data can dramatically reduce the programming time required for the manufacturing process. Chapter 13 discusses the elements of industrial robot off-line programming systems.

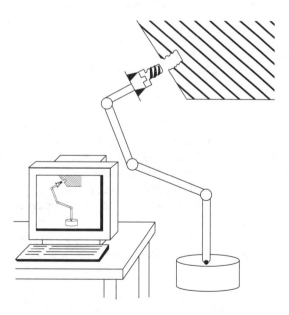

FIGURE 1.16: Off-line programming systems, generally providing a computer graphics interface, allow robots to be programmed without access to the robot itself during programming.

1.3 NOTATION

Notation is always an issue in science and engineering. In this book, we use the following conventions:

1. Usually, variables written in uppercase represent vectors or matrices. Lowercase variables are scalars.

2. Leading subscripts and superscripts identify which coordinate system a quantity is written in. For example, ^AP represents a position vector written in coordinate system {A}, and ^A_BR is a rotation matrix[3] that specifies the relationship between coordinate systems {A} and {B}.

3. Trailing superscripts are used (as widely accepted) for indicating the inverse or transpose of a matrix (e.g., R^{-1}, R^T).

4. Trailing subscripts are not subject to any strict convention but may indicate a vector component (e.g., x, y, or z) or may be used as a description—as in P_{bolt}, the position of a bolt.

5. We will use many trigonometric functions. Our notation for the cosine of an angle θ_1 may take any of the following forms: $\cos\theta_1 = c\theta_1 = c_1$.

Vectors are taken to be column vectors; hence, row vectors will have the transpose indicated explicitly.

A note on vector notation in general: Many mechanics texts treat vector quantities at a very abstract level and routinely use vectors defined relative to different coordinate systems in expressions. The clearest example is that of addition of vectors which are given or known relative to differing reference systems. This is often very convenient and leads to compact and somewhat elegant formulas. For example, consider the angular velocity, $^0\omega_4$, of the last body in a series connection of four rigid bodies (as in the links of a manipulator) relative to the fixed base of the chain. Because angular velocities sum vectorially, we may write a very simple vector equation for the angular velocity of the final link:

$$^0\omega_4 = {}^0\omega_1 + {}^1\omega_2 + {}^2\omega_3 + {}^3\omega_4. \tag{1.1}$$

However, unless these quantities are expressed with respect to a common coordinate system, they cannot be summed, and so, though elegant, equation (1.1) has hidden much of the "work" of the computation. For the particular case of the study of mechanical manipulators, statements like that of (1.1) hide the chore of bookkeeping of coordinate systems, which is often the very idea that we need to deal with in practice.

Therefore, in this book, we carry frame-of-reference information in the notation for vectors, and we do not sum vectors unless they are in the same coordinate system. In this way, we derive expressions that solve the "bookkeeping" problem and can be applied directly to actual numerical computation.

BIBLIOGRAPHY

[1] B. Roth, "Principles of Automation," Future Directions in Manufacturing Technology, Based on the Unilever Research and Engineering Division Symposium held at Port Sunlight, April 1983, Published by Unilever Research, UK.

---

[3]This term will be introduced in Chapter 2.

[2] R. Brooks, "Flesh and Machines," Pantheon Books, New York, 2002.

[3] The International Federation of Robotics, and the United Nations, "World Robotics 2001," Statistics, Market Analysis, Forecasts, Case Studies and Profitability of Robot Investment, United Nations Publication, New York and Geneva, 2001.

General-reference books

[4] R. Paul, *Robot Manipulators*, MIT Press, Cambridge, MA, 1981.

[5] M. Brady et al., *Robot Motion*, MIT Press, Cambridge, MA, 1983.

[6] W. Synder, *Industrial Robots: Computer Interfacing and Control*, Prentice-Hall, Englewood Cliffs, NJ, 1985.

[7] Y. Koren, *Robotics for Engineers*, McGraw-Hill, New York, 1985.

[8] H. Asada and J.J. Slotine, *Robot Analysis and Control*, Wiley, New York, 1986.

[9] K. Fu, R. Gonzalez, and C.S.G. Lee, *Robotics: Control, Sensing, Vision, and Intelligence*, McGraw-Hill, New York, 1987.

[10] E. Riven, *Mechanical Design of Robots*, McGraw-Hill, New York, 1988.

[11] J.C. Latombe, *Robot Motion Planning*, Kluwer Academic Publishers, Boston, 1991.

[12] M. Spong, *Robot Control: Dynamics, Motion Planning, and Analysis*, IEEE Press, New York, 1992.

[13] S.Y. Nof, *Handbook of Industrial Robotics*, 2nd Edition, Wiley, New York, 1999.

[14] L.W. Tsai, *Robot Analysis: The Mechanics of Serial and Parallel Manipulators*, Wiley, New York, 1999.

[15] L. Sciavicco and B. Siciliano, *Modelling and Control of Robot Manipulators*, 2nd Edition, Springer-Verlag, London, 2000.

[16] G. Schmierer and R. Schraft, *Service Robots*, A.K. Peters, Natick, MA, 2000.

General-reference journals and magazines

[17] *Robotics World.*

[18] *IEEE Transactions on Robotics and Automation.*

[19] *International Journal of Robotics Research (MIT Press).*

[20] *ASME Journal of Dynamic Systems, Measurement, and Control.*

[21] *International Journal of Robotics & Automation (IASTED).*

EXERCISES

1.1 [20] Make a chronology of major events in the development of industrial robots over the past 40 years. See Bibliography and general references.

1.2 [20] Make a chart showing the major applications of industrial robots (e.g., spot welding, assembly, etc.) and the percentage of installed robots in use in each application area. Base your chart on the most recent data you can find. See Bibliography and general references.

1.3 [40] Figure 1.3 shows how the cost of industrial robots has declined over the years. Find data on the cost of human labor in various specific industries (e.g., labor in the auto industry, labor in the electronics assembly industry, labor in agriculture, etc.) and create a graph showing how these costs compare to the use of robotics. You should see that the robot cost curve "crosses" various the human cost curves of different industries at different times. From this, derive approximate dates when robotics first became cost effective for use in various industries.

1.4 [10] In a sentence or two, define kinematics, workspace, and trajectory.

1.5 [10] In a sentence or two, define frame, degree of freedom, and position control.

1.6 [10] In a sentence or two, define force control, and robot programming language.

1.7 [10] In a sentence or two, define nonlinear control, and off-line programming.

1.8 [20] Make a chart indicating how labor costs have risen over the past 20 years.

1.9 [20] Make a chart indicating how the computer performance–price ratio has increased over the past 20 years.

1.10 [20] Make a chart showing the major users of industrial robots (e.g., aerospace, automotive, etc.) and the percentage of installed robots in use in each industry. Base your chart on the most recent data you can find. (See reference section.)

PROGRAMMING EXERCISE (PART 1)

Familiarize yourself with the computer you will use to do the programming exercises at the end of each chapter. Make sure you can create and edit files and can compile and execute programs.

MATLAB EXERCISE 1

At the end of most chapters in this textbook, a MATLAB exercise is given. Generally, these exercises ask the student to program the pertinent robotics mathematics in MATLAB and then check the results of the MATLAB Robotics Toolbox. The textbook assumes familiarity with MATLAB and linear algebra (matrix theory). Also, the student must become familiar with the MATLAB Robotics Toolbox. For MATLAB Exercise 1,

a) Familiarize yourself with the MATLAB programming environment if necessary. At the MATLAB software prompt, try typing *demo* and *help*. Using the color-coded MATLAB editor, learn how to create, edit, save, run, and debug m-files (ASCII files with series of MATLAB statements). Learn how to create arrays (matrices and vectors), and explore the built-in MATLAB linear-algebra functions for matrix and vector multiplication, dot and cross products, transposes, determinants, and inverses, and for the solution of linear equations. MATLAB is based on the language C, but is generally much easier to use. Learn how to program logical constructs and loops in MATLAB. Learn how to use subprograms and functions. Learn how to use comments (%) for explaining your programs and tabs for easy readability. Check out www.mathworks.com for more information and tutorials. Advanced MATLAB users should become familiar with Simulink, the graphical interface of MATLAB, and with the MATLAB Symbolic Toolbox.

b) Familiarize yourself with the MATLAB Robotics Toolbox, a third-party toolbox developed by Peter I. Corke of CSIRO, Pinjarra Hills, Australia. This product can be downloaded for free from www.cat.csiro.au/cmst/staff/pic/robot. The source code is readable and changeable, and there is an international community of users, at robot-toolbox@lists.msa.cmst.csiro.au. Download the MATLAB Robotics Toolbox, and install it on your computer by using the *.zip* file and following the instructions. Read the *README* file, and familiarize yourself with the various functions available to the user. Find the *robot.pdf* file—this is the user manual giving background information and detailed usage of all of the Toolbox functions. Don't worry if you can't understand the purpose of these functions yet; they deal with robotics mathematics concepts covered in Chapters 2 through 7 of this book.

C H A P T E R   2

Spatial descriptions and transformations

2.1 INTRODUCTION

Robotic manipulation, by definition, implies that parts and tools will be moved around in space by some sort of mechanism. This naturally leads to a need for representing positions and orientations of parts, of tools, and of the mechanism itself. To define and manipulate mathematical quantities that represent position and orientation, we must define coordinate systems and develop conventions for representation. Many of the ideas developed here in the context of position and orientation will form a basis for our later consideration of linear and rotational velocities, forces, and torques.

We adopt the philosophy that somewhere there is a **universe coordinate system** to which everything we discuss can be referenced. We will describe all positions and orientations with respect to the universe coordinate system or with respect to other Cartesian coordinate systems that are (or could be) defined relative to the universe system.

2.2 DESCRIPTIONS: POSITIONS, ORIENTATIONS, AND FRAMES

A **description** is used to specify attributes of various objects with which a manipulation system deals. These objects are parts, tools, and the manipulator itself. In this section, we discuss the description of positions, of orientations, and of an entity that contains both of these descriptions: the frame.

Description of a position

Once a coordinate system is established, we can locate any point in the universe with a 3×1 **position vector**. Because we will often define many coordinate systems in addition to the universe coordinate system, vectors must be tagged with information identifying which coordinate system they are defined within. In this book, vectors are written with a leading superscript indicating the coordinate system to which they are referenced (unless it is clear from context)—for example, $^A P$. This means that the components of $^A P$ have numerical values that indicate distances along the axes of $\{A\}$. Each of these distances along an axis can be thought of as the result of projecting the vector onto the corresponding axis.

Figure 2.1 pictorially represents a coordinate system, $\{A\}$, with three mutually orthogonal unit vectors with solid heads. A point $^A P$ is represented as a vector and can equivalently be thought of as a position in space, or simply as an ordered set of three numbers. Individual elements of a vector are given the subscripts x, y, and z:

$$^A P = \begin{bmatrix} p_x \\ p_y \\ p_z \end{bmatrix}. \tag{2.1}$$

In summary, we will describe the position of a point in space with a position vector. Other 3-tuple descriptions of the position of points, such as spherical or cylindrical coordinate representations, are discussed in the exercises at the end of the chapter.

Description of an orientation

Often, we will find it necessary not only to represent a point in space but also to describe the **orientation** of a body in space. For example, if vector $^A P$ in Fig. 2.2 locates the point directly between the fingertips of a manipulator's hand, the complete location of the hand is still not specified until its orientation is also given. Assuming that the manipulator has a sufficient number of joints,[1] the hand could be *oriented* arbitrarily while keeping the point between the fingertips at the same

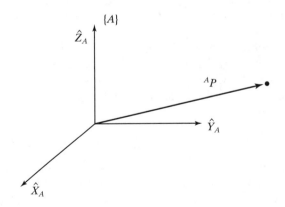

FIGURE 2.1: Vector relative to frame (example).

---

[1] How many are "sufficient" will be discussed in Chapters 3 and 4.

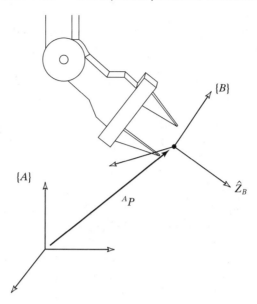

FIGURE 2.2: Locating an object in position and orientation.

position in space. In order to describe the orientation of a body, we will *attach a coordinate system to the body and then give a description of this coordinate system relative to the reference system.* In Fig. 2.2, coordinate system {B} has been attached to the body in a known way. A description of {B} relative to {A} now suffices to give the orientation of the body.

Thus, positions of points are described with vectors and orientations of bodies are described with an attached coordinate system. One way to describe the body-attached coordinate system, {B}, is to write the unit vectors of its three principal axes[2] in terms of the coordinate system {A}.

We denote the unit vectors giving the principal directions of coordinate system {B} as $\hat{X}_B$, $\hat{Y}_B$, and $\hat{Z}_B$. When written in terms of coordinate system {A}, they are called $^A\hat{X}_B$, $^A\hat{Y}_B$, and $^A\hat{Z}_B$. It will be convenient if we stack these three unit vectors together as the columns of a 3×3 matrix, in the order $^A\hat{X}_B$, $^A\hat{Y}_B$, $^A\hat{Z}_B$. We will call this matrix a **rotation matrix**, and, because this particular rotation matrix describes {B} relative to {A}, we name it with the notation $^A_B R$ (the choice of leading sub- and superscripts in the definition of rotation matrices will become clear in following sections):

$$^A_B R = \begin{bmatrix} ^A\hat{X}_B & ^A\hat{Y}_B & ^A\hat{Z}_B \end{bmatrix} = \begin{bmatrix} r_{11} & r_{12} & r_{13} \\ r_{21} & r_{22} & r_{23} \\ r_{31} & r_{32} & r_{33} \end{bmatrix}. \tag{2.2}$$

In summary, a set of three vectors may be used to specify an orientation. For convenience, we will construct a 3×3 matrix that has these three vectors as its columns. Hence, whereas the position of a point is represented with a vector, the

---

[2]It is often convenient to use three, although any two would suffice. (The third can always be recovered by taking the cross product of the two given.)

orientation of a body is represented with a matrix. In Section 2.8, we will consider some other descriptions of orientation that require only three parameters.

We can give expressions for the scalars r_{ij} in (2.2) by noting that the components of any vector are simply the projections of that vector onto the unit directions of its reference frame. Hence, each component of ^A_BR in (2.2) can be written as the dot product of a pair of unit vectors:

$$^A_BR = \begin{bmatrix} ^A\hat{X}_B & ^A\hat{Y}_B & ^A\hat{Z}_B \end{bmatrix} = \begin{bmatrix} \hat{X}_B \cdot \hat{X}_A & \hat{Y}_B \cdot \hat{X}_A & \hat{Z}_B \cdot \hat{X}_A \\ \hat{X}_B \cdot \hat{Y}_A & \hat{Y}_B \cdot \hat{Y}_A & \hat{Z}_B \cdot \hat{Y}_A \\ \hat{X}_B \cdot \hat{Z}_A & \hat{Y}_B \cdot \hat{Z}_A & \hat{Z}_B \cdot \hat{Z}_A \end{bmatrix}. \tag{2.3}$$

For brevity, we have omitted the leading superscripts in the rightmost matrix of (2.3). In fact, the choice of frame in which to describe the unit vectors is arbitrary as long as it is the same for each pair being dotted. The dot product of two unit vectors yields the cosine of the angle between them, so it is clear why the components of rotation matrices are often referred to as **direction cosines**.

Further inspection of (2.3) shows that the rows of the matrix are the unit vectors of $\{A\}$ expressed in $\{B\}$; that is,

$$^A_BR = \begin{bmatrix} ^A\hat{X}_B & ^A\hat{Y}_B & ^A\hat{Z}_B \end{bmatrix} = \begin{bmatrix} ^B\hat{X}^T_A \\ ^B\hat{Y}^T_A \\ ^B\hat{Z}^T_A \end{bmatrix}. \tag{2.4}$$

Hence, ^B_AR, the description of frame $\{A\}$ relative to $\{B\}$, is given by the transpose of (2.3); that is,

$$^B_AR = {}^A_BR^T. \tag{2.5}$$

This suggests that the inverse of a rotation matrix is equal to its transpose, a fact that can be easily verified as

$$^A_BR^T \, {}^A_BR = \begin{bmatrix} ^A\hat{X}^T_B \\ ^A\hat{Y}^T_B \\ ^A\hat{Z}^T_B \end{bmatrix} \begin{bmatrix} ^A\hat{X}_B & ^A\hat{Y}_B & ^A\hat{Z}_B \end{bmatrix} = I_3, \tag{2.6}$$

where I_3 is the 3×3 identity matrix. Hence,

$$^A_BR = {}^B_AR^{-1} = {}^B_AR^T. \tag{2.7}$$

Indeed, from linear algebra [1], we know that the inverse of a matrix with orthonormal columns is equal to its transpose. We have just shown this geometrically.

Description of a frame

The information needed to completely specify the whereabouts of the manipulator hand in Fig. 2.2 is a position and an orientation. The point on the body whose position we describe could be chosen arbitrarily, however. *For convenience, the point whose position we will describe is chosen as the origin of the body-attached frame.* The situation of a position and an orientation pair arises so often in robotics that we define an entity called a **frame**, which is a set of four vectors giving position and orientation information. For example, in Fig. 2.2, one vector locates the fingertip position and three more describe its orientation. Equivalently, the description of a frame can be thought of as a position vector and a rotation matrix. Note that a frame is a coordinate system where, in addition to the orientation, we give a position vector which locates its origin relative to some other embedding frame. For example, frame $\{B\}$ is described by $^A_B R$ and $^A P_{BORG}$, where $^A P_{BORG}$ is the vector that locates the origin of the frame $\{B\}$:

$$\{B\} = \{^A_B R, ^A P_{BORG}\}. \tag{2.8}$$

In Fig. 2.3, there are three frames that are shown along with the universe coordinate system. Frames $\{A\}$ and $\{B\}$ are known relative to the universe coordinate system, and frame $\{C\}$ is known relative to frame $\{A\}$.

In Fig. 2.3, we introduce a *graphical representation* of frames, which is convenient in visualizing frames. A frame is depicted by three arrows representing unit vectors defining the principal axes of the frame. An arrow representing a vector is drawn from one origin to another. This vector represents the position of the origin at the head of the arrow in terms of the frame at the tail of the arrow. The direction of this locating arrow tells us, for example, in Fig. 2.3, that $\{C\}$ is known relative to $\{A\}$ and not vice versa.

In summary, a frame can be used as a description of one coordinate system relative to another. A frame encompasses two ideas by representing both position and orientation and so may be thought of as a generalization of those two ideas. Positions could be represented by a frame whose rotation-matrix part is the identity matrix and whose position-vector part locates the point being described. Likewise, an orientation could be represented by a frame whose position-vector part was the zero vector.

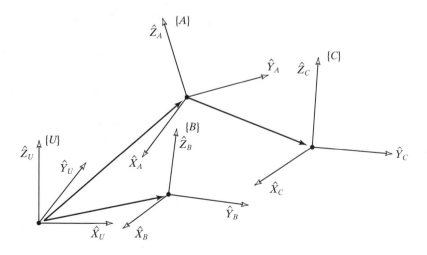

FIGURE 2.3: Example of several frames.

2.3   MAPPINGS: CHANGING DESCRIPTIONS FROM FRAME TO FRAME

In a great many of the problems in robotics, we are concerned with expressing the same quantity in terms of various reference coordinate systems. The previous section introduced descriptions of positions, orientations, and frames; we now consider the mathematics of **mapping** in order to change descriptions from frame to frame.

Mappings involving translated frames

In Fig. 2.4, we have a position defined by the vector ^{B}P. We wish to express this point in space in terms of frame $\{A\}$, when $\{A\}$ has the same orientation as $\{B\}$. In this case, $\{B\}$ differs from $\{A\}$ only by a *translation*, which is given by $^{A}P_{BORG}$, a vector that locates the origin of $\{B\}$ relative to $\{A\}$.

Because both vectors are defined relative to frames of the same orientation, we calculate the description of point P relative to $\{A\}$, ^{A}P, by vector addition:

$$^{A}P = {}^{B}P + {}^{A}P_{BORG}. \tag{2.9}$$

Note that only in the special case of equivalent orientations may we add vectors that are defined in terms of different frames.

In this simple example, we have illustrated **mapping** a vector from one frame to another. This idea of mapping, or changing the description from one frame to another, is an extremely important concept. The quantity itself (here, a point in space) is not changed; only its description is changed. This is illustrated in Fig. 2.4, where the point described by ^{B}P is not translated, but remains the same, and instead we have computed a new description of the same point, but now with respect to system $\{A\}$.

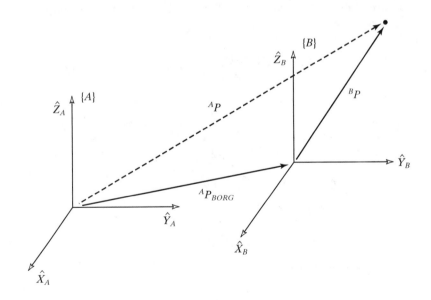

FIGURE 2.4: Translational mapping.

We say that the vector $^A P_{BORG}$ defines this mapping because all the information needed to perform the change in description is contained in $^A P_{BORG}$ (along with the knowledge that the frames had equivalent orientation).

Mappings involving rotated frames

Section 2.2 introduced the notion of describing an orientation by three unit vectors denoting the principal axes of a body-attached coordinate system. For convenience, we stack these three unit vectors together as the columns of a 3×3 matrix. We will call this matrix a rotation matrix, and, if this particular rotation matrix describes $\{B\}$ relative to $\{A\}$, we name it with the notation $^A_B R$.

Note that, by our definition, the columns of a rotation matrix all have unit magnitude, and, further, that these unit vectors are orthogonal. As we saw earlier, a consequence of this is that

$$^A_B R = {}^B_A R^{-1} = {}^B_A R^T. \tag{2.10}$$

Therefore, because the columns of $^A_B R$ are the unit vectors of $\{B\}$ written in $\{A\}$, the *rows* of $^A_B R$ are the unit vectors of $\{A\}$ written in $\{B\}$.

So a rotation matrix can be interpreted as a set of three column vectors or as a set of three row vectors, as follows:

$$^A_B R = \begin{bmatrix} {}^A\hat{X}_B & {}^A\hat{Y}_B & {}^A\hat{Z}_B \end{bmatrix} = \begin{bmatrix} {}^B\hat{X}^T_A \\ {}^B\hat{Y}^T_A \\ {}^B\hat{Z}^T_A \end{bmatrix}. \tag{2.11}$$

As in Fig. 2.5, the situation will arise often where we know the definition of a vector with respect to some frame, $\{B\}$, and we would like to know its definition with respect to another frame, $\{A\}$, where the origins of the two frames are coincident.

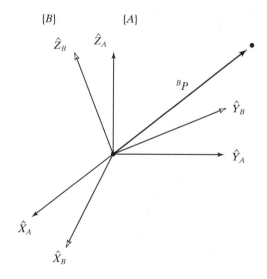

FIGURE 2.5: Rotating the description of a vector.

This computation is possible when a description of the orientation of $\{B\}$ is known relative to $\{A\}$. This orientation is given by the rotation matrix $_B^A R$, whose columns are the unit vectors of $\{B\}$ written in $\{A\}$.

In order to calculate $^A P$, we note that the components of any vector are simply the projections of that vector onto the unit directions of its frame. The projection is calculated as the vector dot product. Thus, we see that the components of $^A P$ may be calculated as

$$^A p_x = {}^B \hat{X}_A \cdot {}^B P,$$

$$^A p_y = {}^B \hat{Y}_A \cdot {}^B P,$$ (2.12)

$$^A p_z = {}^B \hat{Z}_A \cdot {}^B P.$$

In order to express (2.13) in terms of a rotation matrix multiplication, we note from (2.11) that the *rows* of $_B^A R$ are $^B \hat{X}_A$, $^B \hat{Y}_A$, and $^B \hat{Z}_A$. So (2.13) may be written compactly, by using a rotation matrix, as

$$^A P = {}_B^A R \, {}^B P.$$ (2.13)

Equation 2.13 implements a mapping—that is, it changes the description of a vector—from $^B P$, which describes a point in space relative to $\{B\}$, into $^A P$, which is a description of the same point, but expressed relative to $\{A\}$.

We now see that our notation is of great help in keeping track of mappings and frames of reference. A helpful way of viewing the notation we have introduced is to imagine that leading subscripts cancel the leading superscripts of the following entity, for example the Bs in (2.13).

EXAMPLE 2.1

Figure 2.6 shows a frame $\{B\}$ that is rotated relative to frame $\{A\}$ about $\hat{Z}$ by 30 degrees. Here, $\hat{Z}$ is pointing out of the page.

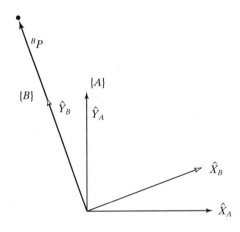

FIGURE 2.6: $\{B\}$ rotated 30 degrees about $\hat{Z}$.

Writing the unit vectors of $\{B\}$ in terms of $\{A\}$ and stacking them as the columns of the rotation matrix, we obtain

$$_B^A R = \begin{bmatrix} 0.866 & -0.500 & 0.000 \\ 0.500 & 0.866 & 0.000 \\ 0.000 & 0.000 & 1.000 \end{bmatrix}. \tag{2.14}$$

Given

$$^B P = \begin{bmatrix} 0.0 \\ 2.0 \\ 0.0 \end{bmatrix}, \tag{2.15}$$

we calculate $^A P$ as

$$^A P = {_B^A R}\, ^B P = \begin{bmatrix} -1.000 \\ 1.732 \\ 0.000 \end{bmatrix}. \tag{2.16}$$

Here, $_B^A R$ acts as a mapping that is used to describe $^B P$ relative to frame $\{A\}$, $^A P$. As was introduced in the case of translations, it is important to remember that, viewed as a mapping, the original vector P is not changed in space. Rather, we compute a new description of the vector relative to another frame.

Mappings involving general frames

Very often, we know the description of a vector with respect to some frame $\{B\}$, and we would like to know its description with respect to another frame, $\{A\}$. We now consider the general case of mapping. Here, the origin of frame $\{B\}$ is not coincident with that of frame $\{A\}$ but has a general vector offset. The vector that locates $\{B\}$'s origin is called $^A P_{BORG}$. Also $\{B\}$ is rotated with respect to $\{A\}$, as described by $_B^A R$. Given $^B P$, we wish to compute $^A P$, as in Fig. 2.7.

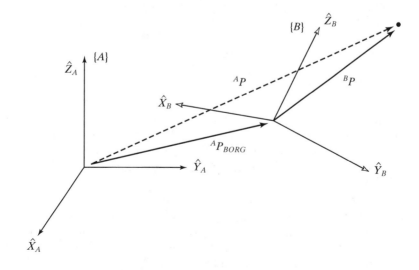

FIGURE 2.7: General transform of a vector.

We can first change $^B P$ to its description relative to an intermediate frame that has the same orientation as $\{A\}$, but whose origin is coincident with the origin of $\{B\}$. This is done by premultiplying by $^A_B R$ as in the last section. We then account for the translation between origins by simple vector addition, as before, and obtain

$$^A P = {}^A_B R \, {}^B P + {}^A P_{BORG}.\tag{2.17}$$

Equation 2.17 describes a general transformation mapping of a vector from its description in one frame to a description in a second frame. Note the following interpretation of our notation as exemplified in (2.17): the B's cancel, leaving all quantities as vectors written in terms of A, which may then be added.

The form of (2.17) is not as appealing as the conceptual form

$$^A P = {}^A_B T \, {}^B P.\tag{2.18}$$

That is, we would like to think of a mapping from one frame to another as an operator in matrix form. This aids in writing compact equations and is conceptually clearer than (2.17). In order that we may write the mathematics given in (2.17) in the matrix operator form suggested by (2.18), we define a 4×4 matrix operator and use 4×1 position vectors, so that (2.18) has the structure

$$\left[\begin{array}{c} {}^A P \\ 1 \end{array}\right] = \left[\begin{array}{ccc|c} & {}^A_B R & & {}^A P_{BORG} \\ \hline 0 & 0 & 0 & 1 \end{array}\right] \left[\begin{array}{c} {}^B P \\ 1 \end{array}\right].\tag{2.19}$$

In other words,

1. a "1" is added as the last element of the 4×1 vectors;
2. a row "$[0\,0\,0\,1]$" is added as the last row of the 4×4 matrix.

We adopt the convention that a position vector is 3×1 or 4×1, depending on whether it appears multiplied by a 3×3 matrix or by a 4×4 matrix. It is readily seen that (2.19) implements

$$^A P = {}^A_B R \, {}^B P + {}^A P_{BORG}$$
$$1 = 1.\tag{2.20}$$

The 4×4 matrix in (2.19) is called a **homogeneous transform**. For our purposes, it can be regarded purely as a construction used to cast the rotation and translation of the general transform into a single matrix form. In other fields of study, it can be used to compute perspective and scaling operations (when the last row is other than "$[0\,0\,0\,1]$" or the rotation matrix is not orthonormal). The interested reader should see [2].

Often, we will write an equation like (2.18) without any notation indicating that it is a homogeneous representation, because it is obvious from context. Note that, although homogeneous transforms are useful in writing compact equations, a computer program to transform vectors would generally not use them, because of time wasted multiplying ones and zeros. Thus, this representation is mainly for our convenience when thinking and writing equations down on paper.

Just as we used rotation matrices to specify an orientation, we will use transforms (usually in homogeneous representation) to specify a frame. Observe that, although we have introduced homogeneous transforms in the context of mappings, they also serve as descriptions of frames. The description of frame {*B*} relative to {*A*} is $^A_B T$.

EXAMPLE 2.2

Figure 2.8 shows a frame {*B*}, which is rotated relative to frame {*A*} about $\hat{Z}$ by 30 degrees, translated 10 units in $\hat{X}_A$, and translated 5 units in $\hat{Y}_A$. Find $^A P$, where $^B P = [3.07.00.0]^T$.

The definition of frame {*B*} is

$$^A_B T = \begin{bmatrix} 0.866 & -0.500 & 0.000 & 10.0 \\ 0.500 & 0.866 & 0.000 & 5.0 \\ 0.000 & 0.000 & 1.000 & 0.0 \\ 0 & 0 & 0 & 1 \end{bmatrix}. \tag{2.21}$$

Given

$$^B P = \begin{bmatrix} 3.0 \\ 7.0 \\ 0.0 \end{bmatrix}, \tag{2.22}$$

we use the definition of {*B*} just given as a transformation:

$$^A P = {}^A_B T \; {}^B P = \begin{bmatrix} 9.098 \\ 12.562 \\ 0.000 \end{bmatrix}. \tag{2.23}$$

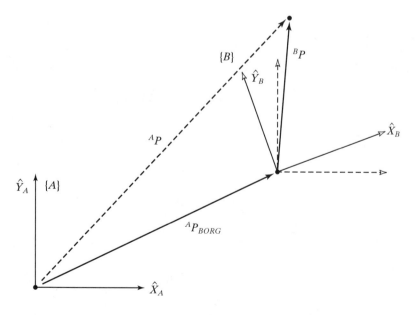

FIGURE 2.8: Frame {*B*} rotated and translated.

2.4   OPERATORS: TRANSLATIONS, ROTATIONS, AND TRANSFORMATIONS

The same mathematical forms used to map points between frames can also be interpreted as operators that translate points, rotate vectors, or do both. This section illustrates this interpretation of the mathematics we have already developed.

Translational operators

A translation moves a point in space a finite distance along a given vector direction. With this interpretation of actually translating the point in space, only one coordinate system need be involved. It turns out that translating the point in space is accomplished with the same mathematics as mapping the point to a second frame. Almost always, it is very important to understand which interpretation of the mathematics is being used. The distinction is as simple as this: When a vector is moved "forward" relative to a frame, we may consider either that the vector moved "forward" or that the frame moved "backward." The mathematics involved in the two cases is identical; only our view of the situation is different. Figure 2.9 indicates pictorially how a vector $^A P_1$ is translated by a vector $^A Q$. Here, the vector $^A Q$ gives the information needed to perform the translation.

The result of the operation is a new vector $^A P_2$, calculated as

$$^A P_2 = {}^A P_1 + {}^A Q. \tag{2.24}$$

To write this translation operation as a matrix operator, we use the notation

$$^A P_2 = D_Q(q) \, {}^A P_1, \tag{2.25}$$

where q is the signed magnitude of the translation along the vector direction $\hat{Q}$. The D_Q operator may be thought of as a homogeneous transform of a special

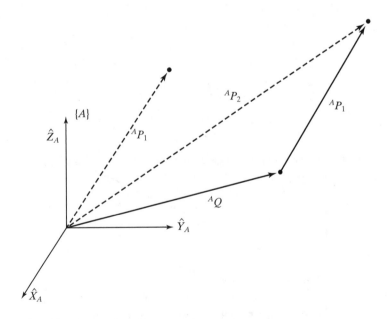

FIGURE 2.9: Translation operator.

simple form:

$$D_Q(q) = \begin{bmatrix} 1 & 0 & 0 & q_x \\ 0 & 1 & 0 & q_y \\ 0 & 0 & 1 & q_z \\ 0 & 0 & 0 & 1 \end{bmatrix},$$

(2.26)

where q_x, q_y, and q_z are the components of the translation vector Q and $q = \sqrt{q_x^2 + q_y^2 + q_z^2}$. Equations (2.9) and (2.24) implement the same mathematics. Note that, if we had defined $^B P_{AORG}$ (instead of $^A P_{BORG}$) in Fig. 2.4 and had used it in (2.9), then we would have seen a sign change between (2.9) and (2.24). This sign change would indicate the difference between moving the vector "forward" and moving the coordinate system "backward." By defining the location of {B} relative to {A} (with $^A P_{BORG}$), we cause the mathematics of the two interpretations to be the same. Now that the "D_Q" notation has been introduced, we may also use it to describe frames and as a mapping.

Rotational operators

Another interpretation of a rotation matrix is as a *rotational operator* that operates on a vector $^A P_1$ and changes that vector to a new vector, $^A P_2$, by means of a rotation, R. Usually, when a rotation matrix is shown as an operator, no sub- or superscripts appear, because it is not viewed as relating two frames. That is, we may write

$$^A P_2 = R\, ^A P_1.$$

(2.27)

Again, as in the case of translations, the mathematics described in (2.13) and in (2.27) is the same; only our interpretation is different. This fact also allows us to see *how to obtain* rotational matrices that are to be used as operators:

   The rotation matrix that rotates vectors through some rotation, R, is the same as the rotation matrix that describes a frame rotated by R relative to the reference frame.

   Although a rotation matrix is easily viewed as an operator, we will also define another notation for a rotational operator that clearly indicates which axis is being rotated about:

$$^A P_2 = R_K(\theta)\, ^A P_1.$$

(2.28)

In this notation, "$R_K(\theta)$" is a rotational operator that performs a rotation about the axis direction $\hat{K}$ by θ degrees. This operator can be written as a homogeneous transform whose position-vector part is zero. For example, substitution into (2.11) yields the operator that rotates about the $\hat{Z}$ axis by θ as

$$R_z(\Theta) = \begin{bmatrix} \cos\theta & -\sin\theta & 0 & 0 \\ \sin\theta & \cos\theta & 0 & 0 \\ 0 & 0 & 1 & 0 \\ 0 & 0 & 0 & 1 \end{bmatrix}.$$

(2.29)

Of course, to rotate a position vector, we could just as well use the 3 × 3 rotation-matrix part of the homogeneous transform. The "R_K" notation, therefore, may be considered to represent a 3 × 3 or a 4 × 4 matrix. Later in this chapter, we will see how to write the rotation matrix for a rotation about a general axis $\hat{K}$.

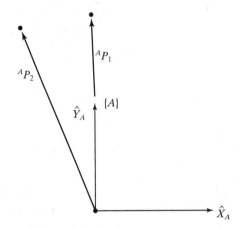

FIGURE 2.10: The vector $^A P_1$ rotated 30 degrees about $\hat{Z}$.

EXAMPLE 2.3

Figure 2.10 shows a vector $^A P_1$. We wish to compute the vector obtained by rotating this vector about $\hat{Z}$ by 30 degrees. Call the new vector $^A P_2$.

The rotation matrix that rotates vectors by 30 degrees about $\hat{Z}$ is the same as the rotation matrix that describes a frame rotated 30 degrees about $\hat{Z}$ relative to the reference frame. Thus, the correct rotational operator is

$$R_z(30.0) = \begin{bmatrix} 0.866 & -0.500 & 0.000 \\ 0.500 & 0.866 & 0.000 \\ 0.000 & 0.000 & 1.000 \end{bmatrix}. \tag{2.30}$$

Given

$$^A P_1 = \begin{bmatrix} 0.0 \\ 2.0 \\ 0.0 \end{bmatrix}, \tag{2.31}$$

we calculate $^A P_2$ as

$$^A P_2 = R_z(30.0) \, ^A P_1 = \begin{bmatrix} -1.000 \\ 1.732 \\ 0.000 \end{bmatrix}. \tag{2.32}$$

Equations (2.13) and (2.27) implement the same mathematics. Note that, if we had defined $^B_A R$ (instead of $^A_B R$) in (2.13), then the inverse of R would appear in (2.27). This change would indicate the difference between rotating the vector "forward" versus rotating the coordinate system "backward." By defining the location of $\{B\}$ relative to $\{A\}$ (by $^A_B R$), we cause the mathematics of the two interpretations to be the same.

Transformation operators

As with vectors and rotation matrices, a frame has another interpretation as a *transformation operator*. In this interpretation, only one coordinate system is involved, and so the symbol T is used without sub- or superscripts. The operator T rotates and translates a vector $^A P_1$ to compute a new vector,

$$^A P_2 = T\ ^A P_1. \tag{2.33}$$

Again, as in the case of rotations, the mathematics described in (2.18) and in (2.33) is the same, only our interpretation is different. This fact also allows us to see how to obtain homogeneous transforms that are to be used as operators:

The transform that rotates by R and translates by Q is the same as the transform that describes a frame rotated by R and translated by Q relative to the reference frame.

A transform is usually thought of as being in the form of a homogeneous transform with general rotation-matrix and position-vector parts.

EXAMPLE 2.4

Figure 2.11 shows a vector $^A P_1$. We wish to rotate it about $\hat{Z}$ by 30 degrees and translate it 10 units in $\hat{X}_A$ and 5 units in $\hat{Y}_A$. Find $^A P_2$, where $^A P_1 = [3.0\ 7.0\ 0.0]^T$.

The operator T, which performs the translation and rotation, is

$$T = \begin{bmatrix} 0.866 & -0.500 & 0.000 & 10.0 \\ 0.500 & 0.866 & 0.000 & 5.0 \\ 0.000 & 0.000 & 1.000 & 0.0 \\ 0 & 0 & 0 & 1 \end{bmatrix}. \tag{2.34}$$

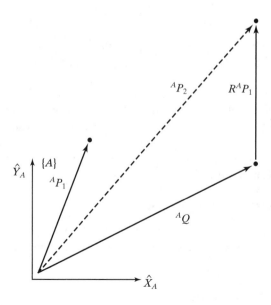

FIGURE 2.11: The vector $^A P_1$ rotated and translated to form $^A P_2$.

Given

$$
{}^A P_1 = \begin{bmatrix} 3.0 \\ 7.0 \\ 0.0 \end{bmatrix}, \tag{2.35}
$$

we use T as an operator:

$$
{}^A P_2 = T \; {}^A P_1 = \begin{bmatrix} 9.098 \\ 12.562 \\ 0.000 \end{bmatrix}. \tag{2.36}
$$

Note that this example is numerically exactly the same as Example 2.2, but the interpretation is quite different.

2.5 SUMMARY OF INTERPRETATIONS

We have introduced concepts first for the case of translation only, then for the case of rotation only, and finally for the general case of rotation about a point and translation of that point. Having understood the general case of rotation and translation, we will not need to explicitly consider the two simpler cases since they are contained within the general framework.

As a general tool to represent frames, we have introduced the *homogeneous transform*, a 4×4 matrix containing orientation and position information. We have introduced three interpretations of this homogeneous transform:

1. It is a *description of a frame*. ${}^A_B T$ describes the frame $\{B\}$ relative to the frame $\{A\}$. Specifically, the columns of ${}^A_B R$ are unit vectors defining the directions of the principal axes of $\{B\}$, and ${}^A P_{BORG}$ locates the position of the origin of $\{B\}$.
2. It is a *transform mapping*. ${}^A_B T$ maps ${}^B P \rightarrow {}^A P$.
3. It is a *transform operator*. T operates on ${}^A P_1$ to create ${}^A P_2$.

From this point on, the terms *frame* and *transform* will both be used to refer to a position vector plus an orientation. *Frame* is the term favored in speaking of a description, and *transform* is used most frequently when function as a mapping or operator is implied. Note that transformations are generalizations of (and subsume) translations and rotations; we will often use the term *transform* when speaking of a pure rotation (or translation).

2.6 TRANSFORMATION ARITHMETIC

In this section, we look at the multiplication of transforms and the inversion of transforms. These two elementary operations form a functionally complete set of transform operators.

Compound transformations

In Fig. 2.12, we have ${}^C P$ and wish to find ${}^A P$.

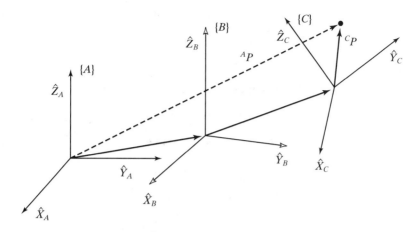

FIGURE 2.12: Compound frames: Each is known relative to the previous one.

Frame $\{C\}$ is known relative to frame $\{B\}$, and frame $\{B\}$ is known relative to frame $\{A\}$. We can transform $^C P$ into $^B P$ as

$$^B P = {}^B_C T \, {}^C P; \tag{2.37}$$

then we can transform $^B P$ into $^A P$ as

$$^A P = {}^A_B T \, {}^B P. \tag{2.38}$$

Combining (2.37) and (2.38), we get the (not unexpected) result

$$^A P = {}^A_B T {}^B_C T {}^C P, \tag{2.39}$$

from which we could define

$$^A_C T = {}^A_B T {}^B_C T. \tag{2.40}$$

Again, note that familiarity with the sub- and superscript notation makes these manipulations simple. In terms of the known descriptions of $\{B\}$ and $\{C\}$, we can give the expression for $^A_C T$ as

$$^A_C T = \left[\begin{array}{ccc|c} & {}^A_B R \, {}^B_C R & & {}^A_B R \, {}^B P_{CORG} + {}^A P_{BORG} \\ \hline 0 & 0 & 0 & 1 \end{array} \right] \tag{2.41}$$

Inverting a transform

Consider a frame $\{B\}$ that is known with respect to a frame $\{A\}$—that is, we know the value of $^A_B T$. Sometimes we will wish to invert this transform, in order to get a description of $\{A\}$ relative to $\{B\}$—that is, $^B_A T$. A straightforward way of calculating the inverse is to compute the inverse of the 4×4 homogeneous transform. However, if we do so, we are not taking full advantage of the structure inherent in the transform. It is easy to find a computationally simpler method of computing the inverse, one that does take advantage of this structure.

To find ^B_AT, we must compute ^B_AR and $^BP_{AORG}$ from ^A_BR and $^AP_{BORG}$. First, recall from our discussion of rotation matrices that

$$^B_AR = {^A_BR}^T.$$ (2.42)

Next, we change the description of $^AP_{BORG}$ into {B} by using (2.13):

$$^B(^AP_{BORG}) = {^B_AR}\,{^AP_{BORG}} + {^BP_{AORG}}.$$ (2.43)

The left-hand side of (2.43) must be zero, so we have

$$^BP_{AORG} = -{^B_AR}\,{^AP_{BORG}} = -{^A_BR}^T\,{^AP_{BORG}}.$$ (2.44)

Using (2.42) and (2.44), we can write the form of ^B_AT as

$$^B_AT = \left[\begin{array}{ccc|c} & {^A_BR}^T & & -{^A_BR}^T\,{^AP_{BORG}} \\ \hline 0 & 0 & 0 & 1 \end{array} \right].$$ (2.45)

Note that, with our notation,

$$^B_AT = {^A_BT}^{-1}.$$

Equation (2.45) is a general and extremely useful way of computing the inverse of a homogeneous transform.

EXAMPLE 2.5

Figure 2.13 shows a frame {B} that is rotated relative to frame {A} about $\hat{Z}$ by 30 degrees and translated four units in $\hat{X}_A$ and three units in $\hat{Y}_A$. Thus, we have a description of ^A_BT. Find ^B_AT.

The frame defining {B} is

$$^A_BT = \begin{bmatrix} 0.866 & -0.500 & 0.000 & 4.0 \\ 0.500 & 0.866 & 0.000 & 3.0 \\ 0.000 & 0.000 & 1.000 & 0.0 \\ 0 & 0 & 0 & 1 \end{bmatrix}.$$ (2.46)

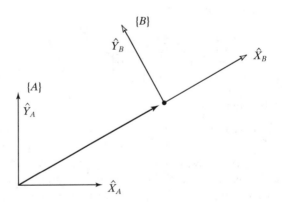

FIGURE 2.13: {B} relative to {A}.

Using (2.45), we compute

$$
{}_{A}^{B}T = \begin{bmatrix} 0.866 & 0.500 & 0.000 & -4.964 \\ -0.500 & 0.866 & 0.000 & -0.598 \\ 0.000 & 0.000 & 1.000 & 0.0 \\ 0 & 0 & 0 & 1 \end{bmatrix}. \tag{2.47}
$$

2.7   TRANSFORM EQUATIONS

Figure 2.14 indicates a situation in which a frame $\{D\}$ can be expressed as products of transformations in two different ways. First,

$$
{}_{D}^{U}T = {}_{A}^{U}T \, {}_{D}^{A}T; \tag{2.48}
$$

second;

$$
{}_{D}^{U}T = {}_{B}^{U}T \, {}_{C}^{B}T \, {}_{D}^{C}T. \tag{2.49}
$$

We can set these two descriptions of ${}_{D}^{U}T$ equal to construct a **transform equation**:

$$
{}_{A}^{U}T \, {}_{D}^{A}T = {}_{B}^{U}T \, {}_{C}^{B}T \, {}_{D}^{C}T. \tag{2.50}
$$

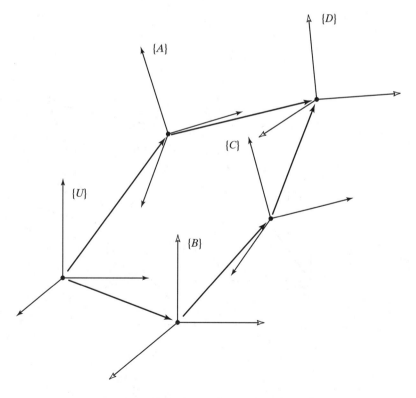

FIGURE 2.14: Set of transforms forming a loop.

Transform equations can be used to solve for transforms in the case of n unknown transforms and n transform equations. Consider (2.50) in the case that all transforms are known except $_C^B T$. Here, we have one transform equation and one unknown transform; hence, we easily find its solution to be

$$_C^B T = {_B^U T}^{-1} \, _A^U T \, _D^A T \, _D^C T^{-1}. \tag{2.51}$$

Figure 2.15 indicates a similar situation.

Note that, in all figures, we have introduced a *graphical* representation of frames as an arrow pointing from one origin to another origin. The arrow's direction indicates which way the frames are defined: In Fig. 2.14, frame $\{D\}$ is defined relative to $\{A\}$; in Fig. 2.15, frame $\{A\}$ is defined relative to $\{D\}$. In order to compound frames when the arrows line up, we simply compute the product of the transforms. If an arrow points the opposite way in a chain of transforms, we simply compute its inverse first. In Fig. 2.15, two possible descriptions of $\{C\}$ are

$$_C^U T = {_A^U T} \, _A^D T^{-1} \, _C^D T \tag{2.52}$$

and

$$_C^U T = {_B^U T} \, _C^B T. \tag{2.53}$$

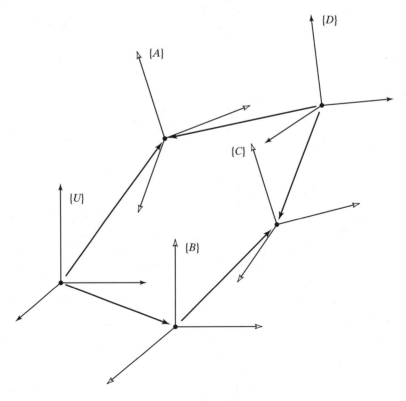

FIGURE 2.15: Example of a transform equation.

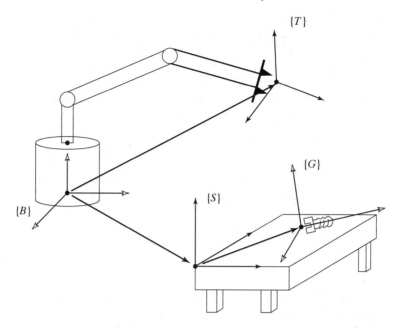

FIGURE 2.16: Manipulator reaching for a bolt.

Again, we might equate (2.52) and (2.53) to solve for, say, $^U_A T$:

$$^U_A T = {}^U_B T \, {}^B_C T \, {}^D_C T^{-1} \, {}^D_A T. \tag{2.54}$$

EXAMPLE 2.6

Assume that we know the transform $^B_T T$ in Fig. 2.16, which describes the frame at the manipulator's fingertips {T} relative to the base of the manipulator, {B}, that we know where the tabletop is located in space relative to the manipulator's base (because we have a description of the frame {S} that is attached to the table as shown, $^B_S T$), and that we know the location of the frame attached to the bolt lying on the table relative to the table frame—that is, $^S_G T$. Calculate the position and orientation of the bolt relative to the manipulator's hand, $^T_G T$.

Guided by our notation (and, it is hoped, our understanding), we compute the bolt frame relative to the hand frame as

$$^T_G T = {}^B_T T^{-1} \, {}^B_S T \, {}^S_G T. \tag{2.55}$$

2.8   MORE ON REPRESENTATION OF ORIENTATION

So far, our only means of representing an orientation is by giving a 3×3 rotation matrix. As shown, rotation matrices are special in that all columns are mutually orthogonal and have unit magnitude. Further, we will see that the determinant of a rotation matrix is always equal to $+1$. Rotation matrices may also be called **proper orthonormal matrices**, where "proper" refers to the fact that the determinant is $+1$ (nonproper orthonormal matrices have the determinant -1).

It is natural to ask whether it is possible to describe an orientation with fewer than nine numbers. A result from linear algebra (known as **Cayley's formula for orthonormal matrices** [3]) states that, for any proper orthonormal matrix R, there exists a skew-symmetric matrix S such that

$$R = (I_3 - S)^{-1}(I_3 + S), \tag{2.56}$$

where I_3 is a 3×3 unit matrix. Now a skew-symmetric matrix (i.e., $S = -S^T$) of dimension 3 is specified by three parameters (s_x, s_y, s_z) as

$$S = \begin{bmatrix} 0 & -s_x & s_y \\ s_x & 0 & -s_x \\ -s_y & s_x & 0 \end{bmatrix}. \tag{2.57}$$

Therefore, an immediate consequence of formula (2.56) is that any 3×3 rotation matrix can be specified by just three parameters.

Clearly, the nine elements of a rotation matrix are not all independent. In fact, given a rotation matrix, R, it is easy to write down the six dependencies between the elements. Imagine R as three columns, as originally introduced:

$$R = [\hat{X} \ \hat{Y} \ \hat{Z}]. \tag{2.58}$$

As we know from Section 2.2, these three vectors are the unit axes of some frame written in terms of the reference frame. Each is a unit vector, and all three must be mutually perpendicular, so we see that there are six constraints on the nine matrix elements:

$$|\hat{X}| = 1,$$
$$|\hat{Y}| = 1,$$
$$|\hat{Z}| = 1, \tag{2.59}$$
$$\hat{X} \cdot \hat{Y} = 0,$$
$$\hat{X} \cdot \hat{Z} = 0,$$
$$\hat{Y} \cdot \hat{Z} = 0.$$

It is natural then to ask whether representations of orientation can be devised such that the representation is *conveniently* specified with three parameters. This section will present several such representations.

Whereas translations along three mutually perpendicular axes are quite easy to visualize, rotations seem less intuitive. Unfortunately, people have a hard time describing and specifying orientations in three-dimensional space. One difficulty is that rotations don't generally commute. That is, ${}^A_B R \ {}^B_C R$ is not the same as ${}^B_C R \ {}^A_B R$.

EXAMPLE 2.7

Consider two rotations, one about $\hat{Z}$ by 30 degrees and one about $\hat{X}$ by 30 degrees:

$$R_z(30) = \begin{bmatrix} 0.866 & -0.500 & 0.000 \\ 0.500 & 0.866 & 0.000 \\ 0.000 & 0.000 & 1.000 \end{bmatrix} \tag{2.60}$$

$$R_x(30) = \begin{bmatrix} 1.000 & 0.000 & 0.000 \\ 0.000 & 0.866 & -0.500 \\ 0.000 & 0.500 & 0.866 \end{bmatrix} \tag{2.61}$$

$$R_z(30)R_x(30) = \begin{bmatrix} 0.87 & -0.43 & 0.25 \\ 0.50 & 0.75 & -0.43 \\ 0.00 & 0.50 & 0.87 \end{bmatrix}$$

$$\neq R_x(30)R_z(30) = \begin{bmatrix} 0.87 & -0.50 & 0.00 \\ 0.43 & 0.75 & -0.50 \\ 0.25 & 0.43 & 0.87 \end{bmatrix} \tag{2.62}$$

The fact that the order of rotations is important should not be surprising; furthermore, it is captured in the fact that we use matrices to represent rotations, because multiplication of matrices is not commutative in general.

Because rotations can be thought of either as operators or as descriptions of orientation, it is not surprising that different representations are favored for each of these uses. Rotation matrices are useful as operators. Their matrix form is such that, when multiplied by a vector, they perform the rotation operation. However, rotation matrices are somewhat unwieldy when used to specify an orientation. A human operator at a computer terminal who wishes to type in the specification of the desired orientation of a robot's hand would have a hard time inputting a nine-element matrix with orthonormal columns. A representation that requires only three numbers would be simpler. The following sections introduce several such representations.

X–Y–Z fixed angles

One method of describing the orientation of a frame $\{B\}$ is as follows:

Start with the frame coincident with a known reference frame $\{A\}$. Rotate $\{B\}$ first about $\hat{X}_A$ by an angle γ, then about $\hat{Y}_A$ by an angle β, and, finally, about $\hat{Z}_A$ by an angle α.

Each of the three rotations takes place about an axis in the fixed reference frame $\{A\}$. We will call this convention for specifying an orientation **X–Y–Z fixed angles**. The word "fixed" refers to the fact that the rotations are specified about the fixed (i.e., nonmoving) reference frame (Fig. 2.17). Sometimes this convention is referred to as **roll, pitch, yaw angles**, but care must be used, as this name is often given to other related but different conventions.

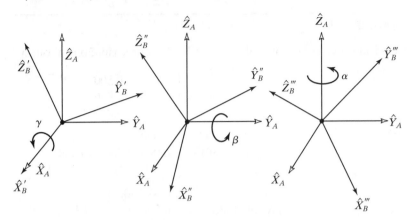

FIGURE 2.17: X–Y–Z fixed angles. Rotations are performed in the order $R_X(\gamma)$, $R_Y(\beta)$, $R_Z(\alpha)$.

The derivation of the equivalent rotation matrix, $^A_B R_{XYZ}(\gamma, \beta, \alpha)$, is straight-forward, because all rotations occur about axes of the reference frame; that is,

$$
^A_B R_{XYZ}(\gamma, \beta, \alpha) = R_Z(\alpha) R_Y(\beta) R_X(\gamma)
$$

$$
= \begin{bmatrix} c\alpha & -s\alpha & 0 \\ s\alpha & c\alpha & 0 \\ 0 & 0 & 1 \end{bmatrix} \begin{bmatrix} c\beta & 0 & s\beta \\ 0 & 1 & 0 \\ -s\beta & 0 & c\beta \end{bmatrix} \begin{bmatrix} 1 & 0 & 0 \\ 0 & c\gamma & -s\gamma \\ 0 & s\gamma & c\gamma \end{bmatrix}, \quad (2.63)
$$

where $c\alpha$ is shorthand for $\cos\alpha$, $s\alpha$ for $\sin\alpha$, and so on. It is extremely important to understand the order of rotations used in (2.63). Thinking in terms of rotations as operators, we have applied the rotations (from the *right*) of $R_X(\gamma)$, then $R_Y(\beta)$, and then $R_Z(\alpha)$. Multiplying (2.63) out, we obtain

$$
^A_B R_{XYZ}(\gamma, \beta, \alpha) = \begin{bmatrix} c\alpha c\beta & c\alpha s\beta s\gamma - s\alpha c\gamma & c\alpha s\beta c\gamma + s\alpha s\gamma \\ s\alpha c\beta & s\alpha s\beta s\gamma + c\alpha c\gamma & s\alpha s\beta c\gamma - c\alpha s\gamma \\ -s\beta & c\beta s\gamma & c\beta c\gamma \end{bmatrix}. \quad (2.64)
$$

Keep in mind that the definition given here specifies the order of the three rotations. Equation (2.64) is correct only for rotations performed in the order: about $\hat{X}_A$ by γ, about $\hat{Y}_A$ by β, about $\hat{Z}_A$ by α.

The inverse problem, that of extracting equivalent X–Y–Z fixed angles from a rotation matrix, is often of interest. The solution depends on solving a set of transcendental equations: there are nine equations and three unknowns if (2.64) is equated to a given rotation matrix. Among the nine equations are six dependencies, so, essentially, we have three equations and three unknowns. Let

$$
^A_B R_{XYZ}(\gamma, \beta, \alpha) = \begin{bmatrix} r_{11} & r_{12} & r_{13} \\ r_{21} & r_{22} & r_{23} \\ r_{31} & r_{32} & r_{33} \end{bmatrix}. \quad (2.65)
$$

From (2.64), we see that, by taking the square root of the sum of the squares of r_{11} and r_{21}, we can compute $\cos\beta$. Then, we can solve for β with the arc tangent of $-r_{31}$ over the computed cosine. Then, as long as $c\beta \neq 0$, we can solve for α by taking the arc tangent of $r_{21}/c\beta$ over $r_{11}/c\beta$ and we can solve for γ by taking the arc tangent of $r_{32}/c\beta$ over $r_{33}/c\beta$.

In summary,

$$\beta = \text{Atan2}(-r_{31}, \sqrt{r_{11}^2 + r_{21}^2}),$$

$$\alpha = \text{Atan2}(r_{21}/c\beta, r_{11}/c\beta), \tag{2.66}$$

$$\gamma = \text{Atan2}(r_{32}/c\beta, r_{33}/c\beta),$$

where $\text{Atan2}(y, x)$ is a two-argument arc tangent function.[3]

Although a second solution exists, by using the positive square root in the formula for β, we always compute the single solution for which $-90.0° \leq \beta \leq 90.0°$. This is usually a good practice, because we can then define one-to-one mapping functions between various representations of orientation. However, in some cases, calculating all solutions is important (more on this in Chapter 4). If $\beta = \pm90.0°$ (so that $c\beta = 0$), the solution of (2.67) degenerates. In those cases, only the sum or the difference of α and γ can be computed. One possible convention is to choose $\alpha = 0.0$ in these cases, which has the results given next.

If $\beta = 90.0°$, then a solution can be calculated to be

$$\beta = 90.0°,$$

$$\alpha = 0.0, \tag{2.67}$$

$$\gamma = \text{Atan2}(r_{12}, r_{22}).$$

If $\beta = -90.0°$, then a solution can be calculated to be

$$\beta = -90.0°,$$

$$\alpha = 0.0, \tag{2.68}$$

$$\gamma = -\text{Atan2}(r_{12}, r_{22}).$$

Z–Y–X Euler angles

Another possible description of a frame $\{B\}$ is as follows:

Start with the frame coincident with a known frame $\{A\}$. Rotate $\{B\}$ first about $\hat{Z}_B$ by an angle α, then about $\hat{Y}_B$ by an angle β, and, finally, about $\hat{X}_B$ by an angle γ.

In this representation, each rotation is performed about an axis of the moving system $\{B\}$ rather than one of the fixed reference $\{A\}$. Such sets of three rotations

---

[3] $\text{Atan2}(y, x)$ computes $\tan^{-1}(\frac{y}{x})$ but uses the signs of both x and y to identify the quadrant in which the resulting angle lies. For example, $\text{Atan 2}(-2.0, -2.0) = -135°$, whereas $\text{Atan 2}(2.0, 2.0) = 45°$, a distinction which would be lost with a single-argument arc tangent function. We are frequently computing angles that can range over a full 360°, so we will make use of the Atan2 function regularly. Note that Atan2 becomes undefined when both arguments are zero. It is sometimes called a "4-quadrant arc tangent," and some programming-language libraries have it predefined.

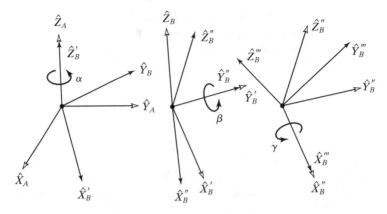

FIGURE 2.18: Z–Y–X Euler angles.

are called **Euler angles**. Note that each rotation takes place about an axis whose location depends upon the preceding rotations. Because the three rotations occur about the axes $\hat{Z}$, $\hat{Y}$, and $\hat{X}$, we will call this representation **Z–Y–X Euler angles**.

Figure 2.18 shows the axes of $\{B\}$ after each Euler-angle rotation is applied. Rotation α about $\hat{Z}$ causes $\hat{X}$ to rotate into $\hat{X}'$, $\hat{Y}$ to rotate into $\hat{Y}'$, and so on. An additional "prime" gets added to each axis with each rotation. A rotation matrix which is parameterized by Z–Y–X Euler angles will be indicated by the notation ${}^A_B R_{Z'Y'X'}(\alpha, \beta, \gamma)$. Note that we have added "primes" to the subscripts to indicate that this rotation is described by Euler angles.

With reference to Fig. 2.18, we can use the intermediate frames $\{B'\}$ and $\{B''\}$ in order to give an expression for ${}^A_B R_{Z'Y'X'}(\alpha, \beta, \gamma)$. Thinking of the rotations as descriptions of these frames, we can immediately write

$$
{}^A_B R = {}^A_{B'} R \; {}^{B'}_{B''} R \; {}^{B''}_{B} R, \tag{2.69}
$$

where each of the relative descriptions on the right-hand side of (2.69) is given by the statement of the Z–Y–X-Euler-angle convention. Namely, the final orientation of $\{B\}$ is given relative to $\{A\}$ as

$$
{}^A_B R_{Z'Y'X'} = R_Z(\alpha) R_Y(\beta) R_X(\gamma)
$$

$$
= \begin{bmatrix} c\alpha & -s\alpha & 0 \\ s\alpha & c\alpha & 0 \\ 0 & 0 & 1 \end{bmatrix} \begin{bmatrix} c\beta & 0 & s\beta \\ 0 & 1 & 0 \\ -s\beta & 0 & c\beta \end{bmatrix} \begin{bmatrix} 1 & 0 & 0 \\ 0 & c\gamma & -s\gamma \\ 0 & s\gamma & c\gamma \end{bmatrix}, \tag{2.70}
$$

where $c\alpha = \cos\alpha$, $s\alpha = \sin\alpha$, and so on. Multiplying out, we obtain

$$
{}^A_B R_{Z'Y'X'}(\alpha, \beta, \gamma) = \begin{bmatrix} c\alpha c\beta & c\alpha s\beta s\gamma - s\alpha c\gamma & c\alpha s\beta c\gamma + s\alpha s\gamma \\ s\alpha c\beta & s\alpha s\beta s\gamma + c\alpha c\gamma & s\alpha s\beta c\gamma - c\alpha s\gamma \\ -s\beta & c\beta s\gamma & c\beta c\gamma \end{bmatrix}. \tag{2.71}
$$

Note that the result is exactly the same as that obtained for the same three rotations taken in the opposite order about fixed axes! This somewhat nonintuitive result holds in general: three rotations taken about fixed axes yield the same final orientation as the same three rotations taken in opposite order about the axes of the moving frame.

Because (2.71) is equivalent to (2.64), there is no need to repeat the solution for extracting Z–Y–X Euler angles from a rotation matrix. That is, (2.66) can also be used to solve for Z–Y–X Euler angles that correspond to a given rotation matrix.

Z–Y–Z Euler angles

Another possible description of a frame $\{B\}$ is

> Start with the frame coincident with a known frame $\{A\}$. Rotate $\{B\}$ first about $\hat{Z}_B$ by an angle α, then about $\hat{Y}_B$ by an angle β, and, finally, about Z_b by an angle γ.

Rotations are described relative to the frame we are moving, namely, $\{B\}$, so this is an Euler-angle description. Because the three rotations occur about the axes $\hat{Z}$, $\hat{Y}$, and $\hat{Z}$, we will call this representation **Z–Y–Z Euler angles**.

Following the development exactly as in the last section, we arrive at the equivalent rotation matrix

$$
{}_{B}^{A}R_{Z'Y'Z'}(\alpha, \beta, \gamma) = \begin{bmatrix} c\alpha c\beta c\gamma - s\alpha s\gamma & -c\alpha c\beta s\gamma - s\alpha c\gamma & c\alpha s\beta \\ s\alpha c\beta c\gamma + c\alpha s\gamma & -s\alpha c\beta s\gamma + c\alpha c\gamma & s\alpha s\beta \\ -s\beta c\gamma & s\beta s\gamma & c\beta \end{bmatrix}. \tag{2.72}
$$

The solution for extracting Z–Y–Z Euler angles from a rotation matrix is stated next.

Given

$$
{}_{B}^{A}R_{Z'Y'Z'}(\alpha, \beta, \gamma) = \begin{bmatrix} r_{11} & r_{12} & r_{13} \\ r_{21} & r_{22} & r_{23} \\ r_{31} & r_{32} & r_{33} \end{bmatrix}, \tag{2.73}
$$

then, if $\sin\beta \neq 0$, it follows that

$$
\beta = \text{Atan2}(\sqrt{r_{31}^2 + r_{32}^2}, r_{33}),
$$
$$
\alpha = \text{Atan2}(r_{23}/s\beta, r_{13}/s\beta), \tag{2.74}
$$
$$
\gamma = \text{Atan2}(r_{32}/s\beta, -r_{31}/s\beta).
$$

Although a second solution exists (which we find by using the positive square root in the formula for β), we always compute the single solution for which $0.0 \leq \beta \leq 180.0°$. If $\beta = 0.0$ or $180.0°$, the solution of (2.74) degenerates. In those cases, only the sum or the difference of α and γ may be computed. One possible convention is to choose $\alpha = 0.0$ in these cases, which has the results given next.

If $\beta = 0.0$, then a solution can be calculated to be

$$
\beta = 0.0,
$$
$$
\alpha = 0.0, \tag{2.75}
$$
$$
\gamma = \text{Atan2}(-r_{12}, r_{11}).
$$

If $\beta = 180.0°$, then a solution can be calculated to be

$$\beta = 180.0°,$$

$$\alpha = 0.0, \tag{2.76}$$

$$\gamma = \text{Atan2}(r_{12}, -r_{11}).$$

Other angle-set conventions

In the preceding subsections we have seen three conventions for specifying orientation: X–Y–Z fixed angles, Z–Y–X Euler angles, and Z–Y–Z Euler angles. Each of these conventions requires performing three rotations about principal axes in a certain order. These conventions are examples of a set of 24 conventions that we will call **angle-set conventions**. Of these, 12 conventions are for fixed-angle sets, and 12 are for Euler-angle sets. Note that, because of the duality of fixed-angle sets with Euler-angle sets, there are really only 12 unique parameterizations of a rotation matrix by using successive rotations about principal axes. There is often no particular reason to favor one convention over another, but various authors adopt different ones, so it is useful to list the equivalent rotation matrices for all 24 conventions. Appendix B (in the back of the book) gives the equivalent rotation matrices for all 24 conventions.

Equivalent angle–axis representation

With the notation $R_X(30.0)$ we give the description of an orientation by giving an axis, $\hat{X}$, and an angle, 30.0 degrees. This is an example of an **equivalent angle–axis** representation. If the axis is a *general* direction (rather than one of the unit directions) any orientation may be obtained through proper axis and angle selection. Consider the following description of a frame $\{B\}$:

> Start with the frame coincident with a known frame $\{A\}$; then rotate $\{B\}$ about the vector $^A\hat{K}$ by an angle θ according to the right-hand rule.

Vector $\hat{K}$ is sometimes called the equivalent axis of a finite rotation. A general orientation of $\{B\}$ relative to $\{A\}$ may be written as $^A_B R(\hat{K}, \theta)$ or $R_K(\theta)$ and will be called the equivalent angle–axis representation.[4] The specification of the vector $^A\hat{K}$ requires only two parameters, because its length is always taken to be one. The angle specifies a third parameter. Often, we will multiply the unit direction, $\hat{K}$, with the amount of rotation, θ, to form a compact 3×1 vector description of orientation, denoted by K (no "hat"). See Fig. 2.19.

When the axis of rotation is chosen from among the principal axes of $\{A\}$, then the equivalent rotation matrix takes on the familiar form of planar rotations:

$$R_X(\theta) = \begin{bmatrix} 1 & 1 & 0 \\ 0 & \cos\theta & -\sin\theta \\ 0 & \sin\theta & \cos\theta \end{bmatrix}, \tag{2.77}$$

---

[4]That such a $\hat{K}$ and θ exist for any orientation of $\{B\}$ relative to $\{A\}$ was shown originally by Euler and is known as Euler's theorem on rotation [3].

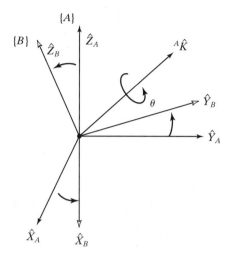

FIGURE 2.19: Equivalent angle–axis representation.

$$R_Y(\theta) = \begin{bmatrix} \cos\theta & 0 & \sin\theta \\ 0 & 1 & 0 \\ -\sin\theta & 0 & \cos\theta \end{bmatrix}, \tag{2.78}$$

$$R_Z(\theta) = \begin{bmatrix} \cos\theta & -\sin\theta & 0 \\ \sin\theta & \cos\theta & 0 \\ 0 & 0 & 1 \end{bmatrix}. \tag{2.79}$$

If the axis of rotation is a general axis, it can be shown (as in Exercise 2.6) that the equivalent rotation matrix is

$$R_K(\theta) = \begin{bmatrix} k_x k_x v\theta + c\theta & k_x k_y v\theta - k_z s\theta & k_x k_z v\theta + k_y s\theta \\ k_x k_y v\theta + k_z s\theta & k_y k_y v\theta + c\theta & k_y k_z v\theta - k_x s\theta \\ k_x k_z v\theta - k_y s\theta & k_y k_z v\theta + k_x s\theta & k_z k_z v\theta + c\theta \end{bmatrix}, \tag{2.80}$$

where $c\theta = \cos\theta$, $s\theta = \sin\theta$, $v\theta = 1 - \cos\theta$, and $^A\hat{K} = [k_x k_y k_z]^T$. The sign of θ is determined by the right-hand rule, with the thumb pointing along the positive sense of $^A\hat{K}$.

Equation (2.80) converts from angle–axis representation to rotation-matrix representation. Note that, given any axis of rotation and any angular amount, we can easily construct an equivalent rotation matrix.

The inverse problem, namely, that of computing $\hat{K}$ and θ from a given rotation matrix, is mostly left for the exercises (Exercises 2.6 and 2.7), but a partial result is given here [3]. If

$$^A_B R_K(\theta) = \begin{bmatrix} r_{11} & r_{12} & r_{13} \\ r_{21} & r_{22} & r_{23} \\ r_{31} & r_{32} & r_{33} \end{bmatrix}, \tag{2.81}$$

then

$$\theta = \text{Acos}\left(\frac{r_{11} + r_{22} + r_{33} - 1}{2}\right)$$

and

$$\hat{K} = \frac{1}{2\sin\theta} \begin{bmatrix} r_{32} - r_{23} \\ r_{13} - r_{31} \\ r_{21} - r_{12} \end{bmatrix}. \tag{2.82}$$

This solution always computes a value of θ between 0 and 180 degrees. For any axis–angle pair $({}^A\hat{K}, \theta)$, there is another pair, namely, $(-{}^A\hat{K}, -\theta)$, which results in the same orientation in space, with the same rotation matrix describing it. Therefore, in converting from a rotation-matrix into an angle–axis representation, we are faced with choosing between solutions. A more serious problem is that, for small angular rotations, the axis becomes ill-defined. Clearly, if the amount of rotation goes to zero, the axis of rotation becomes completely undefined. The solution given by (2.82) fails if $\theta = 0°$ or $\theta = 180°$.

EXAMPLE 2.8

A frame $\{B\}$ is described as initially coincident with $\{A\}$. We then rotate $\{B\}$ about the vector ${}^A\hat{K} = [0.7070\ 7070\ 0]^T$ (passing through the origin) by an amount $\theta = 30$ degrees. Give the frame description of $\{B\}$.

Substituting into (2.80) yields the rotation-matrix part of the frame description. There was no translation of the origin, so the position vector is $[0, 0, 0]^T$. Hence,

$$ {}^A_B T = \begin{bmatrix} 0.933 & 0.067 & 0.354 & 0.0 \\ 0.067 & 0.933 & -0.354 & 0.0 \\ -0.354 & 0.354 & 0.866 & 0.0 \\ 0.0 & 0.0 & 0.0 & 1.0 \end{bmatrix}. \tag{2.83}$$

Up to this point, all rotations we have discussed have been about axes that pass through the origin of the reference system. If we encounter a problem for which this is not true, we can reduce the problem to the "axis through the origin" case by defining additional frames whose origins lie on the axis and then solving a transform equation.

EXAMPLE 2.9

A frame $\{B\}$ is described as initially coincident with $\{A\}$. We then rotate $\{B\}$ about the vector ${}^A\hat{K} = [0.707\ 0.707\ 0.0]^T$ (passing through the point ${}^A P = [1.0\ 2.0\ 3.0]$) by an amount $\theta = 30$ degrees. Give the frame description of $\{B\}$.

Before the rotation, $\{A\}$ and $\{B\}$ are coincident. As is shown in Fig. 2.20, we define two new frames, $\{A'\}$ and $\{B'\}$, which are coincident with each other and have the same orientation as $\{A\}$ and $\{B\}$ respectively, but are translated relative to $\{A\}$ by an offset that places their origins on the axis of rotation. We will choose

$$ {}^A_{A'} T = \begin{bmatrix} 1.0 & 0.0 & 0.0 & 1.0 \\ 0.0 & 1.0 & 0.0 & 2.0 \\ 0.0 & 0.0 & 1.0 & 3.0 \\ 0.0 & 0.0 & 0.0 & 1.0 \end{bmatrix}. \tag{2.84}$$

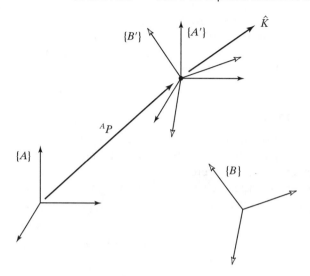

FIGURE 2.20: Rotation about an axis that does not pass through the origin of $\{A\}$. Initially, $\{B\}$ was coincident with $\{A\}$.

Similarly, the description of $\{B\}$ in terms of $\{B'\}$ is

$$
{}^{B'}_{B}T = \begin{bmatrix} 1.0 & 0.0 & 0.0 & -1.0 \\ 0.0 & 1.0 & 0.0 & -2.0 \\ 0.0 & 0.0 & 1.0 & -3.0 \\ 0.0 & 0.0 & 0.0 & 1.0 \end{bmatrix}.
\tag{2.85}
$$

Now, keeping other relationships fixed, we can rotate $\{B'\}$ relative to $\{A'\}$. This is a rotation about an axis that passes through the origin, so we can use (2.80) to compute $\{B'\}$ relative to $\{A'\}$. Substituting into (2.80) yields the rotation-matrix part of the frame description. There was no translation of the origin, so the position vector is $[0, 0, 0]^T$. Thus, we have

$$
{}^{A'}_{B'}T = \begin{bmatrix} 0.933 & 0.067 & 0.354 & 0.0 \\ 0.067 & 0.933 & -0.354 & 0.0 \\ -0.354 & 0.354 & 0.866 & 0.0 \\ 0.0 & 0.0 & 0.0 & 1.0 \end{bmatrix}.
\tag{2.86}
$$

Finally, we can write a transform equation to compute the desired frame,

$$
{}^{A}_{B}T = {}^{A}_{A'}T \, {}^{A'}_{B'}T \, {}^{B'}_{B}T,
\tag{2.87}
$$

which evaluates to

$$
{}^{A}_{B}T = \begin{bmatrix} 0.933 & 0.067 & 0.354 & -1.13 \\ 0.067 & 0.933 & -0.354 & 1.13 \\ -0.354 & 0.354 & 0.866 & 0.05 \\ 0.000 & 0.000 & 0.000 & 1.00 \end{bmatrix}.
\tag{2.88}
$$

A rotation about an axis that does not pass through the origin causes a change in position, plus the same final orientation as if the axis had passed through the origin.

Note that we could have used any definition of $\{A'\}$ and $\{B'\}$ such that their origins were on the axis of rotation. Our particular choice of orientation was arbitrary, and our choice of the position of the origin was one of an infinity of possible choices lying along the axis of rotation. (See also Exercise 2.14.)

Euler parameters

Another representation of orientation is by means of four numbers called the **Euler parameters**. Although complete discussion is beyond the scope of the book, we state the convention here for reference.

In terms of the equivalent axis $\hat{K} = [k_x \ k_y \ k_z]^T$ and the equivalent angle θ, the Euler parameters are given by

$$\epsilon_1 = k_x \sin\frac{\theta}{2},$$

$$\epsilon_2 = k_y \sin\frac{\theta}{2}, \tag{2.89}$$

$$\epsilon_3 = k_z \sin\frac{\theta}{2},$$

$$\epsilon_4 = \cos\frac{\theta}{2}.$$

It is then clear that these four quantities are not independent:

$$\epsilon_1^2 + \epsilon_2^2 + \epsilon_3^2 + \epsilon_4^2 = 1 \tag{2.90}$$

must always hold. Hence, an orientation might be visualized as a point on a unit hypersphere in four-dimensional space.

Sometimes, the Euler parameters are viewed as a 3×1 vector plus a scalar. However, as a 4×1 vector, the Euler parameters are known as a **unit quaternion**.

The rotation matrix R_ϵ that is equivalent to a set of Euler parameters is

$$R_\epsilon = \begin{bmatrix} 1 - 2\epsilon_2^2 - 2\epsilon_3^2 & 2(\epsilon_1\epsilon_2 - \epsilon_3\epsilon_4) & 2(\epsilon_1\epsilon_3 + \epsilon_2\epsilon_4) \\ 2(\epsilon_1\epsilon_2 + \epsilon_3\epsilon_4) & 1 - 2\epsilon_1^2 - 2\epsilon_3^2 & 2(\epsilon_2\epsilon_3 - \epsilon_1\epsilon_4) \\ 2(\epsilon_1\epsilon_3 - \epsilon_2\epsilon_4) & 2(\epsilon_2\epsilon_3 + \epsilon_1\epsilon_4) & 1 - 2\epsilon_1^2 - 2\epsilon_2^2 \end{bmatrix}. \tag{2.91}$$

Given a rotation matrix, the equivalent Euler parameters are

$$\epsilon_1 = \frac{r_{32} - r_{23}}{4\epsilon_4},$$

$$\epsilon_2 = \frac{r_{13} - r_{31}}{4\epsilon_4}, \tag{2.92}$$

$$\epsilon_3 = \frac{r_{21} - r_{12}}{4\epsilon_4},$$

$$\epsilon_4 = \frac{1}{2}\sqrt{1 + r_{11} + r_{22} + r_{33}}.$$

Note that (2.92) is not useful in a computational sense if the rotation matrix represents a rotation of 180 degrees about some axis, because ϵ_4 goes to zero. However, it can be shown that, in the limit, all the expressions in (2.92) remain finite even for this case. In fact, from the definitions in (2.88), it is clear that all ϵ_i remain in the interval $[-1, 1]$.

Taught and predefined orientations

In many robot systems, it will be possible to "teach" positions and orientations by using the robot itself. The manipulator is moved to a desired location, and this position is recorded. A frame taught in this manner need not necessarily be one to which the robot will be commanded to return; it could be a part location or a fixture location. In other words, the robot is used as a measuring tool having six degrees of freedom. Teaching an orientation like this completely obviates the need for the human programmer to deal with orientation representation at all. In the computer, the taught point is stored as a rotation matrix (or however), but the user never has to see or understand it. Robot systems that allow teaching of frames by using the robot are thus highly recommended.

Besides teaching frames, some systems have a set of predefined orientations, such as "pointing down" or "pointing left." These specifications are very easy for humans to deal with. However, if this were the only means of describing and specifying orientation, the system would be very limited.

2.9   TRANSFORMATION OF FREE VECTORS

We have been concerned mostly with position vectors in this chapter. In later chapters, we will discuss velocity and force vectors as well. These vectors will transform differently because they are a different *type* of vector.

In mechanics, one makes a distinction between the equality and the equivalence of vectors. *Two vectors are equal if they have the same dimensions, magnitude, and direction.* Two vectors that are considered *equal* could have different lines of action—for example, the three equal vectors in Fig 2.21. These velocity vectors have the same dimensions, magnitude, and direction and so are equal according to our definition.

Two vectors are equivalent in a certain capacity if each produces the very same effect in this capacity. Thus, if the criterion in Fig. 2.21 is distance traveled, all three vectors give the same result and are thus equivalent in this capacity. If the criterion is height above the xy plane, then the vectors are not equivalent despite their equality. Thus, relationships between vectors and notions of equivalence *depend entirely on the situation at hand*. Furthermore, vectors that are not equal might cause equivalent effects in certain cases.

We will define two basic classes of vector quantities that might be helpful.

The term **line vector** refers to a vector that is dependent on its **line of action**, along with direction and magnitude, for causing its effects. Often, the effects of a force vector depend upon its line of action (or point of application), so it would then be considered a line vector.

A **free vector** refers to a vector that may be positioned anywhere in space without loss or change of meaning, provided that magnitude and direction are preserved.

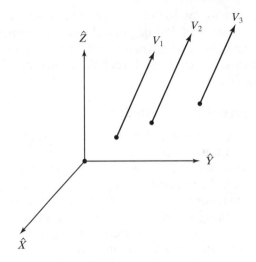

FIGURE 2.21: Equal velocity vectors.

For example, a pure moment vector is always a free vector. If we have a moment vector ^{B}N that is known in terms of $\{B\}$, then we calculate the same moment in terms of frame $\{A\}$ as

$$^{A}N = {}_{B}^{A}R \, {}^{B}N. \tag{2.93}$$

In other words, all that counts is the magnitude and direction (in the case of a free vector), so only the rotation matrix relating the two systems is used in transforming. The relative locations of the origins do not enter into the calculation.

Likewise, a velocity vector written in $\{B\}$, ^{B}V, is written in $\{A\}$ as

$$^{A}V = {}_{B}^{A}R \, {}^{B}V. \tag{2.94}$$

The velocity of a point is a free vector, so all that is important is its direction and magnitude. The operation of rotation (as in (2.94)) does not affect the magnitude, yet accomplishes the rotation that changes the description of the vector from $\{B\}$ to $\{A\}$. Note that $^{A}P_{BORG}$, which would appear in a position-vector transformation, does not appear in a velocity transform. For example, in Fig. 2.22, if $^{B}V = 5\hat{X}$, then $^{A}V = 5\hat{Y}$.

Velocity vectors and force and moment vectors will be introduced more fully in Chapter 5.

2.10 COMPUTATIONAL CONSIDERATIONS

The availability of inexpensive computing power is largely responsible for the growth of the robotics industry; yet, for some time to come, efficient computation will remain an important issue in the design of a manipulation system.

The homogeneous representation is useful as a conceptual entity, but transformation software typically used in industrial manipulation systems does not make use of it directly, because the time spent multiplying by zeros and ones is wasteful.

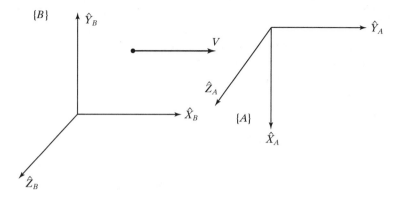

FIGURE 2.22: Transforming velocities.

Usually, the computations shown in (2.41) and (2.45) are performed, rather than the direct multiplication or inversion of 4×4 matrices.

The *order* in which transformations are applied can make a large difference in the amount of computation required to compute the same quantity. Consider performing multiple rotations of a vector, as in

$$^{A}P = {}^{A}_{B}R \, {}^{B}_{C}R \, {}^{C}_{D}R \, {}^{D}P. \qquad (2.95)$$

One choice is to first multiply the three rotation matrices together, to form ${}^{A}_{D}R$ in the expression

$$^{A}P = {}^{A}_{D}R \, {}^{D}P. \qquad (2.96)$$

Forming ${}^{A}_{D}R$ from its three constituents requires 54 multiplications and 36 additions. Performing the final matrix-vector multiplication of (2.96) requires an additional 9 multiplications and 6 additions, bringing the totals to 63 multiplications and 42 additions.

If, instead, we transform the vector through the matrices one at a time, that is,

$$^{A}P = {}^{A}_{B}R \, {}^{B}_{C}R \, {}^{C}_{D}R \, {}^{D}P$$
$$^{A}P = {}^{A}_{B}R \, {}^{B}_{C}R \, {}^{C}P \qquad (2.97)$$
$$^{A}P = {}^{A}_{B}R \, {}^{B}P$$
$$^{A}P = {}^{A}P,$$

then the total computation requires only 27 multiplications and 18 additions, fewer than half the computations required by the other method.

Of course, in some cases, the relationships ${}^{A}_{B}R$, ${}^{B}_{C}R$, and ${}^{C}_{D}R$ are constant, while there are many ${}^{D}P_i$ that need to be transformed into ${}^{A}P_i$. In such a case, it is more efficient to calculate ${}^{A}_{D}R$ once, and then use it for all future mappings. See also Exercise 2.16.

EXAMPLE 2.10

Give a method of computing the product of two rotation matrices, $^A_B R \, ^B_C R$, that uses fewer than 27 multiplications and 18 additions.

Where $\hat{L}_i$ are the columns of $^B_C R$ and $\hat{C}_i$ are the three columns of the result, compute

$$
\hat{C}_1 = {^A_B R} \hat{L}_i,
$$
$$
\hat{C}_2 = {^A_B R} \hat{L}_2, \tag{2.98}
$$
$$
\hat{C}_3 = \hat{C}_1 \times \hat{C}_2,
$$

which requires 24 multiplications and 15 additions.

BIBLIOGRAPHY

[1] B. Noble, *Applied Linear Algebra*, Prentice-Hall, Englewood Cliffs, NJ, 1969.

[2] D. Ballard and C. Brown, *Computer Vision*, Prentice-Hall, Englewood Cliffs, NJ, 1982.

[3] O. Bottema and B. Roth, *Theoretical Kinematics*, North Holland, Amsterdam, 1979.

[4] R.P. Paul, *Robot Manipulators*, MIT Press, Cambridge, MA, 1981.

[5] I. Shames, *Engineering Mechanics*, 2nd edition, Prentice-Hall, Englewood Cliffs, NJ, 1967.

[6] Symon, *Mechanics*, 3rd edition, Addison-Wesley, Reading, MA, 1971.

[7] B. Gorla and M. Renaud, *Robots Manipulateurs*, Cepadues-Editions, Toulouse, 1984.

EXERCISES

2.1 [15] A vector $^A P$ is rotated about $\hat{Z}_A$ by θ degrees and is subsequently rotated about $\hat{X}_A$ by ϕ degrees. Give the rotation matrix that accomplishes these rotations in the given order.

2.2 [15] A vector $^A P$ is rotated about $\hat{Y}_A$ by 30 degrees and is subsequently rotated about $\hat{X}_A$ by 45 degrees. Give the rotation matrix that accomplishes these rotations in the given order.

2.3 [16] A frame $\{B\}$ is located initially coincident with a frame $\{A\}$. We rotate $\{B\}$ about $\hat{Z}_B$ by θ degrees, and then we rotate the resulting frame about $\hat{X}_B$ by ϕ degrees. Give the rotation matrix that will change the descriptions of vectors from $^B P$ to $^A P$.

2.4 [16] A frame $\{B\}$ is located initially coincident with a frame $\{A\}$. We rotate $\{B\}$ about $\hat{Z}_B$ by 30 degrees, and then we rotate the resulting frame about $\hat{X}_B$ by 45 degrees. Give the rotation matrix that will change the description of vectors from $^B P$ to $^A P$.

2.5 [13] $^A_B R$ is a 3×3 matrix with eigenvalues 1, e^{+ai}, and e^{-ai}, where $i = \sqrt{-1}$. What is the physical meaning of the eigenvector of $^A_B R$ associated with the eigenvalue 1?

2.6 [21] Derive equation (2.80).

2.7 [24] Describe (or program) an algorithm that extracts the equivalent angle and axis of a rotation matrix. Equation (2.82) is a good start, but make sure that your algorithm handles the special cases $\theta = 0°$ and $\theta = 180°$.

2.8 [29] Write a subroutine that changes representation of orientation from rotation-matrix form to equivalent angle–axis form. A Pascal-style procedure declaration would begin

```
Procedure RMTOAA (VAR R:mat33; VAR K:vec3; VAR theta: real);
```

Write another subroutine that changes from equivalent angle–axis representation to rotation-matrix representation:

```
Procedure AATORM(VAR K:vec3; VAR theta: real: VAR R:mat33);
```

Write the routines in C if you prefer.

Run these procedures on several cases of test data back-to-back and verify that you get back what you put in. Include some of the difficult cases!

2.9 [27] Do Exercise 2.8 for roll, pitch, yaw angles about fixed axes.

2.10 [27] Do Exercise 2.8 for Z–Y–Z Euler angles.

2.11 [10] Under what condition do two rotation matrices representing finite rotations commute? A proof is not required.

2.12 [14] A velocity vector is given by

$$
{}^B V = \begin{bmatrix} 10.0 \\ 20.0 \\ 30.0 \end{bmatrix}.
$$

Given

$$
{}^A_B T = \begin{bmatrix} 0.866 & -0.500 & 0.000 & 11.0 \\ 0.500 & 0.866 & 0.000 & -3.0 \\ 0.000 & 0.000 & 1.000 & 9.0 \\ 0 & 0 & 0 & 1 \end{bmatrix},
$$

compute ${}^A V$.

2.13 [21] The following frame definitions are given as known:

$$
{}^U_A T = \begin{bmatrix} 0.866 & -0.500 & 0.000 & 11.0 \\ 0.500 & 0.866 & 0.000 & -1.0 \\ 0.000 & 0.000 & 1.000 & 8.0 \\ 0 & 0 & 0 & 1 \end{bmatrix},
$$

$$
{}^B_A T = \begin{bmatrix} 1.000 & 0.000 & 0.000 & 0.0 \\ 0.000 & 0.866 & -0.500 & 10.0 \\ 0.000 & 0.500 & 0.866 & -20.0 \\ 0 & 0 & 0 & 1 \end{bmatrix},
$$

$$
{}^C_U T = \begin{bmatrix} 0.866 & -0.500 & 0.000 & -3.0 \\ 0.433 & 0.750 & -0.500 & -3.0 \\ 0.250 & 0.433 & 0.866 & 3.0 \\ 0 & 0 & 0 & 1 \end{bmatrix}.
$$

Draw a frame diagram (like that of Fig. 2.15) to show their arrangement qualitatively, and solve for ${}^B_C T$.

2.14 [31] Develop a general formula to obtain ${}^A_B T$, where, starting from initial coincidence, $\{B\}$ is rotated by θ about $\hat{K}$ where $\hat{K}$ passes through the point ${}^A P$ (not through the origin of $\{A\}$ in general).

2.15 [34] $\{A\}$ and $\{B\}$ are frames differing only in orientation. $\{B\}$ is attained as follows: starting coincident with $\{A\}$, $\{B\}$ is rotated by θ radians about unit vector $\hat{K}$—that is,

$$
{}^A_B R = {}^A_B R_K(\theta).
$$

Show that

$$^A_B R = e^{k\theta},$$

where

$$K = \begin{bmatrix} 0 & -k_x & k_y \\ k_z & 0 & -k_x \\ -k_y & k_x & 0 \end{bmatrix}.$$

2.16 [22] A vector must be mapped through three rotation matrices:

$$^A P = {}^A_B R\, {}^B_C R\, {}^C_D R\, {}^D P.$$

One choice is to first multiply the three rotation matrices together, to form $^A_D R$ in the expression

$$^A P = {}^A_D R\, {}^D P.$$

Another choice is to transform the vector through the matrices one at a time—that is,

$$^A P = {}^A_B R\, {}^B_C R\, {}^C_D R\, {}^D P,$$

$$^A P = {}^A_B R\, {}^B_C R\, {}^C P,$$

$$^A P = {}^A_B R\, {}^B P,$$

$$^A P = {}^A P.$$

If $^D P$ is changing at 100 Hz, we would have to recalculate $^A P$ at the same rate. However, the three rotation matrices are also changing, as reported by a vision system that gives us new values for $^A_B R$, $^B_C R$, and $^C_D R$ at 30 Hz. What is the best way to organize the computation to minimize the calculation effort (multiplications and additions)?

2.17 [16] Another familiar set of three coordinates that can be used to describe a point in space is cylindrical coordinates. The three coordinates are defined as illustrated in Fig. 2.23. The coordinate θ gives a direction in the xy plane along which to translate radially by an amount r. Finally, z is given to specify the height above the xy plane. Compute the Cartesian coordinates of the point $^A P$ in terms of the cylindrical coordinates θ, r, and z.

2.18 [18] Another set of three coordinates that can be used to describe a point in space is spherical coordinates. The three coordinates are defined as illustrated in Fig. 2.24. The angles α and β can be thought of as describing azimuth and elevation of a ray projecting into space. The third coordinate, r, is the radial distance along that ray to the point being described. Calculate the Cartesian coordinates of the point $^A P$ in terms of the spherical coordinates α, β, and r.

2.19 [24] An object is rotated about its $\hat{X}$ axis by an amount ϕ, and then it is rotated about its new $\hat{Y}$ axis by an amount ψ. From our study of Euler angles, we know that the resulting orientation is given by

$$R_x(\phi) R_y(\psi),$$

whereas, if the two rotations had occurred about axes of the fixed reference frame, the result would have been

$$R_y(\psi) R_x(\phi).$$

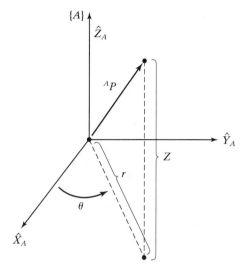

FIGURE 2.23: Cylindrical coordinates.

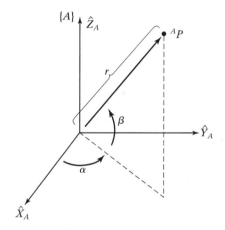

FIGURE 2.24: Spherical coordinates.

It appears that the order of multiplication depends upon whether rotations are described relative to fixed axes or those of the frame being moved. It is more appropriate, however, to realize that, in the case of specifying a rotation about an axis of the frame being moved, we are specifying a rotation in the fixed system given by (for this example)

$$R_x(\phi)R_y(\psi)R_x^{-1}(\phi).$$

This *similarity transform* [1], multiplying the original $R_x(\phi)$ on the left, reduces to the resulting expression in which *it looks as if* the order of matrix multiplication has been reversed. Taking this viewpoint, give a derivation for the form of the rotation matrix that is equivalent to the Z–Y–Z Euler-angle set (α, β, γ). (The result is given by (2.72).)

2.20 [20] Imagine rotating a vector Q about a vector $\hat{K}$ by an amount θ to form a new vector, Q'—that is,

$$Q' = R_K(\theta)Q.$$

Use (2.80) to derive **Rodriques's formula,**

$$Q' = Q\cos\theta + \sin\theta(\hat{K} \times Q) + (1 - \cos\theta)(\hat{K} \cdot \hat{Q})\hat{K}.$$

2.21 [15] For rotations sufficiently small that the approximations $\sin\theta = \theta$, $\cos\theta = 1$, and $\theta^2 = 0$ hold, derive the rotation-matrix equivalent to a rotation of θ about a general axis, $\hat{K}$. Start with (2.80) for your derivation.

2.22 [20] Using the result from Exercise 2.21, show that two infinitesimal rotations commute (i.e., the order in which the rotations are performed is not important).

2.23 [25] Give an algorithm to construct the definition of a frame $^U_A T$ from three points $^U P_1$, $^U P_2$, and $^U P_3$, where the following is known about these points:

    1. $^U P_1$ is at the origin of $\{A\}$;

    2. $^U P_2$ lies somewhere on the positive $\hat{X}$ axis of $\{A\}$;

    3. $^U P_3$ lies near the positive $\hat{Y}$ axis in the XY plane of $\{A\}$.

2.24 [45] Prove Cayley's formula for proper orthonormal matrices.

2.25 [30] Show that the eigenvalues of a rotation matrix are 1, $e^{\alpha i}$, and $e^{-\alpha i}$, where $i = \sqrt{-1}$.

2.26 [33] Prove that any Euler-angle set is sufficient to express all possible rotation matrices.

2.27 [15] Referring to Fig. 2.25, give the value of $^A_B T$.

2.28 [15] Referring to Fig. 2.25, give the value of $^A_C T$.

2.29 [15] Referring to Fig. 2.25, give the value of $^B_C T$.

2.30 [15] Referring to Fig. 2.25, give the value of $^C_A T$.

2.31 [15] Referring to Fig. 2.26, give the value of $^A_B T$.

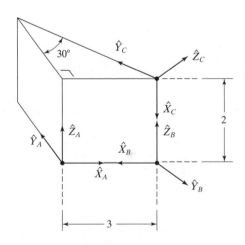

FIGURE 2.25: Frames at the corners of a wedge.

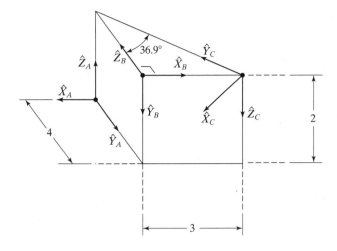

FIGURE 2.26: Frames at the corners of a wedge.

2.32 [15] Referring to Fig. 2.26, give the value of ^A_CT.

2.33 [15] Referring to Fig. 2.26, give the value of ^B_CT.

2.34 [15] Referring to Fig. 2.26, give the value of ^C_AT.

2.35 [20] Prove that the determinant of any rotation matrix is always equal to 1.

2.36 [36] A rigid body moving in a plane (i.e., in 2-space) has three degrees of freedom. A rigid body moving in 3-space has six degrees of freedom. Show that a body in N-space has $\frac{1}{2}(N^2 + N)$ degrees of freedom.

2.37 [15] Given

$$^A_BT = \begin{bmatrix} 0.25 & 0.43 & 0.86 & 5.0 \\ 0.87 & -0.50 & 0.00 & -4.0 \\ 0.43 & 0.75 & -0.50 & 3.0 \\ 0 & 0 & 0 & 1 \end{bmatrix},$$

what is the (2,4) element of ^B_AT?

2.38 [25] Imagine two unit vectors, v_1 and v_2, embedded in a rigid body. Note that, no matter how the body is rotated, the geometric angle between these two vectors is preserved (i.e., rigid-body rotation is an "angle-preserving" operation). Use this fact to give a concise (four- or five-line) proof that the inverse of a rotation matrix must equal its transpose and that a rotation matrix is orthonormal.

2.39 [37] Give an algorithm (perhaps in the form of a C program) that computes the unit quaternion corresponding to a given rotation matrix. Use (2.91) as starting point.

2.40 [33] Give an algorithm (perhaps in the form of a C program) that computes the Z–X–Z Euler angles corresponding to a given rotation matrix. See Appendix B.

2.41 [33] Give an algorithm (perhaps in the form of a C program) that computes the X–Y–X fixed angles corresponding to a given rotation matrix. See Appendix B.

PROGRAMMING EXERCISE (PART 2)

1. If your function library does not include an Atan2 function subroutine, write one.

2. To make a friendly user interface, we wish to describe orientations in the planar world by a single angle, θ, instead of by a 2×2 rotation matrix. The user will always communicate in terms of angle θ, but internally we will need the rotation-matrix form. For the position-vector part of a frame, the user will specify an x and a y value. So, we want to allow the user to specify a *frame* as a 3-tuple: (x, y, θ). Internally, we wish to use a 2×1 position vector and a 2×2 rotation matrix, so we need conversion routines. Write a subroutine whose Pascal definition would begin

```
Procedure UTOI (VAR uform: vec3; VAR iform: frame);
```

where "UTOI" stands for "User form TO Internal form." The first argument is the 3-tuple (x, y, θ), and the second argument is of type "frame," consists of a (2×1) position vector and a (2×2) rotation matrix. If you wish, you may represent the frame with a (3×3) homogeneous transform in which the third row is $[0\ 0\ 1]$. The inverse routine will also be necessary:

```
Procedure ITOU (VAR iform: frame; VAR uform: vec3);
```

3. Write a subroutine to multiply two transforms together. Use the following procedure heading:

```
Procedure TMULT (VAR brela, crelb, crela: frame);
```

The first two arguments are inputs, and the third is an output. Note that the names of the arguments document what the program does (brela = $^A_B T$).

4. Write a subroutine to invert a transform. Use the following procedure heading:

```
Procedure TINVERT (VAR brela, arelb: frame);
```

The first argument is the input, the second the output. Note that the names of the arguments document what the program does (brela = $^A_B T$).

5. The following frame definitions are given as known:

$$^U_A T = [x\ y\ \theta] = [11.0\ -1.0\ 30.0],$$

$$^B_A T = [x\ y\ \theta] = [0.0\ 7.0\ 45.0],$$

$$^C_U T = [x\ y\ \theta] = [-3.0\ -3.0\ -30.0].$$

These frames are input in the user representation $[x, y, \theta]$ (where θ is in degrees). Draw a frame diagram (like Fig. 2.15, only in 2-D) that qualitatively shows their arrangement. Write a program that calls TMULT and TINVERT (defined in programming exercises 3 and 4) as many times as needed to solve for $^B_C T$. Then print out $^B_C T$ in both internal and user representation.

MATLAB EXERCISE 2A

a) Using the $Z-Y-X$ ($\alpha-\beta-\gamma$) Euler angle convention, write a MATLAB program to calculate the rotation matrix $^A_B R$ when the user enters the Euler angles $\alpha-\beta-\gamma$. Test for two examples:

i) $\alpha = 10°,\quad \beta = 20°,\quad \gamma = 30°.$
ii) $\alpha = 30°,\quad \beta = 90°,\quad \gamma = -55°.$

For case (i), demonstrate the six constraints for unitary orthonormal rotation matrices (i.e., there are nine numbers in a 3×3 matrix, but only three are independent). Also, demonstrate the *beautiful* property, $^B_A R = {^A_B R}^{-1} = {^A_B R}^T$, for case i.

b) Write a MATLAB program to calculate the Euler angles $\alpha-\beta-\gamma$ when the user enters the rotation matrix ^A_BR (the inverse problem). Calculate both possible solutions. Demonstrate this inverse solution for the two cases from part (a). Use a circular check to verify your results (i.e., enter Euler angles in code a from part (a); take the resulting rotation matrix ^A_BR, and use this as the input to code b; you get two sets of answers—one should be the original user input, and the second can be verified by once again using the code in part (a).

c) For a simple rotation of β about the Y axis only, for $\beta = 20°$ and $^BP = \{1\ 0\ 1\ \}^T$, calculate ^AP; demonstrate with a sketch that your results are correct.

d) Check all results, by means of the Corke MATLAB Robotics Toolbox. Try the functions *rpy2tr()*, *tr2rpy()*, *rotx()*, *roty()*, and *rotz()*.

MATLAB EXERCISE 2B

a) Write a MATLAB program to calculate the homogeneous transformation matrix ^A_BT when the user enters $Z-Y-X$ Euler angles $\alpha - \beta - \gamma$ and the position vector ^AP_B. Test for two examples:

  i) $\alpha = 10°$, $\beta = 20°$, $\gamma = 30°$, and $^AP_B = \{1\ 2\ 3\ \}^T$.
  ii) For $\beta = 20°$ $(\alpha = \gamma = 0°)$, $^AP_B = \{3\ 0\ 1\ \}^T$.

b) For $\beta = 20°$ $(\alpha = \gamma = 0°)$, $^AP_B = \{3\ 0\ 1\}^T$, and $^BP = \{1\ 0\ 1\}^T$, use MATLAB to calculate ^AP; demonstrate with a sketch that your results are correct. Also, using the same numbers, demonstrate all three interpretations of the homogeneous transformation matrix—the (b) assignment is the second interpretation, transform mapping.

c) Write a MATLAB program to calculate the inverse homogeneous transformation matrix $^A_BT^{-1} = {}^B_AT$, using the symbolic formula. Compare your result with a numerical MATLAB function (e.g., *inv*). Demonstrate that both methods yield correct results (i.e., $^A_BT\ {}^A_BT^{-1} = {}^A_BT^{-1}\ {}^A_BT = I_4$). Demonstrate this for examples (i) and (ii) from (a) above.

d) Define ^A_BT to be the result from (a)(i) and ^B_CT to be the result from (a)(ii).

  i) Calculate ^A_CT, and show the relationship via a transform graph. Do the same for ^C_AT.
  ii) Given ^A_CT and ^B_CT from (d)(i)—assume you don't know ^A_BT, calculate it, and compare your result with the answer you know.
  iii) Given ^A_CT and ^A_BT from (d)(i)—assume you don't know ^B_CT, calculate it, and compare your result with the answer you know.

e) Check all results by means of the Corke MATLAB Robotics Toolbox. Try functions *rpy2tr()* and *transl()*.

CHAPTER 3

Manipulator kinematics

3.1 INTRODUCTION

Kinematics is the science of motion that treats the subject without regard to the forces that cause it. Within the science of kinematics, one studies the position, the velocity, the acceleration, and all higher order derivatives of the position variables (with respect to time or any other variable(s)). Hence, the study of the kinematics of manipulators refers to all the geometrical and time-based properties of the motion. The relationships between these motions and the forces and torques that cause them constitute the problem of dynamics, which is the subject of Chapter 6.

In this chapter, we consider position and orientation of the manipulator linkages in static situations. In Chapters 5 and 6, we will consider the kinematics when velocities and accelerations are involved.

In order to deal with the complex geometry of a manipulator, we will affix frames to the various parts of the mechanism and then describe the relationships between these frames. The study of manipulator kinematics involves, among other things, how the locations of these frames change as the mechanism articulates. The central topic of this chapter is a method to compute the position and orientation of the manipulator's end-effector relative to the base of the manipulator as a function of the joint variables.

3.2 LINK DESCRIPTION

A manipulator may be thought of as a set of bodies connected in a chain by joints. These bodies are called links. Joints form a connection between a neighboring pair of links. The term **lower pair** is used to describe the connection between a pair of

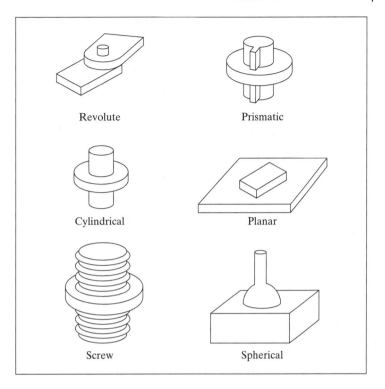

Revolute

Prismatic

Cylindrical

Planar

Screw

Spherical

FIGURE 3.1: The six possible lower-pair joints.

bodies when the relative motion is characterized by two surfaces sliding over one another. Figure 3.1 shows the six possible lower pair joints.

Mechanical-design considerations favor manipulators' generally being constructed from joints that exhibit just one degree of freedom. Most manipulators have **revolute joints** or have sliding joints called **prismatic joints**. In the rare case that a mechanism is built with a joint having n degrees of freedom, it can be modeled as n joints of one degree of freedom connected with $n-1$ links of zero length. Therefore, without loss of generality, we will consider only manipulators that have joints with a single degree of freedom.

The links are numbered starting from the immobile base of the arm, which might be called link 0. The first moving body is link 1, and so on, out to the free end of the arm, which is link n. In order to position an end-effector generally in 3-space, a minimum of six joints is required.[1] Typical manipulators have five or six joints. Some robots are not actually as simple as a single kinematic chain—these have parallelogram linkages or other closed kinematic structures. We will consider one such manipulator later in this chapter.

A single link of a typical robot has many attributes that a mechanical designer had to consider during its design: the type of material used, the strength and stiffness

---

[1]This makes good intuitive sense, because the description of an object in space requires six parameters—three for position and three for orientation.

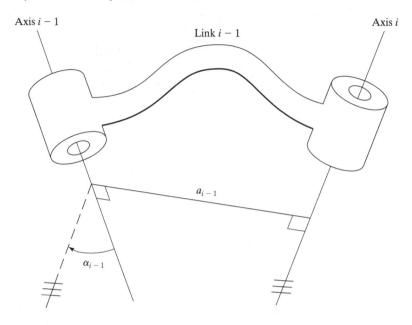

FIGURE 3.2: The kinematic function of a link is to maintain a fixed relationship between the two joint axes it supports. This relationship can be described with two parameters: the link length, a, and the link twist, α.

of the link, the location and type of the joint bearings, the external shape, the weight and inertia, and more. However, for the purposes of obtaining the kinematic equations of the mechanism, *a link is considered only as a rigid body that defines the relationship between two neighboring joint axes of a manipulator.* Joint axes are defined by lines in space. Joint axis i is defined by a line in space, or a vector direction, about which link i rotates relative to link $i - 1$. It turns out that, for kinematic purposes, a link can be specified with two numbers, which define the relative location of the two axes in space.

For any two axes in 3-space, there exists a well-defined measure of distance between them. This distance is measured along a line that is mutually perpendicular to both axes. This mutual perpendicular always exists; it is unique except when both axes are parallel, in which case there are many mutual perpendiculars of equal length. Figure 3.2 shows link $i - 1$ and the mutually perpendicular line along which the **link length**, a_{i-1}, is measured. Another way to visualize the link parameter a_{i-1} is to imagine an expanding cylinder whose axis is the joint $i - 1$ axis—when it just touches joint axis i, the radius of the cylinder is equal to a_{i-1}.

The second parameter needed to define the relative location of the two axes is called the **link twist**. If we imagine a plane whose normal is the mutually perpendicular line just constructed, we can project the axes $i - 1$ and i onto this plane and measure the angle between them. This angle is measured from axis $i - 1$ to axis i in the right-hand sense about a_{i-1}.[2] We will use this definition of the twist

[2] In this case, a_{i-1} is given the direction pointing from axis $i - 1$ to axis i.

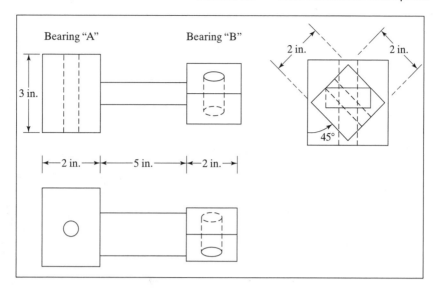

FIGURE 3.3: A simple link that supports two revolute axes.

of link $i - 1$, α_{i-1}. In Fig. 3.2, α_{i-1} is indicated as the angle between axis $i - 1$ and axis i. (The lines with the triple hash marks are parallel.) In the case of intersecting axes, twist is measured in the plane containing both axes, but the sense of α_{i-1} is lost. In this special case, one is free to assign the sign of α_{i-1} arbitrarily.

You should convince yourself that these two parameters, length and twist, as defined above, can be used to define the relationship between any two lines (in this case axes) in space.

EXAMPLE 3.1

Figure 3.3 shows the mechanical drawings of a robot link. If this link is used in a robot, with bearing "A" used for the lower-numbered joint, give the length and twist of this link. Assume that holes are centered in each bearing.

By inspection, the common perpendicular lies right down the middle of the metal bar connecting the bearings, so the link length is 7 inches. The end view actually shows a projection of the bearings onto the plane whose normal is the mutual perpendicular. Link twist is measured in the right-hand sense about the common perpendicular from axis $i - 1$ to axis i, so, in this example, it is clearly +45 degrees.

3.3   LINK-CONNECTION DESCRIPTION

The problem of connecting the links of a robot together is again one filled with many questions for the mechanical designer to resolve. These include the strength of the joint, its lubrication, and the bearing and gearing mounting. However, for the investigation of kinematics, we need only worry about two quantities, which will completely specify the way in which links are connected together.

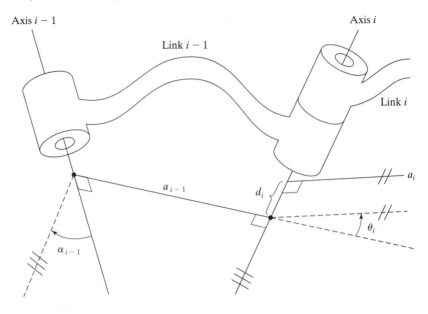

FIGURE 3.4: The link offset, d, and the joint angle, θ, are two parameters that may be used to describe the nature of the connection between neighboring links.

Intermediate links in the chain

Neighboring links have a common joint axis between them. One parameter of interconnection has to do with the distance along this common axis from one link to the next. This parameter is called the **link offset**. The offset at joint axis i is called d_i. The second parameter describes the amount of rotation about this common axis between one link and its neighbor. This is called the **joint angle**, θ_i.

Figure 3.4 shows the interconnection of link $i - 1$ and link i. Recall that a_{i-1} is the mutual perpendicular between the two axes of link $i - 1$. Likewise, a_i is the mutual perpendicular defined for link i. The first parameter of interconnection is the link offset, d_i, which is the signed distance measured along the axis of joint i from the point where a_{i-1} intersects the axis to the point where a_i intersects the axis. The offset d_i is indicated in Fig. 3.4. The link offset d_i is variable if joint i is prismatic. The second parameter of interconnection is the angle made between an extension of a_{i-1} and a_i measured about the axis of joint i. This is indicated in Fig. 3.4, where the lines with the double hash marks are parallel. This parameter is named θ_i and is variable for a revolute joint.

First and last links in the chain

Link length, a_i, and link twist, α_i, depend on joint axes i and $i+1$. Hence, a_1 through a_{n-1} and α_1 through α_{n-1} are defined as was discussed in this section. At the ends of the chain, it will be our convention to assign zero to these quantities. That is, $a_0 = a_n = 0.0$ and $\alpha_0 = \alpha_n = 0.0$.[3] Link offset, d_i, and joint angle, θ_i, are well defined

---

[3]In fact, a_n and α_n do not need to be defined at all.

for joints 2 through $n - 1$ according to the conventions discussed in this section. If joint 1 is revolute, the zero position for θ_1 may be chosen arbitrarily; $d_1 = 0.0$ will be our convention. Similarly, if joint 1 is prismatic, the zero position of d_1 may be chosen arbitrarily; $\theta_1 = 0.0$ will be our convention. Exactly the same statements apply to joint n.

These conventions have been chosen so that, in a case where a quantity could be assigned arbitrarily, a zero value is assigned so that later calculations will be as simple as possible.

Link parameters

Hence, any robot can be described kinematically by giving the values of four quantities for each link. Two describe the link itself, and two describe the link's connection to a neighboring link. In the usual case of a revolute joint, θ_i is called the **joint variable**, and the other three quantities would be fixed **link parameters**. For prismatic joints, d_i is the joint variable, and the other three quantities are fixed link parameters. The definition of mechanisms by means of these quantities is a convention usually called the **Denavit–Hartenberg notation** [1].[4] Other methods of describing mechanisms are available, but are not presented here.

At this point, we could inspect any mechanism and determine the Denavit–Hartenberg parameters that describe it. For a six-jointed robot, 18 numbers would be required to describe the fixed portion of its kinematics completely. In the case of a six-jointed robot with all revolute joints, the 18 numbers are in the form of six sets of $(\alpha_i, \alpha_i, d_i)$.

EXAMPLE 3.2

Two links, as described in Fig. 3.3, are connected as links 1 and 2 of a robot. Joint 2 is composed of a "B" bearing of link 1 and an "A" bearing of link 2, arranged so that the flat surfaces of the "A" and "B" bearings lie flush against each other. What is d_2?

The link offset d_2 is the offset at joint 2, which is the distance, measured along the joint 2 axis, between the mutual perpendicular of link 1 and that of link 2. From the drawings in Fig. 3.3, this is 2.5 inches.

Before introducing more examples, we will define a convention for attaching a frame to each link of the manipulator.

3.4   CONVENTION FOR AFFIXING FRAMES TO LINKS

In order to describe the location of each link relative to its neighbors, we define a frame attached to each link. The link frames are named by number according to the link to which they are attached. That is, frame $\{i\}$ is attached rigidly to link i.

---

[4]Note that many related conventions go by the name Denavit–Hartenberg, but differ in a few details. For example, the version used in this book differs from some of the robotic literature in the manner of frame numbering. Unlike some other conventions, in this book frame $\{i\}$ is attached to link i and has its origin lying on joint axis i.

Intermediate links in the chain

The convention we will use to locate frames on the links is as follows: The $\hat{Z}$-axis of frame $\{i\}$, called $\hat{Z}_i$, is coincident with the joint axis i. The origin of frame $\{i\}$ is located where the a_i perpendicular intersects the joint i axis. $\hat{X}_i$ points along a_i in the direction from joint i to joint $i + 1$.

In the case of $a_i = 0$, $\hat{X}_i$ is normal to the plane of $\hat{Z}_i$ and $\hat{Z}_{i+1}$. We define α_i as being measured in the right-hand sense about $\hat{X}_i$, and so we see that the freedom of choosing the sign of α_i in this case corresponds to two choices for the direction of $\hat{X}_i$. $\hat{Y}_i$ is formed by the right-hand rule to complete the ith frame. Figure 3.5 shows the location of frames $\{i - 1\}$ and $\{i\}$ for a general manipulator.

First and last links in the chain

We attach a frame to the base of the robot, or link 0, called frame $\{0\}$. This frame does not move; for the problem of arm kinematics, it can be considered the reference frame. We may describe the position of all other link frames in terms of this frame.

Frame $\{0\}$ is arbitrary, so it always simplifies matters to choose $\hat{Z}_0$ along axis 1 and to locate frame $\{0\}$ so that it coincides with frame $\{1\}$ when joint variable 1 is zero. Using this convention, we will always have $a_0 = 0.0$, $\alpha_0 = 0.0$. Additionally, this ensures that $d_1 = 0.0$ if joint 1 is revolute, or $\theta_1 = 0.0$ if joint 1 is prismatic.

For joint n revolute, the direction of $\hat{X}_N$ is chosen so that it aligns with $\hat{X}_{N-1}$ when $\theta_n = 0.0$, and the origin of frame $\{N\}$ is chosen so that $d_n = 0.0$. For joint n prismatic, the direction of $\hat{X}_N$ is chosen so that $\theta_n = 0.0$, and the origin of frame $\{N\}$ is chosen at the intersection of $\hat{X}_{N-1}$ and joint axis n when $d_n = 0.0$.

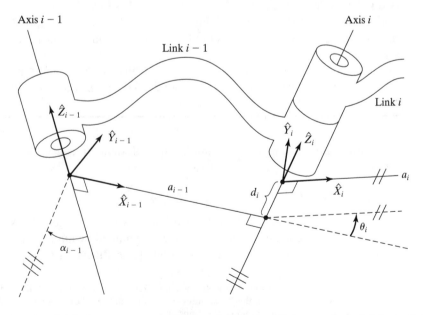

FIGURE 3.5: Link frames are attached so that frame $\{i\}$ is attached rigidly to link i.

Summary of the link parameters in terms of the link frames

If the link frames have been attached to the links according to our convention, the following definitions of the link parameters are valid:

$$a_i = the\ distance\ from\ \hat{Z}_i\ to\ \hat{Z}_{i+1}\ measured\ along\ \hat{X}_i;$$

$$\alpha_i = the\ angle\ from\ \hat{Z}_i\ to\ \hat{Z}_{i+1}\ measured\ about\ \hat{X}_i;$$

$$d_i = the\ distance\ from\ \hat{X}_{i-1}\ to\ \hat{X}_i\ measured\ along\ \hat{Z}_i;\ and$$

$$\theta_i = the\ angle\ from\ \hat{X}_{i-1}\ to\ \hat{X}_i\ measured\ about\ \hat{Z}_i.$$

We usually choose $a_i > 0$, because it corresponds to a distance; however, α_i, d_i, and θ_i are signed quantities.

A final note on uniqueness is warranted. The convention outlined above does not result in a unique attachment of frames to links. First of all, when we first align the $\hat{Z}_i$ axis with joint axis i, there are two choices of direction in which to point $\hat{Z}_i$. Furthermore, in the case of intersecting joint axes (i.e., $a_i = 0$), there are two choices for the direction of $\hat{X}_i$, corresponding to the choice of signs for the normal to the plane containing $\hat{Z}_i$ and $\hat{Z}_{i+1}$. When axes i and $i + 1$ are parallel, the choice of origin location for $\{i\}$ is arbitrary (though generally chosen in order to cause d_i to be zero). Also, when prismatic joints are present, there is quite a bit of freedom in frame assignment. (See also Example 3.5.)

Summary of link-frame attachment procedure

The following is a summary of the procedure to follow when faced with a new mechanism, in order to properly attach the link frames:

1. Identify the joint axes and imagine (or draw) infinite lines along them. For steps 2 through 5 below, consider two of these neighboring lines (at axes i and $i + 1$).
2. Identify the common perpendicular between them, or point of intersection. At the point of intersection, or at the point where the common perpendicular meets the ith axis, assign the link-frame origin.
3. Assign the $\hat{Z}_i$ axis pointing along the ith joint axis.
4. Assign the $\hat{X}_i$ axis pointing along the common perpendicular, or, if the axes intersect, assign $\hat{X}_i$ to be normal to the plane containing the two axes.
5. Assign the $\hat{Y}_i$ axis to complete a right-hand coordinate system.
6. Assign $\{0\}$ to match $\{1\}$ when the first joint variable is zero. For $\{N\}$, choose an origin location and $\hat{X}_N$ direction freely, but generally so as to cause as many linkage parameters as possible to become zero.

EXAMPLE 3.3

Figure 3.6(a) shows a three-link planar arm. Because all three joints are revolute, this manipulator is sometimes called an **RRR** (or **3R**) **mechanism**. Fig. 3.6(b) is a schematic representation of the same manipulator. Note the double hash marks

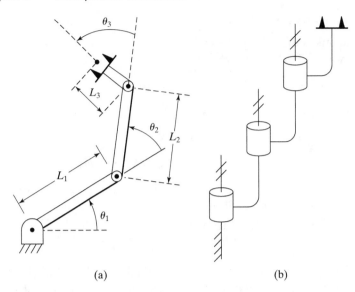

(a)   (b)

FIGURE 3.6: A three-link planar arm. On the right, we show the same manipulator by means of a simple schematic notation. Hash marks on the axes indicate that they are mutually parallel.

indicated on each of the three axes, which indicate that these axes are parallel. Assign link frames to the mechanism and give the Denavit–Hartenberg parameters.

We start by defining the reference frame, frame {0}. It is fixed to the base and aligns with frame {1} when the first joint variable (θ_1) is zero. Therefore, we position frame {0} as shown in Fig. 3.7 with $\hat{Z}_0$ aligned with the joint-1 axis. For this arm, all joint axes are oriented perpendicular to the plane of the arm. Because the arm

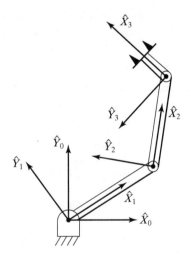

FIGURE 3.7: Link-frame assignments.

i	α_{i-1}	a_{i-1}	d_i	θ_i
1	0	0	0	θ_1
2	0	L_1	0	θ_2
3	0	L_2	0	θ_3

FIGURE 3.8: Link parameters of the three-link planar manipulator.

lies in a plane with all $\hat{Z}$ axes parallel, there are no link offsets—all d_i are zero. All joints are rotational, so when they are at zero degrees, all $\hat{X}$ axes must align.

With these comments in mind, it is easy to find the frame assignments shown in Fig. 3.7. The corresponding link parameters are shown in Fig. 3.8.

Note that, because the joint axes are all parallel and all the $\hat{Z}$ axes are taken as pointing out of the paper, all α_i are zero. This is obviously a very simple mechanism. Note also that our kinematic analysis always ends at a frame whose origin lies on the last joint axis; therefore, l_3 does not appear in the link parameters. Such final offsets to the end-effector are dealt with separately later.

EXAMPLE 3.4

Figure 3.9(a) shows a robot having three degrees of freedom and one prismatic joint. This manipulator can be called an "*RPR* mechanism," in a notation that specifies the type and order of the joints. It is a "cylindrical" robot whose first two joints are analogous to polar coordinates when viewed from above. The last joint (joint 3) provides "roll" for the hand. Figure 3.9(b) shows the same manipulator in schematic

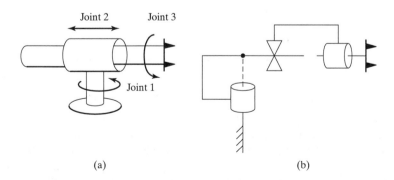

(a)   (b)

FIGURE 3.9: Manipulator having three degrees of freedom and one prismatic joint.

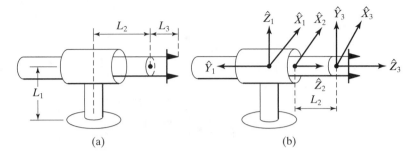

FIGURE 3.10: Link-frame assignments.

form. Note the symbol used to represent prismatic joints, and note that a "dot" is used to indicate the point at which two adjacent axes intersect. Also, the fact that axes 1 and 2 are orthogonal has been indicated.

Figure 3.10(a) shows the manipulator with the prismatic joint at minimum extension; the assignment of link frames is shown in Fig. 3.10(b).

Note that frame {0} and frame {1} are shown as exactly coincident in this figure, because the robot is drawn for the position $\theta_1 = 0$. Note that frame {0}, although not at the bottom of the flanged base of the robot, is nonetheless rigidly affixed to link 0, the nonmoving part of the robot. Just as our link frames are not used to describe the kinematics all the way out to the hand, they need not be attached all the way back to the lowest part of the base of the robot. It is sufficient that frame {0} be attached anywhere to the nonmoving link 0, and that frame {N}, the final frame, be attached anywhere to the last link of the manipulator. Other offsets can be handled later in a general way.

Note that rotational joints rotate about the $\hat{Z}$ axis of the associated frame, but prismatic joints slide along $\hat{Z}$. In the case where joint i is prismatic, θ_i is a fixed constant, and d_i is the variable. If d_i is zero at minimum extension of the link, then frame {2} should be attached where shown, so that d_2 will give the true offset. The link parameters are shown in Fig. 3.11.

Note that θ_2 is zero for this robot and that d_2 is a variable. Axes 1 and 2 intersect, so a_1 is zero. Angle α_1 must be 90 degrees in order to rotate $\hat{Z}_1$ so as to align with $\hat{Z}_2$ (about $\hat{X}_1$).

EXAMPLE 3.5

Figure 3.12(a) shows a three-link, $3R$ manipulator for which joint axes 1 and 2 intersect and axes 2 and 3 are parallel. Figure 3.12(b) shows the kinematic schematic of the manipulator. Note that the schematic includes annotations indicating that the first two axes are orthogonal and that the last two are parallel.

Demonstrate the nonuniqueness of frame assignments and of the Denavit–Hartenberg parameters by showing several possible correct assignments of frames {1} and {2}.

i	α_{i-1}	a_{i-1}	d_i	θ_i
1	0	0	0	θ_1
2	90°	0	d_2	0
3	0	0	L_2	θ_3

FIGURE 3.11: Link parameters for the *RPR* manipulator of Example 3.4.

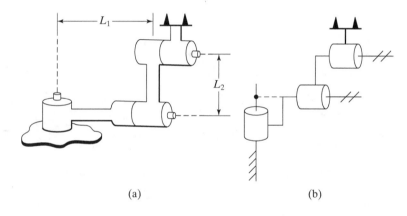

(a)   (b)

FIGURE 3.12: Three-link, nonplanar manipulator.

Figure 3.13 shows two possible frame assignments and corresponding parameters for the two possible choices of direction of $\hat{Z}_2$.

In general, when $\hat{Z}_i$ and $\hat{Z}_{i+1}$ intersect, there are two choices for $\hat{X}_i$. In this example, joint axes 1 and 2 intersect, so there are two choices for the direction of $\hat{X}_1$. Figure 3.14 shows two more possible frame assignments, corresponding to the second choice of $\hat{X}_1$.

In fact, there are four more possibilities, corresponding to the preceding four choices, but with $\hat{Z}_1$ pointing downward.

3.5   MANIPULATOR KINEMATICS

In this section, we derive the general form of the transformation that relates the frames attached to neighboring links. We then concatenate these individual transformations to solve for the position and orientation of link n relative to link 0.

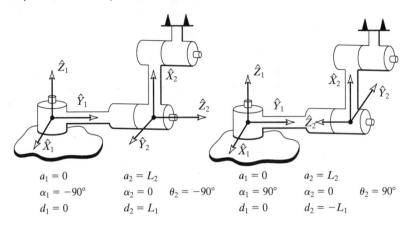

$$a_1 = 0 \qquad a_2 = L_2$$
$$\alpha_1 = -90° \qquad \alpha_2 = 0 \qquad \theta_2 = -90°$$
$$d_1 = 0 \qquad d_2 = L_1$$

$$a_1 = 0 \qquad a_2 = L_2$$
$$\alpha_1 = 90° \qquad \alpha_2 = 0 \qquad \theta_2 = 90°$$
$$d_1 = 0 \qquad d_2 = -L_1$$

FIGURE 3.13: Two possible frame assignments.

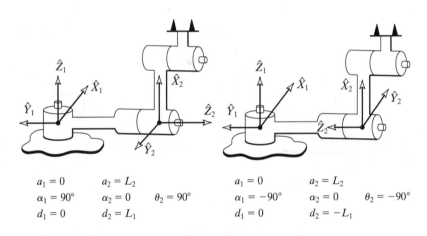

$$a_1 = 0 \qquad a_2 = L_2$$
$$\alpha_1 = 90° \qquad \alpha_2 = 0 \qquad \theta_2 = 90°$$
$$d_1 = 0 \qquad d_2 = L_1$$

$$a_1 = 0 \qquad a_2 = L_2$$
$$\alpha_1 = -90° \qquad \alpha_2 = 0 \qquad \theta_2 = -90°$$
$$d_1 = 0 \qquad d_2 = -L_1$$

FIGURE 3.14: Two more possible frame assignments.

Derivation of link transformations

We wish to construct the transform that defines frame $\{i\}$ relative to the frame $\{i-1\}$. In general, this transformation will be a function of the four link parameters. For any *given* robot, this transformation will be a function of only one variable, the other three parameters being fixed by mechanical design. By defining a frame for each link, we have broken the kinematics problem into n subproblems. In order to solve each of these subproblems, namely $^{i-1}_{i}T$, we will further break each subproblem into four subsubproblems. *Each of these four transformations will be a function of one link parameter only and will be simple enough that we can write down its form by inspection.* We begin by defining three intermediate frames for each link—$\{P\}$, $\{Q\}$, and $\{R\}$.

Figure 3.15 shows the same pair of joints as before with frames $\{P\}$, $\{Q\}$, and $\{R\}$ defined. Note that only the $\hat{X}$ and $\hat{Z}$ axes are shown for each frame, to make the drawing clearer. Frame $\{R\}$ differs from frame $\{i-1\}$ only by a rotation of α_{i-1}.

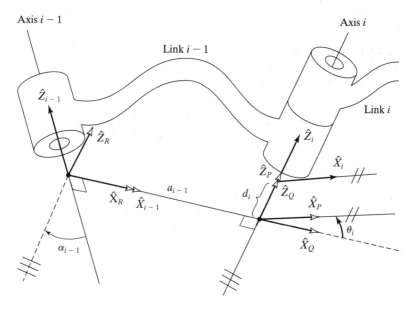

FIGURE 3.15: Location of intermediate frames $\{P\}$, $\{Q\}$, and $\{R\}$.

Frame $\{Q\}$ differs from $\{R\}$ by a translation a_{i-1}. Frame $\{P\}$ differs from $\{Q\}$ by a rotation θ_i, and frame $\{i\}$ differs from $\{P\}$ by a translation d_i. If we wish to write the transformation that transforms vectors defined in $\{i\}$ to their description in $\{i-1\}$, we may write

$$^{i-1}P = {}^{i-1}_{R}T \; {}^{R}_{Q}T \; {}^{Q}_{P}T \; {}^{P}_{i}T \, {}^{i}P, \tag{3.1}$$

or

$$^{i-1}P = {}^{i-1}_{i}T \, {}^{i}P, \tag{3.2}$$

where

$$^{i-1}_{i}T = {}^{i-1}_{R}T \; {}^{R}_{Q}T \; {}^{Q}_{P}T \; {}^{P}_{i}T. \tag{3.3}$$

Considering each of these transformations, we see that (3.3) may be written

$$^{i-1}_{i}T = R_X(\alpha_{i-1})D_X(a_{i-1})R_Z(\theta_i)D_Z(d_i), \tag{3.4}$$

or

$$^{i-1}_{i}T = \text{Screw}_X(a_{i-1}, \alpha_{i-1}) \; \text{Screw}_Z(d_i, \theta_i), \tag{3.5}$$

where the notation $\text{Screw}_Q(r, \phi)$ stands for the combination of a translation along an axis $\hat{Q}$ by a distance r and a rotation about the same axis by an angle ϕ. Multiplying out (3.4), we obtain the general form of $^{i-1}_{i}T$:

$$^{i-1}_{i}T = \begin{bmatrix} c\theta_i & -s\theta_i & 0 & a_{i-1} \\ s\theta_i c\alpha_{i-1} & c\theta_i c\alpha_{i-1} & -s\alpha_{i-1} & -s\alpha_{i-1}d_i \\ s\theta_i s\alpha_{i-1} & c\theta_i s\alpha_{i-1} & c\alpha_{i-1} & c\alpha_{i-1}d_i \\ 0 & 0 & 0 & 1 \end{bmatrix}. \tag{3.6}$$

EXAMPLE 3.6

Using the link parameters shown in Fig. 3.11 for the robot of Fig. 3.9, compute the individual transformations for each link.

Substituting the parameters into (3.6), we obtain

$$
{}^0_1T = \begin{bmatrix} c\theta_1 & -s\theta_1 & 0 & 0 \\ s\theta_1 & c\theta_1 & 0 & 0 \\ 0 & 0 & 1 & 0 \\ 0 & 0 & 0 & 1 \end{bmatrix},
$$

$$
{}^1_2T = \begin{bmatrix} 1 & 0 & 0 & 0 \\ 0 & 0 & -1 & -d_2 \\ 0 & 1 & 0 & 0 \\ 0 & 0 & 0 & 1 \end{bmatrix}, \tag{3.7}
$$

$$
{}^2_3T = \begin{bmatrix} c\theta_3 & -s\theta_3 & 0 & 0 \\ s\theta_3 & c\theta_3 & 0 & 0 \\ 0 & 0 & 1 & l_2 \\ 0 & 0 & 0 & 1 \end{bmatrix}.
$$

Once having derived these link transformations, we will find it a good idea to check them against common sense. For example, the elements of the fourth column of each transform should give the coordinates of the origin of the next higher frame.

Concatenating link transformations

Once the link frames have been defined and the corresponding link parameters found, developing the kinematic equations is straightforward. From the values of the link parameters, the individual link-transformation matrices can be computed. Then, the link transformations can be multiplied together to find the single transformation that relates frame $\{N\}$ to frame $\{0\}$:

$$
{}^0_NT = {}^0_1T\, {}^1_2T\, {}^2_3T \cdots {}^{N-1}_NT. \tag{3.8}
$$

This transformation, ${}^0_NT$, will be a function of all n joint variables. If the robot's joint-position sensors are queried, the Cartesian position and orientation of the last link can be computed by ${}^0_NT$.

3.6   ACTUATOR SPACE, JOINT SPACE, AND CARTESIAN SPACE

The position of all the links of a manipulator of n degrees of freedom can be specified with a set of n joint variables. This set of variables is often referred to as the $n \times 1$ **joint vector**. The space of all such joint vectors is referred to as **joint space**. Thus far in this chapter, we have been concerned with computing the **Cartesian space** description from knowledge of the joint-space description. We use the term *Cartesian space* when position is measured along orthogonal axes and orientation is measured according to any of the conventions outlined in Chapter 2. Sometimes, the terms **task-oriented space** and **operational space** are used for what we will call Cartesian space.

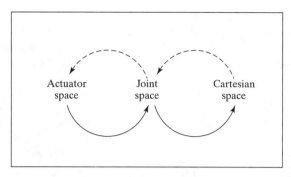

FIGURE 3.16: Mappings between kinematic descriptions.

So far, we have implicitly assumed that each kinematic joint is actuated directly by some sort of actuator. However, in the case of many industrial robots, this is not so. For example, sometimes two actuators work together in a differential pair to move a single joint, or sometimes a linear actuator is used to rotate a revolute joint, through the use of a four-bar linkage. In these cases, it is helpful to consider the notion of *actuator positions*. The sensors that measure the position of the manipulator are often located at the actuators, so some computations must be performed to realize the joint vector as a function of a set of actuator values, or **actuator vector**.

As is indicated in Fig. 3.16, there are three representations of a manipulator's position and orientation: descriptions in **actuator space**, in **joint space**, and in **Cartesian space**. In this chapter, we are concerned with the mappings between representations, as indicated by the solid arrows in Fig. 3.16. In Chapter 4, we will consider the inverse mappings, indicated by the dashed arrows.

The ways in which actuators might be connected to move a joint are quite varied; they might be catalogued, but we will not do so here. For each robot we design or seek to analyze, the correspondence between actuator positions and joint positions must be solved. In the next section, we will solve an example problem for an industrial robot.

3.7 EXAMPLES: KINEMATICS OF TWO INDUSTRIAL ROBOTS

Current industrial robots are available in many different kinematic configurations [2], [3]. In this section, we work out the kinematics of two typical industrial robots. First we consider the Unimation PUMA 560, a rotary-joint manipulator with six degrees of freedom. We will solve for the kinematic equations as functions of the joint angles. For this example, we will skip the additional problem of the relationship between actuator space and joint space. Second, we consider the Yasukawa Motoman L-3, a robot with five degrees of freedom and rotary joints. This example is done in detail, including the actuator-to-joint transformations. This example may be skipped on first reading of the book.

The PUMA 560

The Unimation PUMA 560 (Fig. 3.17) is a robot with six degrees of freedom and all rotational joints (i.e., it is a 6R mechanism). It is shown in Fig. 3.18, with

FIGURE 3.17: The Unimation PUMA 560. Courtesy of Unimation Incorporated, Shelter Rock Lane, Danbury, Conn.

link-frame assignments in the position corresponding to all joint angles equal to zero.[5] Figure 3.19 shows a detail of the forearm of the robot.

Note that the frame {0} (not shown) is coincident with frame {1} when θ_1 is zero. Note also that, for this robot, as for many industrial robots, the joint axes of joints 4, 5, and 6 all intersect at a common point, and this point of intersection coincides with the origin of frames {4}, {5}, and {6}. Furthermore, the joint axes 4, 5, and 6 are mutually orthogonal. This wrist mechanism is illustrated schematically in Fig. 3.20.

The link parameters corresponding to this placement of link frames are shown in Fig. 3.21. In the case of the PUMA 560, a gearing arrangement in the wrist of the manipulator couples together the motions of joints 4, 5, and 6. What this means is that, for these three joints, we must make a distinction between joint space and actuator space and solve the complete kinematics in two steps. However, in this example, we will consider only the kinematics from joint space to Cartesian space.

---

[5]Unimation has used a slightly different assignment of zero location of the joints, such that $\theta_3^* = \theta_3 - 180°$, where θ_3^* is the position of joint 3 in Unimation's convention.

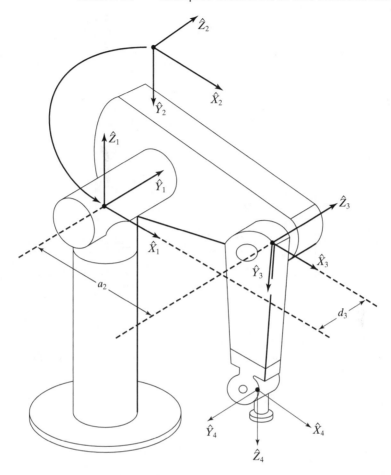

FIGURE 3.18: Some kinematic parameters and frame assignments for the PUMA 560 manipulator.

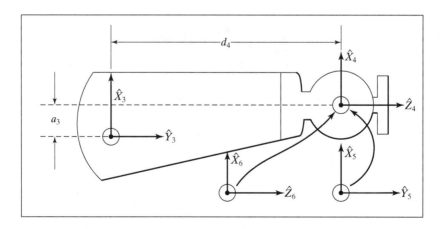

FIGURE 3.19: Kinematic parameters and frame assignments for the forearm of the PUMA 560 manipulator.

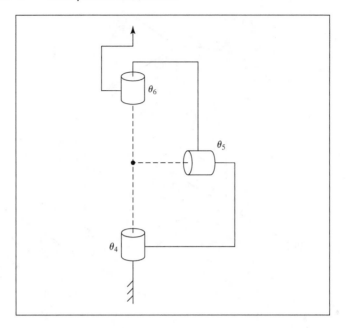

FIGURE 3.20: Schematic of a $3R$ wrist in which all three axes intersect at a point and are mutually orthogonal. This design is used in the PUMA 560 manipulator and many other industrial robots.

i	α_{i-1}	a_{i-1}	d_i	θi
1	0	0	0	θ_1
2	$-90°$	0	0	θ_2
3	0	a_2	d_3	θ_3
4	$-90°$	a_3	d_4	θ_4
5	$90°$	0	0	θ_5
6	$-90°$	0	0	θ_6

FIGURE 3.21: Link parameters of the PUMA 560.

Using (3.6), we compute each of the link transformations:

$$
{}^0_1T = \begin{bmatrix} c\theta_1 & -s\theta_1 & 0 & 0 \\ s\theta_1 & c\theta_1 & 0 & 0 \\ 0 & 0 & 1 & 0 \\ 0 & 0 & 0 & 1 \end{bmatrix},
$$

$$
{}^1_2T = \begin{bmatrix} c\theta_2 & -s\theta_2 & 0 & 0 \\ 0 & 0 & 1 & 0 \\ -s\theta_2 & -c\theta_2 & 0 & 0 \\ 0 & 0 & 0 & 1 \end{bmatrix},
$$

$$
{}^2_3T = \begin{bmatrix} c\theta_3 & -s\theta_3 & 0 & a_2 \\ s\theta_3 & c\theta_3 & 0 & 0 \\ 0 & 0 & 1 & d_3 \\ 0 & 0 & 0 & 1 \end{bmatrix},
$$

$$
{}^3_4T = \begin{bmatrix} c\theta_4 & -s\theta_4 & 0 & a_3 \\ 0 & 0 & 1 & d_4 \\ -s\theta_4 & -c\theta_4 & 0 & 0 \\ 0 & 0 & 0 & 1 \end{bmatrix},
$$

$$
{}^4_5T = \begin{bmatrix} c\theta_5 & -s\theta_5 & 0 & 0 \\ 0 & 0 & -1 & 0 \\ s\theta_5 & c\theta_5 & 0 & 0 \\ 0 & 0 & 0 & 1 \end{bmatrix},
$$

$$
{}^5_6T = \begin{bmatrix} c\theta_6 & -s\theta_6 & 0 & 0 \\ 0 & 0 & 1 & 0 \\ -s\theta_6 & -c\theta_6 & 0 & 0 \\ 0 & 0 & 0 & 1 \end{bmatrix}. \tag{3.9}
$$

We now form ${}^0_6T$ by matrix multiplication of the individual link matrices. While forming this product, we will derive some subresults that will be useful when solving the inverse kinematic problem in Chapter 4. We start by multiplying ${}^4_5T$ and ${}^5_6T$; that is,

$$
{}^4_6T = {}^4_5T\,{}^5_6T = \begin{bmatrix} c_5c_6 & -c_5s_6 & -s_5 & 0 \\ s_6 & c_6 & 0 & 0 \\ s_5c_6 & -s_5s_6 & c_5 & 0 \\ 0 & 0 & 0 & 1 \end{bmatrix}, \tag{3.10}
$$

where c_5 is shorthand for $\cos\theta_5$, s_5 for $\sin\theta_5$, and so on.[6] Then we have

$$
{}^3_6T = {}^3_4T\,{}^4_6T = \begin{bmatrix} c_4c_5c_6 - s_4s_6 & -c_4c_5s_6 - s_4c_6 & -c_4s_5 & a_3 \\ s_5c_6 & -s_5s_6 & c_5 & d_4 \\ -s_4c_5c_6 - c_4s_6 & s_4c_5s_6 - c_4c_6 & s_4s_5 & 0 \\ 0 & 0 & 0 & 1 \end{bmatrix}. \tag{3.11}
$$

---

[6]Depending on the amount of space available to show expressions, we use any of the following three forms: $\cos\theta_5$, $c\theta_5$, or c_5.

Because joints 2 and 3 are always parallel, multiplying ^1_2T and ^2_3T first and then applying sum-of-angle formulas will yield a somewhat simpler final expression. This can be done whenever two rotational joints have parallel axes and we have

$$^1_3T = {}^1_2T\,{}^2_3T = \begin{bmatrix} c_{23} & -s_{23} & 0 & a_2c_2 \\ 0 & 0 & 1 & d_3 \\ -s_{23} & -c_{23} & 0 & -a_2s_2 \\ 0 & 0 & 0 & 1 \end{bmatrix}, \tag{3.12}$$

where we have used the sum-of-angle formulas (from Appendix A):

$$c_{23} = c_2c_3 - s_2s_3,$$

$$s_{23} = c_2s_3 + s_2c_3.$$

Then we have

$$^1_6T = {}^1_3T\,{}^3_6T = \begin{bmatrix} {}^1r_{11} & {}^1r_{12} & {}^1r_{13} & {}^1p_x \\ {}^1r_{21} & {}^1r_{22} & {}^1r_{23} & {}^1p_y \\ {}^1r_{31} & {}^1r_{32} & {}^1r_{33} & {}^1p_z \\ 0 & 0 & 0 & 1 \end{bmatrix},$$

where

$$
\begin{aligned}
{}^1r_{11} &= c_{23}[c_4c_5c_6 - s_4s_6] - s_{23}s_5s_6, \\
{}^1r_{21} &= -s_4c_5c_6 - c_4s_6, \\
{}^1r_{31} &= -s_{23}[c_4c_5c_6 - s_4s_6] - c_{23}s_5c_6, \\
{}^1r_{12} &= -c_{23}[c_4c_5s_6 + s_4c_6] + s_{23}s_5s_6, \\
{}^1r_{22} &= s_4c_5s_6 - c_4c_6, \\
{}^1r_{32} &= s_{23}[c_4c_5s_6 + s_4c_6] + c_{23}s_5s_6, \\
{}^1r_{13} &= -c_{23}c_4s_5 - s_{23}c_5, \\
{}^1r_{23} &= s_4s_5, \\
{}^1r_{33} &= s_{23}c_4s_5 - c_{23}c_5, \\
{}^1p_x &= a_2c_2 + a_3c_{23} - d_4s_{23}, \\
{}^1p_y &= d_3, \\
{}^1p_x &= -a_3s_{23} - a_2s_2 - d_4c_{23}.
\end{aligned} \tag{3.13}
$$

Finally, we obtain the product of all six link transforms:

$$^0_6T = {}^0_1T\,{}^1_6T = \begin{bmatrix} r_{11} & r_{12} & r_{13} & p_x \\ r_{21} & r_{22} & r_{23} & p_y \\ r_{31} & r_{32} & r_{33} & p_z \\ 0 & 0 & 0 & 1 \end{bmatrix}.$$

Here,

$$r_{11} = c_1[c_{23}(c_4c_5c_6 - s_4s_5) - s_{23}s_5c_5] + s_1(s_4c_5c_6 + c_4s_6),$$

$$r_{21} = s_1[c_{23}(c_4c_5c_6 - s_4s_6) - s_{23}s_5c_6 - c_1(s_4c_5c_6 + c_4s_6),$$

$$r_{31} = -s_{23}(c_4c_5c_6 - s_4s_6) - c_{23}s_5c_6,$$

$$r_{12} = c_1[c_{23}(-c_4c_5s_6 - s_4c_6) + s_{23}s_5s_6] + s_1(c_4c_6 - s_4c_5s_6),$$

$$r_{22} = s_1[c_{23}(-c_4c_5s_6 - s_4c_6) + s_{23}s_5s_6] - c_1(c_4c_6 - s_4c_5s_6),$$

$$r_{32} = -s_{23}(-c_4c_5s_6 - s_4c_6) + c_{23}s_5s_6,$$

$$r_{13} = -c_1(c_{23}c_4s_5 + s_{23}c_5) - s_1s_4s_5,$$

$$r_{23} = -s_1(c_{23}c_4s_5 + s_{23}c_5) + c_1s_4s_5,$$

$$r_{33} = s_{23}c_4s_5 - c_{23}c_5,$$

$$p_x = c_1[a_2c_2 + a_3c_{23} - d_4s_{23}] - d_3s_1,$$

$$p_y = s_1[a_2c_2 + a_3c_{23} - d_4s_{23}] + d_3c_1,$$

$$p_z = -a_3s_{23} - a_2s_2 - d_4c_{23}. \tag{3.14}$$

Equations (3.14) constitute the kinematics of the PUMA 560. They specify how to compute the position and orientation of frame {6} relative to frame {0} of the robot. These are the basic equations for all kinematic analysis of this manipulator.

The Yasukawa Motoman L-3

The Yasukawa Motoman L-3 is a popular industrial manipulator with five degrees of freedom (Fig. 3.22). Unlike the examples we have seen thus far, the Motoman is not a simple open kinematic chain, but rather makes use of two linear actuators coupled to links 2 and 3 with four-bar linkages. Also, through a chain drive, joints 4 and 5 are operated by two actuators in a differential arrangement.

   In this example, we will solve the kinematics in two stages. First, we will solve for joint angles from actuator positions; second, we will solve for Cartesian position and orientation of the last link from joint angles. In this second stage, we can treat the system as if it were a simple open-kinematic-chain 5R device.

   Figure 3.23 shows the linkage mechanism that connects actuator number 2 to links 2 and 3 of the robot. The actuator is a linear one that directly controls the length of the segment labeled DC. Triangle ABC is fixed, as is the length BD. Joint 2 pivots about point B, and the actuator pivots slightly about point C as the linkage moves. We give the following names to the constants (lengths and angles) associated with actuator 2:

$$\gamma_2 = AB, \phi_2 = AC, \alpha_2 = BC,$$

$$\beta_2 = BD, \Omega_2 = \angle JBD, l_2 = BJ,$$

FIGURE 3.22: The Yasukawa Motoman L-3. Courtesy of Yasukawa.

we give the following names to the variables:

$$\theta_2 = -\angle JBQ, \psi_2 = \angle CBD, g_2 = DC.$$

Figure 3.24 shows the linkage mechanism that connects actuator number 3 to links 2 and 3 of the robot. The actuator is a linear one that directly controls the length of the segment labeled HG. Triangle EFG is fixed, as is the length FH. Joint 3 pivots about point J, and the actuator pivots slightly about point G as the linkage moves. We give the following names to the constants (lengths and angles) associated with actuator 3:

$$\gamma_3 = EF, \phi_3 = EG, \alpha_3 = GF,$$
$$\beta_3 = HF, l_3 = JK.$$

We give the following names to the variables:

$$\theta_3 = \angle PJK, \ \psi_3 = \angle GFH, \ g_3 = GH.$$

This arrangement of actuators and linkages has the following functional effect. Actuator 2 is used to position joint 2; while it is doing so, link 3 remains in the same orientation relative to the base of the robot. Actuator 3 is used to adjust

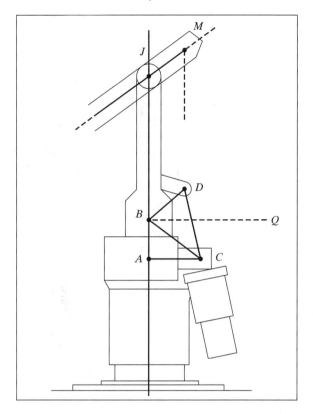

FIGURE 3.23: Kinematic details of the Yasukawa actuator-2 linkage.

the orientation of link 3 relative to the base of the robot (rather than relative to the preceding link as in a serial-kinematic-chain robot). One purpose of such a linkage arrangement is to increase the structural rigidity of the main linkages of the robot. This often pays off in terms of an increased ability to position the robot precisely.

The actuators for joints 4 and 5 are attached to link 1 of the robot with their axes aligned with that of joint 2 (points B and F in Figs. 3.23 and 3.24). They operate the wrist joints through two sets of chains—one set located interior to link 2, the second set interior to link 3. The effect of this transmission system, along with its interaction with the actuation of links 2 and 3, is described functionally as follows: Actuator 4 is used to position joint 4 relative to the base of the robot, rather than relative to the preceding link 3. This means that holding actuator 4 constant will keep link 4 at a constant orientation relative to the base of the robot, regardless of the positions of joints 2 and 3. Finally, actuator 5 behaves as if directly connected to joint 5.

We now state the equations that map a set of actuator values (A_i) to the equivalent set of joint values (θ_i). In this case, these equations were derived by straightforward plane geometry—mostly just application of the "law of cosines."[7]

---

[7]If a triangle's angles are labeled a, b, and c, where angle a is opposite side A, and so on, then $A^2 = B^2 + C^2 - 2BC \cos a$.

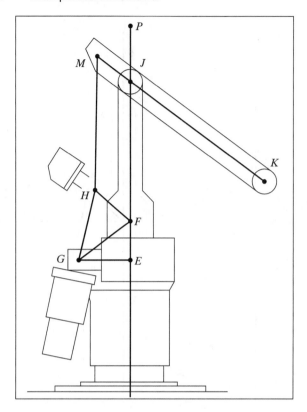

FIGURE 3.24: Kinematic details of the Yasukawa actuator-3 linkage.

Appearing in these equations are scale (k_i) and offset (λ_i) constants for each actuator. For example, actuator 1 is directly connected to joint axis 1, and so the conversion is simple; it is just a matter of a scale factor plus an offset. Thus,

$$\theta_1 = k_1 A_1 + \lambda_1,$$

$$\theta_2 = \cos^{-1}\left(\frac{(k_2 A_2 + \lambda_2)^2 - \alpha_2^2 - \beta_2^2}{-2\alpha_2\beta_2}\right) + \tan^{-1}\left(\frac{\phi_2}{\gamma_2}\right) + \Omega_2 - 270°,$$

$$\theta_3 = \cos^{-1}\left(\frac{(k_3 A_3 + \lambda_3)^2 - \alpha_3^2 - \beta_3^2}{-2\alpha_3\beta_3}\right) - \theta_2 + \tan^{-1}\left(\frac{\phi_3}{\gamma_3}\right) - 90°,$$

$$\theta_4 = -k_4 A_4 - \theta_2 - \theta_3 + \lambda_4 + 180°, \tag{3.15}$$

$$\theta_5 = -k_5 A_5 + \lambda_5.$$

Figure 3.25 shows the attachment of the link frames. In this figure, the manipulator is shown in a position corresponding to the joint vector $\Theta = (0, -90°, 90°, 90°, 0)$. Figure 3.26 shows the link parameters for this manipulator. The resulting link-transformation matrices are

$$
{}_1^0T = \begin{bmatrix} c\theta_1 & -s\theta_1 & 0 & 0 \\ s\theta_1 & c\theta_1 & 0 & 0 \\ 0 & 0 & 1 & 0 \\ 0 & 0 & 0 & 1 \end{bmatrix},
$$

$$
{}_2^1T = \begin{bmatrix} c\theta_2 & -s\theta_2 & 0 & 0 \\ 0 & 0 & 1 & 0 \\ -s\theta_2 & -c\theta_2 & 0 & 0 \\ 0 & 0 & 0 & 1 \end{bmatrix},
$$

$$
{}_3^2T = \begin{bmatrix} c\theta_3 & -s\theta_3 & 0 & l_2 \\ s\theta_3 & c\theta_3 & 0 & 0 \\ 0 & 0 & 1 & 0 \\ 0 & 0 & 0 & 1 \end{bmatrix}, \qquad (3.16)
$$

$$
{}_4^3T = \begin{bmatrix} c\theta_4 & -s\theta_4 & 0 & l_3 \\ s\theta_4 & c\theta_4 & 0 & 0 \\ 0 & 0 & 1 & 0 \\ 0 & 0 & 0 & 1 \end{bmatrix},
$$

$$
{}_5^4T = \begin{bmatrix} c\theta_5 & -s\theta_5 & 0 & 0 \\ 0 & 0 & -1 & 0 \\ s\theta_5 & c\theta_5 & 0 & 0 \\ 0 & 0 & 0 & 1 \end{bmatrix}.
$$

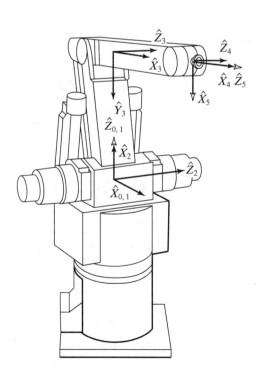

FIGURE 3.25: Assignment of link frames for the Yasukawa L-3.

i	$\alpha_i - 1$	$a_i - 1$	d_i	θ_i
1	0	0	0	θ_1
2	$-90°$	0	0	θ_2
3	0	L_2	0	θ_3
4	0	L_3	0	θ_4
5	$90°$	0	0	θ_5

FIGURE 3.26: Link parameters of the Yasukawa L-3 manipulator.

Forming the product to obtain ${}_5^0T$, we obtain

$$
{}_5^0T = \begin{bmatrix} r_{11} & r_{12} & r_{13} & p_x \\ r_{21} & r_{22} & r_{23} & p_y \\ r_{31} & r_{32} & r_{33} & p_z \\ 0 & 0 & 0 & 1 \end{bmatrix},
$$

where

$$r_{11} = c_1 c_{234} c_5 - s_1 s_5,$$
$$r_{21} = s_1 c_{234} c_5 + c_1 s_5,$$
$$r_{31} = -s_{234} c_5,$$

$$r_{12} = -c_1 c_{234} s_5 - s_1 c_5,$$
$$r_{22} = -s_1 c_{234} s_5 + c_1 c_5,$$
$$r_{32} = s_{234} s_5,$$

$$r_{13} = c_1 s_{234},$$
$$r_{23} = s_1 s_{234},$$
$$r_{33} = c_{234},$$

$$p_x = c_1(l_2c_2 + l_3c_{23}),$$
$$p_y = s_1(l_2c_2 + l_3c_{23}),$$
$$p_z = -l_2s_2 - l_3s_{23}. \tag{3.17}$$

We developed the kinematic equations for the Yasukawa Motoman in two steps. In the first step, we computed a joint vector from an actuator vector; in the second step, we computed a position and orientation of the wrist frame from the joint vector. If we wish to compute only Cartesian position and not joint angles, it is possible to derive equations that map directly from actuator space to Cartesian space. These equations are somewhat simpler computationally than the two-step approach. (See Exercise 3.10.)

3.8   FRAMES WITH STANDARD NAMES

As a matter of convention, it will be helpful if we assign specific names and locations to certain "standard" frames associated with a robot and its workspace. Figure 3.27 shows a typical situation in which a robot has grasped some sort of tool and is to position the tool tip to a user-defined location. The five frames indicated in Fig. 3.27 are so often referred to that we will define names for them. The naming and subsequent use of these five frames in a robot programming and control system facilitates providing general capabilities in an easily understandable way. All robot motions will be described in terms of these frames.

Brief definitions of the frames shown in Fig. 3.27 follow.

The base frame, $\{B\}$

$\{B\}$ is located at the base of the manipulator. It is merely another name for frame $\{0\}$. It is affixed to a nonmoving part of the robot, sometimes called link 0.

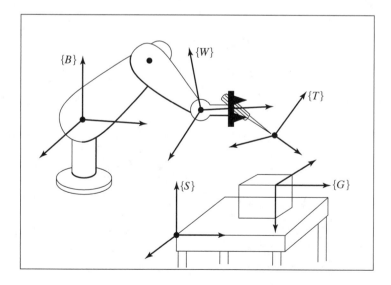

FIGURE 3.27: The standard frames.

The station frame, {S}

{S} is located in a task-relevant location. In Fig. 3.28, it is at the corner of a table upon which the robot is to work. As far as the user of this robot system is concerned, {S} is the universe frame, and all actions of the robot are performed relative to it. It is sometimes called the task frame, the world frame, or the universe frame. The station frame is always specified with respect to the base frame, that is, $_S^B T$.

The wrist frame, {W}

{W} is affixed to the last link of the manipulator. It is another name for frame {N}, the link frame attached to the last link of the robot. Very often, {W} has its origin fixed at a point called the wrist of the manipulator, and {W} moves with the last link of the manipulator. It is defined relative to the base frame—that is, {W} $= _W^B T = _N^0 T$.

The tool frame, {T}

{T} is affixed to the end of any tool the robot happens to be holding. When the hand is empty, {T} is usually located with its origin between the fingertips of the robot. The tool frame is always specified with respect to the wrist frame. In Fig. 3.28, the tool frame is defined with its origin at the tip of a pin that the robot is holding.

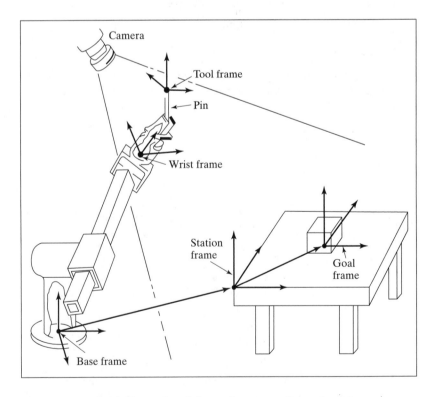

FIGURE 3.28: Example of the assignment of standard frames.

The goal frame, {G}

{G} is a description of the location to which the robot is to move the tool. Specifically this means that, at the end of the motion, the tool frame should be brought to coincidence with the goal frame. {G} is always specified relative to the station frame. In Fig. 3.28, the goal is located at a hole into which we want the pin to be inserted.

All robot motions may be described in terms of these frames without loss of generality. Their use helps to give us a standard language for talking about robot tasks.

3.9   WHERE IS THE TOOL?

One of the first capabilities a robot must have is to be able to calculate the position and orientation of the tool it is holding (or of its empty hand) with respect to a convenient coordinate system. That is, we wish to calculate the value of the tool frame, {T}, relative to the station frame, {S}. Once $^B_W T$ has been computed via the kinematic equations, we can use Cartesian transforms, as studied in Chapter 2, to calculate {T} relative to {S}. Solving a simple transform equation leads to

$$^S_T T = {}^B_S T^{-1}\, {}^B_W T\, {}^W_T T. \tag{3.18}$$

Equation (3.18) implements what is called the **WHERE** function in some robot systems. It computes "where" the arm is. For the situation in Fig. 3.28, the output of **WHERE** would be the position and orientation of the pin relative to the table top.

Equation (3.18) can be thought of as *generalizing* the kinematics. $^S_T T$ computes the kinematics due to the geometry of the linkages, along with a general transform (which might be considered a fixed link) at the base end ($^B_S T$) and another at the end-effector ($^W_T T$). These extra transforms allow us to include tools with offsets and twists and to operate with respect to an arbitrary station frame.

3.10   COMPUTATIONAL CONSIDERATIONS

In many practical manipulator systems, the time required to perform kinematic calculations is a consideration. In this section, we briefly discuss various issues involved in computing manipulator kinematics, as exemplified by (3.14), for the case of the PUMA 560.

One choice to be made is the use of fixed- or floating-point representation of the quantities involved. Many implementations use floating point for ease of software development, because the programmer does not have to be concerned with scaling operations capturing the relative magnitudes of the variables. However, when speed is crucial, fixed-point representation is quite possible, because the variables do not have a large dynamic range, and these ranges are fairly well known. Rough estimations of the number of bits needed in fixed-point representation seem to indicate that 24 are sufficient [4].

By factoring equations such as (3.14), it is possible to reduce the number of multiplications and additions—at the cost of creating local variables (usually a good trade-off). The point is to avoid computing common terms over and over throughout the computation. There has been some application of computer-assisted automatic factorization of such equations [5].

The major expense in calculating kinematics is often the calculation of the transcendental functions (sine and cosine). When these functions are available as part of a standard library, they are often computed from a series expansion at the cost of many multiply times. At the expense of some required memory, many manipulation systems employ table-lookup implementations of the transcendental functions. Depending on the scheme, this reduces the amount of time required to calculate a sine or cosine to two or three multiply times or less [6].

The computation of the kinematics as in (3.14) is redundant, in that nine quantities are calculated to represent orientation. One means that usually reduces computation is to calculate only two columns of the rotation matrix and then to compute a cross product (requiring only six multiplications and three additions) to compute the third column. Obviously, one chooses the two least complicated columns to compute.

BIBLIOGRAPHY

[1] J. Denavit and R.S. Hartenberg, "A Kinematic Notation for Lower-Pair Mechanisms Based on Matrices," *Journal of Applied Mechanics*, pp. 215–221, June 1955.

[2] J. Lenarčič, "Kinematics," in *The International Encyclopedia of Robotics*, R. Dorf and S. Nof, Editors, John C. Wiley and Sons, New York, 1988.

[3] J. Colson and N.D. Perreira, "Kinematic Arrangements Used in Industrial Robots," *13th Industrial Robots Conference Proceedings*, April 1983.

[4] T. Turner, J. Craig, and W. Gruver, "A Microprocessor Architecture for Advanced Robot Control," 14th ISIR, Stockholm, Sweden, October 1984.

[5] W. Schiehlen, "Computer Generation of Equations of Motion," in *Computer Aided Analysis and Optimization of Mechanical System Dynamics*, E.J. Haug, Editor, Springer-Verlag, Berlin & New York, 1984.

[6] C. Ruoff, "Fast Trigonometric Functions for Robot Control," *Robotics Age*, November 1981.

EXERCISES

3.1 [15] Compute the kinematics of the planar arm from Example 3.3.

3.2 [37] Imagine an arm like the PUMA 560, except that joint 3 is replaced with a prismatic joint. Assume the prismatic joint slides along the direction of $\hat{X}_1$ in Fig. 3.18; however, there is still an offset equivalent to d_3 to be accounted for. Make any additional assumptions needed. Derive the kinematic equations.

3.3 [25] The arm with three degrees of freedom shown in Fig. 3.29 is like the one in Example 3.3, except that joint 1's axis is not parallel to the other two. Instead, there is a twist of 90 degrees in magnitude between axes 1 and 2. Derive link parameters and the kinematic equations for $^{B}_{W}T$. Note that no l_3 need be defined.

3.4 [22] The arm with three degrees of freedom shown in Fig. 3.30 has joints 1 and 2 perpendicular and joints 2 and 3 parallel. As pictured, all joints are at their zero location. Note that the positive sense of the joint angle is indicated. Assign link frames {0} through {3} for this arm—that is, sketch the arm, showing the attachment of the frames. Then derive the transformation matrices $^{0}_{1}T$, $^{1}_{2}T$, and $^{2}_{3}T$.

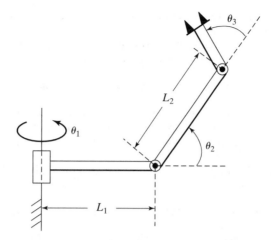

FIGURE 3.29: The $3R$ nonplanar arm (Exercise 3.3).

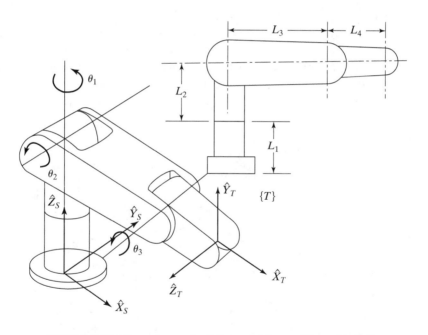

FIGURE 3.30: Two views of a $3R$ manipulator (Exercise 3.4).

3.5 [26] Write a subroutine to compute the kinematics of a PUMA 560. Code for speed, trying to minimize the number of multiplications as much as possible. Use the procedure heading (or equivalent in C)

```
Procedure KIN(VAR theta: vec6; VAR wrelb: frame);
```

Count a sine or cosine evaluation as costing 5 multiply times. Count additions as costing 0.333 multiply times and assignment statements as 0.2 multiply times.

Count a square-root computation as costing 4 multiply times. How many multiply times do you need?

3.6 [20] Write a subroutine to compute the kinematics of the cylindrical arm in Example 3.4. Use the procedure heading (or equivalent in C)

```
Procedure KIN(VAR jointvar: vec3; VAR wrelb: frames);
```

Count a sine or cosine evaluation as costing 5 multiply times. Count additions as costing 0.333 multiply times and assignment statements as 0.2 multiply times. Count a square-root computation as costing 4 multiply times. How many multiply times do you need?

3.7 [22] Write a subroutine to compute the kinematics of the arm in Exercise 3.3. Use the procedure heading (or equivalent in C)

```
Procedure KIN(VAR theta: vec3; VAR wrelb: frame);
```

Count a sine or cosine evaluation as costing 5 multiply times. Count additions as costing 0.333 multiply times and assignment statements as 0.2 multiply times. Count a square-root computation as costing 4 multiply times. How many multiply times do you need?

3.8 [13] In Fig. 3.31, the location of the tool, $^W_T T$, is not accurately known. Using force control, the robot feels around with the tool tip until it inserts it into the socket (or Goal) at location $^S_G T$. Once in this "calibration" configuration (in which $\{G\}$ and $\{T\}$ are coincident), the position of the robot, $^B_W T$, is figured out by reading the joint angle sensors and computing the kinematics. Assuming $^B_S T$ and $^S_G T$ are known, give the transform equation to compute the unknown tool frame, $^W_T T$.

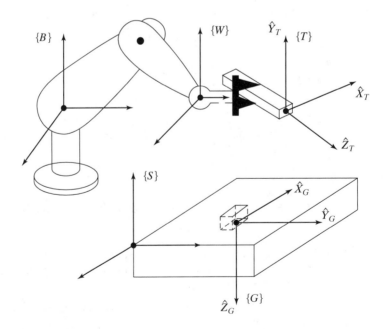

FIGURE 3.31: Determination of the tool frame (Exercise 3.8).

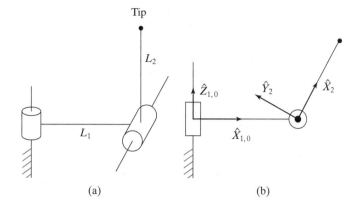

FIGURE 3.32: Two-link arm with frame assignments (Exercise 3.9).

3.9 [11] For the two-link manipulator shown in Fig. 3.32(a), the link-transformation matrices, ^0_1T and ^1_2T, were constructed. Their product is

$$^0_2T = \begin{bmatrix} c\theta_1 c\theta_2 & -c\theta_1 s\theta_2 & s\theta_1 & l_1 c\theta_1 \\ s\theta_1 c\theta_2 & -s\theta_1 s\theta_2 & -c\theta_1 & l_1 s\theta_1 \\ s\theta_2 & c\theta_2 & 0 & 0 \\ 0 & 0 & 0 & 1 \end{bmatrix}.$$

The link-frame assignments used are indicated in Fig. 3.32(b). Note that frame {0} is coincident with frame {1} when $\theta_1 = 0$. The length of the second link is l_2. Find an expression for the vector $^0P_{tip}$, which locates the tip of the arm relative to the {0} frame.

3.10 [39] Derive kinematic equations for the Yasukawa Motoman robot (see Section 3.7) that compute the position and orientation of the wrist frame directly from actuator values, rather than by first computing the joint angles. A solution is possible that requires only 33 multiplications, two square roots, and six sine or cosine evaluations.

3.11 [17] Figure 3.33 shows the schematic of a wrist which has three intersecting axes that are not orthogonal. Assign link frames to this wrist (as if it were a 3-DOF manipulator), and give the link parameters.

3.12 [08] Can an arbitrary rigid-body transformation always be expressed with four parameters (a, α, d, θ) in the form of equation (3.6)?

3.13 [15] Show the attachment of link frames for the 5-DOF manipulator shown schematically in Fig. 3.34.

3.14 [20] As was stated, the relative position of any two lines in space can be given with two parameters, a and α, where a is the length of the common perpendicular joining the two and α is the angle made by the two axes when projected onto a plane normal to the common perpendicular. Given a line defined as passing through point p with unit-vector direction $\hat{m}$ and a second passing through point q with unit-vector direction $\hat{n}$, write expressions for a and α.

3.15 [15] Show the attachment of link frames for the 3-DOF manipulator shown schematically in Fig. 3.35.

3.16 [15] Assign link frames to the RPR planar robot shown in Fig. 3.36, and give the linkage parameters.

3.17 [15] Show the attachment of link frames on the three-link robot shown in Fig. 3.37.

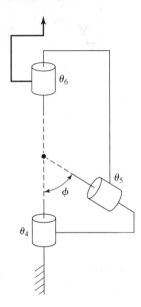

FIGURE 3.33: $3R$ nonorthogonal-axis robot (Exercise 3.11).

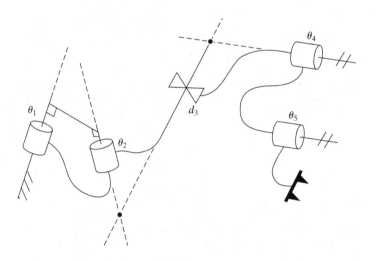

FIGURE 3.34: Schematic of a $2RP2R$ manipulator (Exercise 3.13).

3.18 [15] Show the attachment of link frames on the three-link robot shown in Fig. 3.38.
3.19 [15] Show the attachment of link frames on the three-link robot shown in Fig. 3.39.
3.20 [15] Show the attachment of link frames on the three-link robot shown in Fig. 3.40.
3.21 [15] Show the attachment of link frames on the three-link robot shown in Fig. 3.41.
3.22 [18] Show the attachment of link frames on the $P3R$ robot shown in Fig. 3.42. Given your frame assignments, what are the signs of d_2, d_3, and a_2?

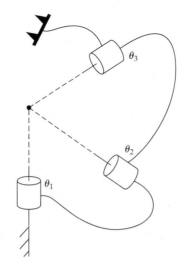

FIGURE 3.35: Schematic of a $3R$ manipulator (Exercise 3.15).

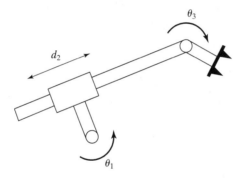

FIGURE 3.36: RPR planar robot (Exercise 3.16).

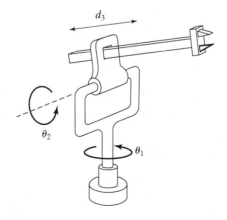

FIGURE 3.37: Three-link RRP manipulator (Exercise 3.17).

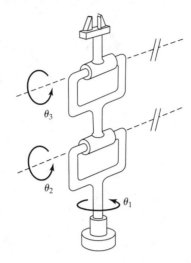

FIGURE 3.38: Three-link RRR manipulator (Exercise 3.18).

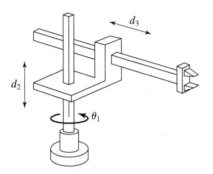

FIGURE 3.39: Three-link RPP manipulator (Exercise 3.19).

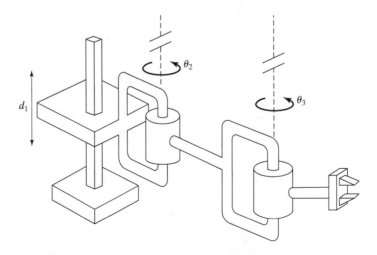

FIGURE 3.40: Three-link PRR manipulator (Exercise 3.20).

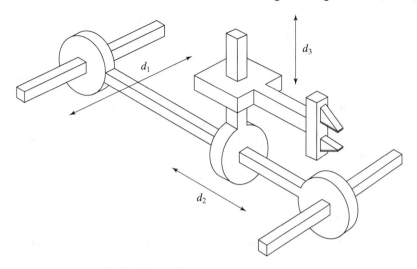

FIGURE 3.41: Three-link PPP manipulator (Exercise 3.21).

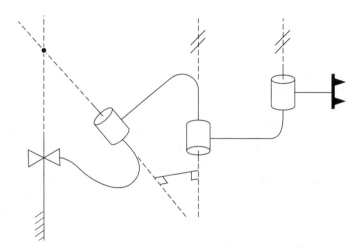

FIGURE 3.42: Schematic of a $P3R$ manipulator (Exercise 3.22).

PROGRAMMING EXERCISE (PART 3)

1. Write a subroutine to compute the kinematics of the planar $3R$ robot in Example 3.3—that is, a routine with the joint angles' values as input, and a frame (the wrist frame relative to the base frame) as output. Use the procedure heading (or equivalent in C)

   ```
   Procedure KIN(VAR theta: vec3; VAR wrelb: frame);
   ```

   where "wrelb" is the wrist frame relative to the base frame, $_W^B T$. The type "frame" consists of a 2×2 rotation matrix and a 2×1 position vector. If desired, you may represent the frame with a 3×3 homogeneous transform in which the third row is [0 0 1]. (The manipulator data are $l_1 = l_2 = 0.5$ meters.)

2. Write a routine that calculates where the tool is, relative to the station frame. The input to the routine is a vector of joint angles:

   Procedure WHERE(VAR theta: vec3; VAR trels: frame);

   Obviously, WHERE must make use of descriptions of the tool frame and the robot base frame in order to compute the location of the tool relative to the station frame. The values of $^W_T T$ and $^S_B T$ should be stored in global memory (or, as a second choice, you may pass them as arguments in WHERE).

3. A tool frame and a station frame for a certain task are defined as follows by the user:

$$^W_T T = [x\ y\ \theta] = [0.1\ 0.2\ 30.0],$$

$$^B_S T = [x\ y\ \theta] = [-0.1\ 0.3\ 0.0].$$

   Calculate the position and orientation of the tool relative to the station frame for the following three configurations (in units of degrees) of the arm:

$$[\theta_1\ \theta_2\ \theta_3] = [0.0\ 90.0\ -90.0],$$

$$[\theta_1\ \theta_2\ \theta_3] = [-23.6\ -30.3\ 48.0],$$

$$[\theta_1\ \theta_2\ \theta_3] = [130.0\ 40.0\ 12.0].$$

MATLAB EXERCISE 3

This exercise focuses on DH parameters and on the forward-pose (position and orientation) kinematics transformation for the planar 3-DOF, 3R robot (of Figures 3.6 and 3.7). The following fixed-length parameters are given: $L_1 = 4$, $L_2 = 3$, and $L_3 = 2$ (m).

a) Derive the DH parameters. You can check your results against Figure 3.8.

b) Derive the neighboring homogeneous transformation matrices $^{i-1}_i T$, $i = 1, 2, 3$. These are functions of the joint-angle variables θ_i, $i = 1, 2, 3$. Also, derive the constant $^3_H T$ by inspection: The origin of $\{H\}$ is in the center of the gripper fingers, and the orientation of $\{H\}$ is always the same as the orientation of $\{3\}$.

c) Use Symbolic MATLAB to derive the forward-pose kinematics solution $^0_3 T$ and $^0_H T$ symbolically (as a function of θ_i). Abbreviate your answer, using $s_i = \sin(\theta_i)$, $c_i = \cos(\theta_i)$, and so on. Also, there is a $(\theta_1 + \theta_2 + \theta_3)$ simplification, by using sum-of-angle formulas, that is due to the parallel $\hat{Z}_i$ axes. Calculate the forward-pose kinematics results (both $^0_3 T$ and $^0_H T$) via MATLAB for the following input cases:

   i) $\Theta = \{\theta_1\ \theta_2\ \theta_3\}^T = \{0\ 0\ 0\}^T$.

   ii) $\Theta = \{10°\ 20°\ 30°\}^T$.

   iii) $\Theta = \{90°\ 90°\ 90°\}^T$.

   For all three cases, check your results by sketching the manipulator configuration and deriving the forward-pose kinematics transformation by inspection. (Think of the definition of $^0_H T$ in terms of a rotation matrix and a position vector.) Include frames $\{H\}$, $\{3\}$, and $\{0\}$ in your sketches.

d) Check all your results by means of the Corke MATLAB Robotics Toolbox. Try functions *link()*, *robot()*, and *fkine()*.

CHAPTER 4

Inverse manipulator kinematics

4.1 INTRODUCTION

In the last chapter, we considered the problem of computing the position and orientation of the tool relative to the user's workstation when given the joint angles of the manipulator. In this chapter, we investigate the more difficult converse problem: Given the desired position and orientation of the tool relative to the station, how do we compute the set of joint angles which will achieve this desired result? Whereas Chapter 3 focused on the **direct kinematics** of manipulators, here the focus is the **inverse kinematics** of manipulators.

Solving the problem of finding the required joint angles to place the tool frame, $\{T\}$, relative to the station frame, $\{S\}$, is split into two parts. First, frame transformations are performed to find the wrist frame, $\{W\}$, relative to the base frame, $\{B\}$, and then the inverse kinematics are used to solve for the joint angles.

4.2 SOLVABILITY

The problem of solving the kinematic equations of a manipulator is a nonlinear one. Given the numerical value of ${}^0_N T$, we attempt to find values of $\theta_1, \theta_2, \ldots, \theta_n$. Consider the equations given in (3.14). In the case of the PUMA 560 manipulator, the precise statement of our current problem is as follows: Given ${}^0_6 T$ as sixteen numeric values (four of which are trivial), solve (3.14) for the six joint angles θ_1 through θ_6.

For the case of an arm with six degrees of freedom (like the one corresponding to the equations in (3.14)), we have 12 equations and six unknowns. However, among the 9 equations arising from the rotation-matrix portion of ${}^0_6 T$, only 3 are independent. These, added to the 3 equations from the position-vector portion of ${}^0_6 T$, give 6 equations with six unknowns. These equations are nonlinear, transcendental equations, which can be quite difficult to solve. The equations of (3.14) are those of a robot that had very simple link parameters—many of the α_i were 0 or ±90 degrees. Many link offsets and lengths were zero. It is easy to imagine that, for the case of a general mechanism with six degrees of freedom (with all link parameters nonzero) the kinematic equations would be much more complex than those of (3.14). As with any nonlinear set of equations, we must concern ourselves with the existence of solutions, with multiple solutions, and with the method of solution.

Existence of solutions

The question of whether any solution exists at all raises the question of the manipulator's **workspace**. Roughly speaking, workspace is that volume of space that the end-effector of the manipulator can reach. For a solution to exist, the specified goal point must lie within the workspace. Sometimes, it is useful to consider two definitions of workspace: **Dextrous workspace** is that volume of space that the robot end-effector can reach with all orientations. That is, at each point in the dextrous workspace, the end-effector can be arbitrarily oriented. The **reachable workspace** is that volume of space that the robot can reach in at least one orientation. Clearly, the dextrous workspace is a subset of the reachable workspace.

   Consider the workspace of the two-link manipulator in Fig. 4.1. If $l_1 = l_2$, then the reachable workspace consists of a disc of radius $2l_1$. The dextrous workspace consists of only a single point, the origin. If $l_1 \neq l_2$, then there is no dextrous workspace, and the reachable workspace becomes a ring of outer radius $l_1 + l_2$ and inner radius $|l_1 - l_2|$. Inside the reachable workspace there are two possible orientations of the end-effector. On the boundaries of the workspace there is only one possible orientation.

   These considerations of workspace for the two-link manipulator have assumed that all the joints can rotate 360 degrees. This is rarely true for actual mechanisms. When joint limits are a subset of the full 360 degrees, then the workspace is obviously correspondingly reduced, either in extent, or in the number of possible orientations

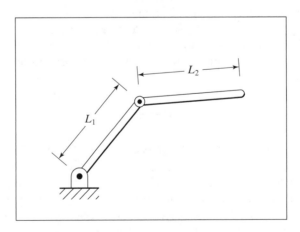

FIGURE 4.1: Two-link manipulator with link lengths l_1 and l_2.

attainable. For example, if the arm in Fig. 4.1 has full 360-degree motion for θ_1, but only $0 \leq \theta_2 \leq 180°$, then the reachable workspace has the same extent, but only one orientation is attainable at each point.

When a manipulator has fewer than six degrees of freedom, it cannot attain general goal positions and orientations in 3-space. Clearly, the planar manipulator in Fig. 4.1 cannot reach out of the plane, so any goal point with a nonzero Z-coordinate value can be quickly rejected as unreachable. In many realistic situations, manipulators with four or five degrees of freedom are employed that operate out of a plane, but that clearly cannot reach general goals. Each such manipulator must be studied to understand its workspace. In general, the workspace of such a robot is a subset of a subspace that can be associated with any particular robot. Given a general goal-frame specification, an interesting problem arises in connection with manipulators having fewer than six degrees of freedom: What is the nearest attainable goal frame?

Workspace also depends on the tool-frame transformation, because it is usually the tool-tip that is discussed when we speak of reachable points in space. Generally, the tool transformation is performed independently of the manipulator kinematics and inverse kinematics, so we are often led to consider the workspace of the wrist frame, $\{W\}$. For a given end-effector, a tool frame, $\{T\}$, is defined; given a goal frame, $\{G\}$, the corresponding $\{W\}$ frame is calculated, and then we ask: Does this desired position and orientation of $\{W\}$ lie in the workspace? In this way, the workspace that we must concern ourselves with (in a computational sense) is different from the one imagined by the user, who is concerned with the workspace of the end-effector (the $\{T\}$ frame).

If the desired position and orientation of the wrist frame is in the workspace, then at least one solution exists.

Multiple solutions

Another possible problem encountered in solving kinematic equations is that of multiple solutions. A planar arm with three revolute joints has a large dextrous workspace in the plane (given "good" link lengths and large joint ranges), because any position in the interior of its workspace can be reached with any orientation. Figure 4.2 shows a three-link planar arm with its end-effector at a certain position

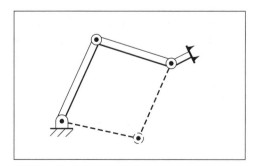

FIGURE 4.2: Three-link manipulator. Dashed lines indicate a second solution.

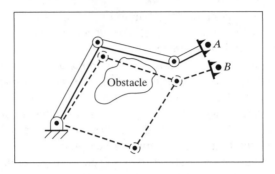

FIGURE 4.3: One of the two possible solutions to reach point B causes a collision.

and orientation. The dashed lines indicate a second possible configuration in which the same end-effector position and orientation are achieved.

The fact that a manipulator has multiple solutions can cause problems, because the system has to be able to choose one. The criteria upon which to base a decision vary, but a very reasonable choice would be the *closest* solution. For example, if the manipulator is at point A, as in Fig. 4.3, and we wish to move it to point B, a good choice would be the solution that minimizes the amount that each joint is required to move. Hence, in the absence of the obstacle, the upper dashed configuration in Fig. 4.3 would be chosen. This suggests that one input argument to our kinematic inverse procedure might be the present position of the manipulator. In this way, if there is a choice, our algorithm can choose the solution closest in joint-space. However, the notion of "close" might be defined in several ways. For example, typical robots could have three large links followed by three smaller, orienting links near the end-effector. In this case, weights might be applied in the calculation of which solution is "closer" so that the selection favors moving smaller joints rather than moving the large joints, when a choice exists. The presence of obstacles might force a "farther" solution to be chosen in cases where the "closer" solution would cause a collision—in general, then, we need to be able to calculate all the possible solutions. Thus, in Fig. 4.3, the presence of the obstacle implies that the lower dashed configuration is to be used to reach point B.

The number of solutions depends upon the number of joints in the manipulator but is also a function of the link parameters (α_i, a_i, and d_i for a rotary joint manipulator) and the allowable ranges of motion of the joints. For example, the PUMA 560 can reach certain goals with eight different solutions. Figure 4.4 shows four solutions; all place the hand with the same position and orientation. For each solution pictured, there is another solution in which the last three joints "flip" to an alternate configuration according to the following formulas:

$$\theta_4' = \theta_4 + 180°,$$
$$\theta_5' = -\theta_5, \tag{4.1}$$
$$\theta_6' = \theta_6 + 180°.$$

So, in total, there can be eight solutions for a single goal. Because of limits on joint ranges, some of these eight could be inaccessible.

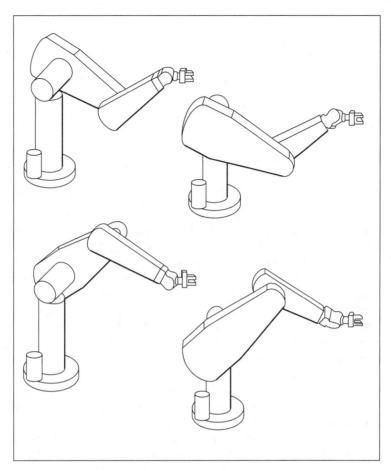

FIGURE 4.4: Four solutions of the PUMA 560.

In general, the more nonzero link parameters there are, the more ways there will be to reach a certain goal. For example, consider a manipulator with six rotational joints. Figure 4.5 shows how the maximum number of solutions is related to how many of the link length parameters (the a_i) are zero. The more that are nonzero, the bigger is the maximum number of solutions. For a completely general rotary-jointed manipulator with six degrees of freedom, there are up to sixteen solutions possible [1, 6].

Method of solution

Unlike linear equations, there are no general algorithms that may be employed to solve a set of nonlinear equations. In considering methods of solution, it will be wise to define what constitutes the "solution" of a given manipulator.

A manipulator will be considered solvable if the joint variables can be determined by an algorithm that allows one to determine *all* the sets of joint variables associated with a given position and orientation [2].

a_i	Number of solutions
$a_1 = a_3 = a_5 = 0$	$\leqslant 4$
$a_3 = a_5 = 0$	$\leqslant 8$
$a_3 = 0$	$\leqslant 16$
All $a_i \neq 0$	$\leqslant 16$

FIGURE 4.5: Number of solutions vs. nonzero a_i.

The main point of this definition is that we require, in the case of multiple solutions, that it be possible to calculate all solutions. Hence, we do not consider some numerical iterative procedures as solving the manipulator—namely, those methods not guaranteed to find all the solutions.

We will split all proposed manipulator solution strategies into two broad classes: **closed-form solutions** and **numerical solutions**. Because of their iterative nature, numerical solutions generally are much slower than the corresponding closed-form solution; so much so, in fact, that, for most uses, we are not interested in the numerical approach to solution of kinematics. Iterative numerical solution to kinematic equations is a whole field of study in itself (see [6,11,12]) and is beyond the scope of this text.

We will restrict our attention to closed-form solution methods. In this context, "closed form" means a solution method based on analytic expressions or on the solution of a polynomial of degree 4 or less, such that noniterative calculations suffice to arrive at a solution. Within the class of closed-form solutions, we distinguish two methods of obtaining the solution: **algebraic** and **geometric**. These distinctions are somewhat hazy: Any geometric methods brought to bear are applied by means of algebraic expressions, so the two methods are similar. The methods differ perhaps in approach only.

A major recent result in kinematics is that, according to our definition of solvability, *all systems with revolute and prismatic joints having a total of six degrees of freedom in a single series chain are solvable.* However, this general solution is a numerical one. Only in special cases can robots with six degrees of freedom be solved analytically. These robots for which an analytic (or closed-form) solution exists are characterized either by having several intersecting joint axes or by having many α_i equal to 0 or ± 90 degrees. Calculating numerical solutions is generally time consuming relative to evaluating analytic expressions; hence, it is considered very important to design a manipulator so that a closed-form solution exists. Manipulator designers discovered this very soon, and now virtually all industrial manipulators are designed sufficiently simply that a closed-form solution can be developed.

A sufficient condition that a manipulator with six revolute joints have a closed-form solution is that three neighboring joint axes intersect at a point. Section 4.6 discusses this condition. Almost every manipulator with six degrees of freedom built today has three axes intersecting. For example, axes 4, 5, and 6 of the PUMA 560 intersect.

4.3   THE NOTION OF MANIPULATOR SUBSPACE WHEN *n* < 6

The set of reachable goal frames for a given manipulator constitutes its reachable workspace. For a manipulator with n degrees of freedom (where $n < 6$), this reachable workspace can be thought of as a portion of an n-degree-of-freedom **subspace**. In the same manner in which the workspace of a six-degree-of-freedom manipulator is a subset of space, the workspace of a simpler manipulator is a subset of its subspace. For example, the subspace of the two-link robot of Fig. 4.1 is a plane, but the workspace is a subset of this plane, namely a circle of radius $l_1 + l_2$ for the case that $l_1 = l_2$.

One way to specify the subspace of an n-degree-of-freedom manipulator is to give an expression for its wrist or tool frame as a function of n variables that locate it. If we consider these n variables to be free, then, as they take on all possible values, the subspace is generated.

EXAMPLE 4.1

Give a description of the subspace of $^B_W T$ for the three-link manipulator from Chapter 3, Fig. 3.6.

The subspace of $^B_W T$ is given by

$$
^B_W T = \begin{bmatrix} c_\phi & -s_\phi & 0.0 & x \\ s_\phi & c_\phi & 0.0 & y \\ 0.0 & 0.0 & 1.0 & 0.0 \\ 0 & 0 & 0 & 1 \end{bmatrix}, \tag{4.2}
$$

where x and y give the position of the wrist and ϕ describes the orientation of the terminal link. As x, y, and ϕ are allowed to take on arbitrary values, the subspace is generated. Any wrist frame that does not have the structure of (4.2) lies outside the subspace (and therefore lies outside the workspace) of this manipulator. Link lengths and joint limits restrict the workspace of the manipulator to be a subset of this subspace.

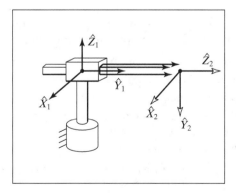

FIGURE 4.6: A polar two-link manipulator.

EXAMPLE 4.2

Give a description of the subspace of ^0_2T for the polar manipulator with two degrees of freedom shown in Fig. 4.6. We have

$$^0P_{2ORG} = \begin{bmatrix} x \\ y \\ 0 \end{bmatrix}, \tag{4.3}$$

where x and y can take any values. The orientation is restricted because the $^0\hat{Z}_2$ axis must point in a direction that depends on x and y. The $^0\hat{Y}_2$ axis always points down, and the $^0\hat{X}_2$ axis can be computed as the cross product $^0\hat{Y}_2 \times ^0\hat{Z}_2$. In terms of x and y, we have

$$^0\hat{Z}_2 = \begin{bmatrix} \dfrac{x}{\sqrt{x^2+y^2}} \\ \dfrac{y}{\sqrt{x^2+y^2}} \\ 0 \end{bmatrix}. \tag{4.4}$$

The subspace can therefore be given as

$$^0_2T = \begin{bmatrix} \dfrac{y}{\sqrt{x^2+y^2}} & 0 & \dfrac{x}{\sqrt{x^2+y^2}} & x \\ \dfrac{-x}{\sqrt{x^2+y^2}} & 0 & \dfrac{y}{\sqrt{x^2+y^2}} & y \\ 0 & -1 & 0 & 0 \\ 0 & 0 & 0 & 1 \end{bmatrix}. \tag{4.5}$$

Usually, in defining a goal for a manipulator with n degrees of freedom, we use n parameters to specify the goal. If, on the other hand, we give a specification of a full six degrees of freedom, we will not in general be able to reach the goal with an $n < 6$ manipulator. In this case, we might be interested instead in reaching a goal that lies in the manipulator's subspace and is as "near" as possible to the original desired goal.

Hence, when specifying *general* goals for a manipulator with fewer than six degrees of freedom, one solution strategy is the following:

1. Given a general goal frame, ^S_GT, compute a modified goal frame, $^S_{G'}T$, such that $^S_{G'}T$ lies in the manipulator's subspace and is as "near" to ^S_GT as possible. A definition of "near" must be chosen.

2. Compute the inverse kinematics to find joint angles using $^S_{G'}T$ as the desired goal. Note that a solution still might not be possible if the goal point is not in the manipulator's workspace.

It generally makes sense to position the tool-frame origin to the desired location and then choose an attainable orientation that is near the desired orientation. As we saw in Examples 4.1 and 4.2, computation of the subspace is dependent on manipulator geometry. Each manipulator must be individually considered to arrive at a method of making this computation.

Section 4.7 gives an example of *projecting* a general goal into the subspace of a manipulator with five degrees of freedom in order to compute joint angles that will result in the manipulator's reaching the attainable frame nearest to the desired one.

4.4 ALGEBRAIC VS. GEOMETRIC

As an introduction to solving kinematic equations, we will consider two different approaches to the solution of a simple planar three-link manipulator.

Algebraic solution

Consider the three-link planar manipulator introduced in Chapter 3. It is shown with its link parameters in Fig. 4.7.

Following the method of Chapter 3, we can use the link parameters easily to find the kinematic equations of this arm:

$$
{}_{W}^{B}T = {}_{3}^{0}T = \begin{bmatrix} c_{123} & -s_{123} & 0.0 & l_1 c_1 + l_2 c_{12} \\ s_{123} & c_{123} & 0.0 & l_1 s_1 + l_2 s_{12} \\ 0.0 & 0.0 & 1.0 & 0.0 \\ 0 & 0 & 0 & 1 \end{bmatrix}. \tag{4.6}
$$

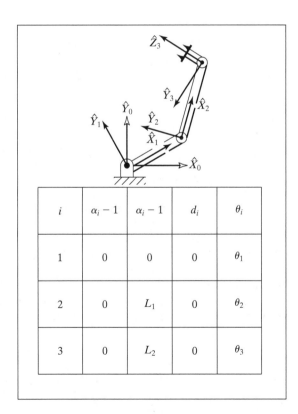

i	$\alpha_i - 1$	$\alpha_i - 1$	d_i	θ_i
1	0	0	0	θ_1
2	0	L_1	0	θ_2
3	0	L_2	0	θ_3

FIGURE 4.7: Three-link planar manipulator and its link parameters.

To focus our discussion on inverse kinematics, we will assume that the necessary transformations have been performed so that the goal point is a specification of the wrist frame relative to the base frame, that is, ${}^{B}_{W}T$. Because we are working with a planar manipulator, specification of these goal points can be accomplished most easily by specifying three numbers: x, y, and ϕ, where ϕ is the orientation of link 3 in the plane (relative to the $+\hat{X}$ axis). Hence, rather than giving a general ${}^{B}_{W}T$ as a goal specification, we will assume a transformation with the structure

$${}^{B}_{W}T = \begin{bmatrix} c_\phi & -s_\phi & 0.0 & x \\ s_\phi & c_\phi & 0.0 & y \\ 0.0 & 0.0 & 1.0 & 0.0 \\ 0 & 0 & 0 & 1 \end{bmatrix}. \tag{4.7}$$

All attainable goals must lie in the subspace implied by the structure of equation (4.7). By equating (4.6) and (4.7), we arrive at a set of four nonlinear equations that must be solved for θ_1, θ_2, and θ_3:

$$c_\phi = c_{123}, \tag{4.8}$$

$$s_\phi = s_{123}, \tag{4.9}$$

$$x = l_1 c_1 + l_2 c_{12}, \tag{4.10}$$

$$y = l_1 s_1 + l_2 s_{12}. \tag{4.11}$$

We now begin our algebraic solution of equations (4.8) through (4.11). If we square both (4.10) and (4.11) and add them, we obtain

$$x^2 + y^2 = l_1^2 + l_2^2 + 2l_1 l_2 c_2, \tag{4.12}$$

where we have made use of

$$c_{12} = c_1 c_2 - s_1 s_2,$$

$$s_{12} = c_1 s_2 + s_1 c_2. \tag{4.13}$$

Solving (4.12) for c_2, we obtain

$$c_2 = \frac{x^2 + y^2 - l_1^2 - l_2^2}{2l_1 l_2}. \tag{4.14}$$

In order for a solution to exist, the right-hand side of (4.14) must have a value between -1 and 1. In the solution algorithm, this constraint would be checked at this time to find out whether a solution exists. Physically, if this constraint is not satisfied, then the goal point is too far away for the manipulator to reach.

Assuming the goal is in the workspace, we write an expression for s_2 as

$$s_2 = \pm\sqrt{1 - c_2^2}. \tag{4.15}$$

Finally, we compute θ_2, using the two-argument arctangent routine[1]:

$$\theta_2 = \text{Atan2}(s_2, c_2). \tag{4.16}$$

---

[1] See Section 2.8.

The choice of signs in (4.15) corresponds to the multiple solution in which we can choose the "elbow-up" or the "elbow-down" solution. In determining θ_2, we have used one of the recurring methods for solving the type of kinematic relationships that often arise, namely, to determine both the sine and cosine of the desired joint angle and then apply the two-argument arctangent. This ensures that we have found all solutions and that the solved angle is in the proper quadrant.

Having found θ_2, we can solve (4.10) and (4.11) for θ_1. We write (4.10) and (4.11) in the form

$$x = k_1 c_1 - k_2 s_1, \tag{4.17}$$

$$y = k_1 s_1 + k_2 c_1, \tag{4.18}$$

where

$$k_1 = l_1 + l_2 c_2,$$

$$k_2 = l_2 s_2. \tag{4.19}$$

In order to solve an equation of this form, we perform a change of variables. Actually, we are changing the way in which we write the constants k_1 and k_2. If

$$r = +\sqrt{k_1^2 + k_2^2} \tag{4.20}$$

and

$$\gamma = \text{Atan2}(k_2, k_1),$$

then

$$k_1 = r \cos \gamma,$$

$$k_2 = r \sin \gamma. \tag{4.21}$$

Equations (4.17) and (4.18) can now be written as

$$\frac{x}{r} = \cos \gamma \cos \theta_1 - \sin \gamma \sin \theta_1, \tag{4.22}$$

$$\frac{y}{r} = \cos \gamma \sin \theta_1 + \sin \gamma \cos \theta_1, \tag{4.23}$$

so

$$\cos(\gamma + \theta_1) = \frac{x}{r}, \tag{4.24}$$

$$\sin(\gamma + \theta_1) = \frac{y}{r}. \tag{4.25}$$

Using the two-argument arctangent, we get

$$\gamma + \theta_1 = \text{Atan2}\left(\frac{y}{r}, \frac{x}{r}\right) = \text{Atan2}(y, x), \tag{4.26}$$

and so

$$\theta_1 = \text{Atan2}(y, x) - \text{Atan2}(k_2, k_1). \tag{4.27}$$

Note that, when a choice of sign is made in the solution of θ_2 above, it will cause a sign change in k_2, thus affecting θ_1. The substitutions used, (4.20) and (4.21), constitute a method of solution of a form appearing frequently in kinematics—namely, that of (4.10) or (4.11). Note also that, if $x = y = 0$, then (4.27) becomes undefined—in this case, θ_1 is arbitrary.

Finally, from (4.8) and (4.9), we can solve for the sum of θ_1 through θ_3:

$$\theta_1 + \theta_2 + \theta_3 = \text{Atan2}(s_\phi, c_\phi) = \phi. \tag{4.28}$$

From this, we can solve for θ_3, because we know the first two angles. It is typical with manipulators that have two or more links moving in a plane that, in the course of solution, expressions for sums of joint angles arise.

In summary, an algebraic approach to solving kinematic equations is basically one of manipulating the given equations into a form for which a solution is known. It turns out that, for many common geometries, several forms of transcendental equations commonly arise. We have encountered a couple of them in this preceding section. In Appendix C, more are listed.

Geometric solution

In a geometric approach to finding a manipulator's solution, we try to decompose the spatial geometry of the arm into several plane-geometry problems. For many manipulators (particularly when the $\alpha_i = 0$ or ± 90) this can be done quite easily. Joint angles can then be solved for by using the tools of plane geometry [7]. For the arm with three degrees of freedom shown in Fig. 4.7, because the arm is planar, we can apply plane geometry directly to find a solution.

Figure 4.8 shows the triangle formed by l_1, l_2, and the line joining the origin of frame {0} with the origin of frame {3}. The dashed lines represent the other possible configuration of the triangle, which would lead to the same position of the frame {3}. Considering the solid triangle, we can apply the "law of cosines" to solve for θ_2:

$$x^2 + y^2 = l_1^2 + l_2^2 - 2l_1 l_2 \cos(180 + \theta_2). \tag{4.29}$$

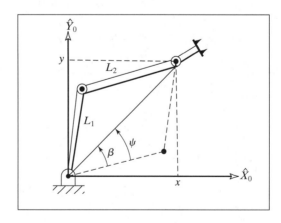

FIGURE 4.8: Plane geometry associated with a three-link planar robot.

Now; $\cos(180 + \theta_2) = -\cos(\theta_2)$, so we have

$$c_2 = \frac{x^2 + y^2 - l_1^2 - l_2^2}{2l_1l_2}. \tag{4.30}$$

In order for this triangle to exist, the distance to the goal point $\sqrt{x^2 + y^2}$ must be less than or equal to the sum of the link lengths, $l_1 + l_2$. This condition would be checked at this point in a computational algorithm to verify existence of solutions. This condition is not satisfied when the goal point is out of reach of the manipulator. Assuming a solution exists, this equation is solved for that value of θ_2 that lies between 0 and -180 degrees, because only for these values does the triangle in Fig. 4.8 exist. The other possible solution (the one indicated by the dashed-line triangle) is found by symmetry to be $\theta_2' = -\theta_2$.

To solve for θ_1, we find expressions for angles ψ and β as indicated in Fig. 4.8. First, β may be in any quadrant, depending on the signs of x and y. So we must use a two-argument arctangent:

$$\beta = \text{Atan2}(y, x). \tag{4.31}$$

We again apply the law of cosines to find ψ:

$$\cos\psi = \frac{x^2 + y^2 + l_1^2 - l_2^2}{2l_1\sqrt{x^2 + y^2}}. \tag{4.32}$$

Here, the arccosine must be solved so that $0 \leq \psi \leq 180°$, in order that the geometry which leads to (4.32) will be preserved. These considerations are typical when using a geometric approach—we must apply the formulas we derive only over a range of variables such that the geometry is preserved. Then we have

$$\theta_1 = \beta \pm \psi, \tag{4.33}$$

where the plus sign is used if $\theta_2 < 0$ and the minus sign if $\theta_2 > 0$.

We know that angles in a plane add, so the sum of the three joint angles must be the orientation of the last link:

$$\theta_1 + \theta_2 + \theta_3 = \phi. \tag{4.34}$$

This equation is solved for θ_3 to complete our solution.

4.5   ALGEBRAIC SOLUTION BY REDUCTION TO POLYNOMIAL

Transcendental equations are often difficult to solve because, even when there is only one variable (say, θ), it generally appears as $\sin\theta$ and $\cos\theta$. Making the following substitutions, however, yields an expression in terms of a single variable, u:

$$u = \tan\frac{\theta}{2},$$

$$\cos\theta = \frac{1 - u^2}{1 + u^2}, \tag{4.35}$$

$$\sin\theta = \frac{2u}{1 + u^2}.$$

This is a very important geometric substitution used often in solving kinematic equations. These substitutions convert transcendental equations into polynomial equations in u. Appendix A lists these and other trigonometric identities.

EXAMPLE 4.3

Convert the transcendental equation

$$a \cos \theta + b \sin \theta = c \tag{4.36}$$

into a polynomial in the tangent of the half angle, and solve for θ.

Substituting from (4.35) and multiplying through by $1 + u^2$, we have

$$a(1 - u^2) + 2bu = c(1 + u^2). \tag{4.37}$$

Collecting powers of u yields

$$(a + c)u^2 - 2bu + (c - a) = 0, \tag{4.38}$$

which is solved by the quadratic formula:

$$u = \frac{b \pm \sqrt{b^2 + a^2 - c^2}}{a + c}. \tag{4.39}$$

Hence,

$$\theta = 2 \tan^{-1} \left(\frac{b \pm \sqrt{b^2 + a^2 - c^2}}{a + c} \right). \tag{4.40}$$

Should the solution for u from (4.39) be complex, there is no real solution to the original transcendental equation. Note that, if $a + c = 0$, the argument of the arctangent becomes infinity and hence $\theta = 180°$. In a computer implementation, this potential division by zero should be checked for ahead of time. This situation results when the quadratic term of (4.38) vanishes, so that the quadratic degenerates into a linear equation.

Polynomials up to degree four possess closed-form solutions [8, 9], so manipulators sufficiently simple that they can be solved by algebraic equations of this degree (or lower) are called **closed-form-solvable** manipulators.

4.6   PIEPER'S SOLUTION WHEN THREE AXES INTERSECT

As mentioned earlier, although a completely general robot with six degrees of freedom does not have a closed-form solution, certain important special cases can be solved. Pieper [3, 4] studied manipulators with six degrees of freedom in which three consecutive axes intersect at a point.[2] In this section, we outline the method he developed for the case of all six joints revolute, with the last three axes intersecting. His method applies to other configurations, which include prismatic

---

[2]Included in this family of manipulators are those with three consecutive parallel axes, because they meet at the point at infinity.

joints, and the interested reader should see [4]. Pieper's work applies to the majority of commercially available industrial robots.

When the last three axes intersect, the origins of link frames {4}, {5}, and {6} are all located at this point of intersection. This point is given in base coordinates as

$$^0P_{4ORG} = {}^0_1T\,{}^1_2T\,{}^2_3T\,{}^3P_{4ORG} = \begin{bmatrix} x \\ y \\ z \\ 1 \end{bmatrix},$$ (4.41)

or, using the fourth column of (3.6) for $i = 4$, as

$$^0P_{4ORG} = {}^0_1T\,{}^1_2T\,{}^2_3T \begin{bmatrix} a_3 \\ -d_4s\alpha_3 \\ d_4c\alpha_3 \\ 1 \end{bmatrix},$$ (4.42)

or as

$$^0P_{4ORG} = {}^0_1T\,{}^1_2T \begin{bmatrix} f_1(\theta_3) \\ f_2(\theta_3) \\ f_3(\theta_3) \\ 1 \end{bmatrix},$$ (4.43)

where

$$\begin{bmatrix} f_1 \\ f_2 \\ f_3 \\ 1 \end{bmatrix} = {}^2_3T \begin{bmatrix} a_3 \\ -d_4s\alpha_3 \\ d_4c\alpha_3 \\ 1 \end{bmatrix}.$$ (4.44)

Using (3.6) for ^2_3T in (4.44) yields the following expressions for f_1:

$$f_1 = a_3c_3 + d_4s\alpha_3s_3 + a_2,$$
$$f_2 = a_3c\alpha_2s_3 - d_4s\alpha_3c\alpha_2c_3 - d_4s\alpha_2c\alpha_3 - d_3s\alpha_2,$$ (4.45)
$$f_3 = a_3s\alpha_2s_3 - d_4s\alpha_3s\alpha_2c_3 + d_4c\alpha_2c\alpha_3 + d_3c\alpha_2.$$

Using (3.6) for ^0_1T and ^1_2T in (4.43), we obtain

$$^0P_{4ORG} = \begin{bmatrix} c_1g_1 - s_1g_2 \\ s_1g_1 + c_1g_2 \\ g_3 \\ 1 \end{bmatrix},$$ (4.46)

where

$$g_1 = c_2f_1 - s_2f_2 + a_1,$$
$$g_2 = s_2c\alpha_1f_1 + c_2c\alpha_1f_2 - s\alpha_1f_3 - d_2s\alpha_1,$$ (4.47)
$$g_3 = s_2s\alpha_1f_1 + c_2s\alpha_1f_2 + c\alpha_1f_3 + d_2c\alpha_1.$$

We now write an expression for the squared magnitude of $^0P_{4ORG}$, which we will denote as $r = x^2 + y^2 + z^2$, and which is seen from (4.46) to be

$$r = g_1^2 + g_2^2 + g_3^2;$$ (4.48)

so, using (4.47) for the g_i, we have

$$r = f_1^2 + f_2^2 + f_3^2 + a_1^2 + d_2^2 + 2d_2f_3 + 2a_1(c_2f_1 - s_2f_2). \qquad (4.49)$$

We now write this equation, along with the Z-component equation from (4.46), as a system of two equations in the form

$$r = (k_1c_2 + k_2s_2)2a_1 + k_3,$$
$$z = (k_1s_2 - k_2c_2)s\alpha_1 + k_4, \qquad (4.50)$$

where

$$k_1 = f_1,$$
$$k_2 = -f_2,$$
$$k_3 = f_1^2 + f_2^2 + f_3^2 + a_1^2 + d_2^2 + 2d_2f_3, \qquad (4.51)$$
$$k_4 = f_3c\alpha_1 + d_2c\alpha_1.$$

Equation (4.50) is useful because dependence on θ_1 has been eliminated and because dependence on θ_2 takes a simple form.

Now let us consider the solution of (4.50) for θ_3. We distinguish three cases:

1. If $a_1 = 0$, then we have $r = k_3$, where r is known. The right-hand side (k_3) is a function of θ_3 only. After the substitution (4.35), a quadratic equation in $\tan \frac{\theta_3}{2}$ may be solved for θ_3.
2. If $s\alpha_1 = 0$, then we have $z = k_4$, where z is known. Again, after substituting via (4.35), a quadratic equation arises that can be solved for θ_3.
3. Otherwise, eliminate s_2 and c_2 from (4.50) to obtain

$$\frac{(r - k_3)^2}{4a_1^2} + \frac{(z - k_4)^2}{s^2\alpha_1} = k_1^2 + k_2^2. \qquad (4.52)$$

This equation, after the (4.35) substitution for θ_3, results in an equation of degree 4, which can be solved for θ_3.[3]

Having solved for θ_3, we can solve (4.50) for θ_2 and (4.46) for θ_1.

To complete our solution, we need to solve for θ_4, θ_5, and θ_6. These axes intersect, so these joint angles affect the orientation of only the last link. We can compute them from nothing more than the rotation portion of the specified goal, ^0_6R. Having obtained θ_1, θ_2, and θ_3, we can compute $^0_4R|_{\theta_4=0}$, by which notation we mean the orientation of link frame {4} relative to the base frame when $\theta_4 = 0$. The desired orientation of {6} differs from this orientation only by the action of the last three joints. Because the problem was specified as given ^0_6R, we can compute

$$^4_6R|_{\theta_4=0} = {^0_4R}^{-1}|_{\theta_4=0} {^0_6R}. \qquad (4.53)$$

---

[3]It is helpful to note that $f_1^2 + f_2^2 + f_3^2 = a_3^2 + d_4^2 + d_3^2 + a_2^2 + 2d_4d_3c\alpha_3 + 2a_2a_3c_3 + 2a_2d_4s\alpha_3s_3$.

For many manipulators, these last three angles can be solved for by using exactly the Z–Y–Z Euler angle solution given in Chapter 2, applied to $^4_6R|_{\theta_4=0}$. For any manipulator (with intersecting axes 4, 5, and 6), the last three joint angles can be solved for as a set of appropriately defined Euler angles. There are always two solutions for these last three joints, so the total number of solutions for the manipulator will be twice the number found for the first three joints.

4.7 EXAMPLES OF INVERSE MANIPULATOR KINEMATICS

In this section, we work out the inverse kinematics of two industrial robots. One manipulator solution is done purely algebraically; the second solution is partially algebraic and partially geometric. The following solutions do not constitute a cookbook method of solving manipulator kinematics, but they do show many of the common manipulations likely to appear in most kinematic solutions. Note that Pieper's method of solution (covered in the preceding section) can be used for these manipulators, but here we choose to approach the solution a different way, to give insight into various available methods.

The Unimation PUMA 560

As an example of the algebraic solution technique applied to a manipulator with six degrees of freedom, we will solve the kinematic equations of the PUMA 560, which were developed in Chapter 3. This solution is in the style of [5].
    We wish to solve

$$
^0_6T = \begin{bmatrix} r_{11} & r_{12} & r_{13} & p_x \\ r_{21} & r_{22} & r_{23} & p_y \\ r_{31} & r_{32} & r_{33} & p_z \\ 0 & 0 & 0 & 1 \end{bmatrix}
$$

$$
= {}^0_1T(\theta_1)\,{}^1_2T(\theta_2)\,{}^2_3T(\theta_3)\,{}^3_4T(\theta_4)\,{}^4_5T(\theta_5)\,{}^5_6T(\theta_6) \tag{4.54}
$$

for θ_i when ^0_6T is given as numeric values.
    A restatement of (4.54) that puts the dependence on θ_1 on the left-hand side of the equation is

$$
[{}^0_1T(\theta_1)]^{-1}\,{}^0_6T = {}^1_2T(\theta_2)\,{}^2_3T(\theta_3)\,{}^3_4T(\theta_4)\,{}^4_5T(\theta_5)\,{}^5_6T(\theta_6). \tag{4.55}
$$

Inverting ^0_1T, we write (4.55) as

$$
\begin{bmatrix} c_1 & s_1 & 0 & 0 \\ -s_1 & c_1 & 0 & 0 \\ 0 & 0 & 1 & 0 \\ 0 & 0 & 0 & 1 \end{bmatrix} \begin{bmatrix} r_{11} & r_{12} & r_{13} & p_x \\ r_{21} & r_{22} & r_{23} & p_y \\ r_{31} & r_{32} & r_{33} & p_z \\ 0 & 0 & 0 & 1 \end{bmatrix} = {}^1_6T, \tag{4.56}
$$

where ^1_6T is given by equation (3.13) developed in Chapter 3. This simple technique of multiplying each side of a transform equation by an inverse is often used to advantage in separating out variables in the search for a solvable equation.
    Equating the (2, 4) elements from both sides of (4.56), we have

$$
-s_1 p_x + c_1 p_y = d_3. \tag{4.57}
$$

To solve an equation of this form, we make the trigonometric substitutions

$$p_x = \rho \cos \phi,$$
$$p_y = \rho \sin \phi, \qquad (4.58)$$

where

$$\rho = \sqrt{p_x^2 + p_y^2},$$
$$\phi = \text{Atan2}(p_y, p_x). \qquad (4.59)$$

Substituting (4.58) into (4.57), we obtain

$$c_1 s_\phi - s_1 c_\phi = \frac{d_3}{\rho}. \qquad (4.60)$$

From the difference-of-angles formula,

$$\sin(\phi - \theta_1) = \frac{d_3}{\rho}. \qquad (4.61)$$

Hence,

$$\cos(\phi - \theta_1) = \pm\sqrt{1 - \frac{d_3^2}{\rho^2}}, \qquad (4.62)$$

and so

$$\phi - \theta_1 = \text{Atan2}\left(\frac{d_3}{\rho}, \pm\sqrt{1 - \frac{d_3^2}{\rho^2}}\right). \qquad (4.63)$$

Finally, the solution for θ_1 may be written as

$$\theta_1 = \text{Atan2}(p_y, p_x) - \text{Atan2}\left(d_3, \pm\sqrt{p_x^2 + p_y^2 - d_3^2}\right). \qquad (4.64)$$

Note that we have found two possible solutions for θ_1, corresponding to the plus-or-minus sign in (4.64). Now that θ_1 is known, the left-hand side of (4.56) is known. If we equate both the (1,4) elements and the (3,4) elements from the two sides of (4.56), we obtain

$$c_1 p_x + s_1 p_y = a_3 c_{23} - d_4 s_{23} + a_2 c_2,$$
$$-p_x = a_3 s_{23} + d_4 c_{23} + a_2 s_2. \qquad (4.65)$$

If we square equations (4.65) and (4.57) and add the resulting equations, we obtain

$$a_3 c_3 - d_4 s_3 = K, \qquad (4.66)$$

where

$$K = \frac{p_x^2 + p_y^2 + p_x^2 - a_2^2 - a_3^2 - d_3^2 - d_4^2}{2a_2}. \qquad (4.67)$$

Note that dependence on θ_1 has been removed from (4.66). Equation (4.66) is of the same form as (4.57) and so can be solved by the same kind of trigonometric substitution to yield a solution for θ_3:

$$\theta_3 = \text{Atan2}(a_3, d_4) - \text{Atan2}(K, \pm\sqrt{a_3^2 + d_4^2 - K^2}). \tag{4.68}$$

The plus-or-minus sign in (4.68) leads to two different solutions for θ_3. If we consider (4.54) again, we can now rewrite it so that all the left-hand side is a function of only knowns and θ_2:

$$[{}_3^0T(\theta_2)]^{-1} {}_6^0T = {}_4^3T(\theta_4){}_5^4T(\theta_5){}_6^5T(\theta_6), \tag{4.69}$$

or

$$\begin{bmatrix} c_1 c_{23} & s_1 c_{23} & -s_{23} & -a_2 c_3 \\ -c_1 s_{23} & -s_1 s_{23} & -c_{23} & a_2 s_3 \\ -s_1 & c_1 & 0 & -d_3 \\ 0 & 0 & 0 & 1 \end{bmatrix} \begin{bmatrix} r_{11} & r_{12} & r_{13} & p_x \\ r_{21} & r_{22} & r_{23} & p_y \\ r_{31} & r_{32} & r_{33} & p_z \\ 0 & 0 & 0 & 1 \end{bmatrix} = {}_6^3T, \tag{4.70}$$

where ${}_6^3T$ is given by equation (3.11) developed in Chapter 3. Equating both the (1,4) elements and the (2,4) elements from the two sides of (4.70), we get

$$c_1 c_{23} p_x + s_1 c_{23} p_y - s_{23} p_z - a_2 c_3 = a_3,$$

$$-c_1 s_{23} p_x - s_1 s_{23} p_y - c_{23} p_z + a_2 s_3 = d_4. \tag{4.71}$$

These equations can be solved simultaneously for s_{23} and c_{23}, resulting in

$$s_{23} = \frac{(-a_3 - a_2 c_3)p_z + (c_1 p_x + s_1 p_y)(a_2 s_3 - d_4)}{p_z^2 + (c_1 p_x + s_1 p_y)^2},$$

$$c_{23} = \frac{(a_2 s_3 - d_4)p_z - (a_3 + a_2 c_3)(c_1 p_x + s_1 p_y)}{p_z^2 + (c_1 p_x + s_1 p_y)^2}. \tag{4.72}$$

The denominators are equal and positive, so we solve for the sum of θ_2 and θ_3 as

$$\theta_{23} = \text{Atan2}[(-a_3 - a_2 c_3)p_z - (c_1 p_x + s_1 p_y)(d_4 - a_2 s_3),$$

$$(a_2 s_3 - d_4)p_z - (a_3 + a_2 c_3)(c_1 p_x + s_1 p_y)]. \tag{4.73}$$

Equation (4.73) computes four values of θ_{23}, according to the four possible combinations of solutions for θ_1 and θ_3; then, four possible solutions for θ_2 are computed as

$$\theta_2 = \theta_{23} - \theta_3, \tag{4.74}$$

where the appropriate solution for θ_3 is used when forming the difference.

Now the entire left side of (4.70) is known. Equating both the (1,3) elements and the (3,3) elements from the two sides of (4.70), we get

$$r_{13} c_1 c_{23} + r_{23} s_1 c_{23} - r_{33} s_{23} = -c_4 s_5,$$

$$-r_{13} s_1 + r_{23} c_1 = s_4 s_5. \tag{4.75}$$

As long as $s_5 \neq 0$, we can solve for θ_4 as

$$\theta_4 = \text{Atan2}(-r_{13} s_1 + r_{23} c_1, -r_{13} c_1 c_{23} - r_{23} s_1 c_{23} + r_{33} s_{23}). \tag{4.76}$$

When $\theta_5 = 0$, the manipulator is in a singular configuration in which joint axes 4 and 6 line up and cause the same motion of the last link of the robot. In this case, all that matters (and all that can be solved for) is the sum or difference of θ_4 and θ_6. This situation is detected by checking whether both arguments of the Atan2 in (4.76) are near zero. If so, θ_4 is chosen arbitrarily,[4] and when θ_6 is computed later, it will be computed accordingly.

If we consider (4.54) again, we can now rewrite it so that all the left-hand side is a function of only knowns and θ_4, by rewriting it as

$$[{}_4^0T(\theta_4)]^{-1}\,{}_6^0T = {}_5^4T(\theta_5){}_6^5T(\theta_6), \tag{4.77}$$

where $[{}_4^0T(\theta4)]^{-1}$ is given by

$$\begin{bmatrix} c_1c_{23}c_4 + s_1s_4 & s_1c_{23}c_4 - c_1s_4 & -s_{23}c_4 & -a_2c_3c_4 + d_3s_4 - a_3c_4 \\ -c_1c_{23}s_4 + s_1c_4 & -s_1c_{23}s_4 - c_1c_4 & s_{23}s_4 & a_2c_3s_4 + d_3c_4 + a_3s_4 \\ -c_1s_{23} & -s_1s_{23} & -c_{23} & a_2s_3 - d_4 \\ 0 & 0 & 0 & 1 \end{bmatrix}, \tag{4.78}$$

and ${}_6^4T$ is given by equation (3.10) developed in Chapter 3. Equating both the (1,3) elements and the (3,3) elements from the two sides of (4.77), we get

$$r_{13}(c_1c_{23}c_4 + s_1s_4) + r_{23}(s_1c_{23}c_4 - c_1s_4) - r_{33}(s_{23}c_4) = -s_5,$$

$$r_{13}(-c_1s_{23}) + r_{23}(-s_1s_{23}) + r_{33}(-c_{23}) = c_5. \tag{4.79}$$

Hence, we can solve for θ_5 as

$$\theta_5 = \text{Atan2}(s_5, c_5), \tag{4.80}$$

where s_5 and c_5 are given by (4.79).

Applying the same method one more time, we compute $({}_5^0T)^{-1}$ and write (4.54) in the form

$$({}_5^0T)^{-1}\,{}_6^0T = {}_6^5T(\theta_6). \tag{4.81}$$

Equating both the (3,1) elements and the (1,1) elements from the two sides of (4.77) as we have done before, we get

$$\theta_6 = \text{Atan2}(s_6, c_6), \tag{4.82}$$

where

$$s_6 = -r_{11}(c_1c_{23}s_4 - s_1c_4) - r_{21}(s_1c_{23}s_4 + c_1c_4) + r_{31}(s_{23}s_4),$$

$$c_6 = r_{11}[(c_1c_{23}c_4 + s_1s_4)c_5 - c_1s_{23}s_5] + r_{21}[(s_1c_{23}c_4 - c_1s_4)c_5 - s_1s_{23}s_5]$$

$$-r_{31}(s_{23}c_4c_5 + c_{23}s_5).$$

Because of the plus-or-minus signs appearing in (4.64) and (4.68), these equations compute four solutions. Additionally, there are four more solutions obtained by

---

[4]It is usually chosen to be equal to the present value of joint 4.

"flipping" the wrist of the manipulator. For each of the four solutions computed above, we obtain the flipped solution by

$$\theta'_4 = \theta_4 + 180°,$$

$$\theta'_5 = -\theta_5, \tag{4.83}$$

$$\theta'_6 = \theta_6 + 180°.$$

After all eight solutions have been computed, some (or even all) of them might have to be discarded because of joint-limit violations. Of any remaining valid solutions, usually the one closest to the present manipulator configuration is chosen.

The Yasukawa Motoman L-3

As the second example, we will solve the kinematic equations of the Yasukawa Motoman L-3, which were developed in Chapter 3. This solution will be partially algebraic and partially geometric. The Motoman L-3 has three features that make the inverse kinematic problem quite different from that of the PUMA. First, the manipulator has only five joints, so it is not able to position and orient its end-effector in order to attain *general* goal frames. Second, the four-bar type of linkages and chain-drive scheme cause one actuator to move two or more joints. Third, the actuator position limits are not constants, but depend on the positions of the other actuators, so finding out whether a computed set of actuator values is in range is not trivial.

If we consider the nature of the subspace of the Motoman manipulator (and the same applies to many manipulators with five degrees of freedom), we quickly realize that this subspace can be described by giving one constraint on the attainable orientation: The pointing direction of the tool, that is, the $\hat{Z}_T$ axis, must lie in the "plane of the arm." This plane is the vertical plane that contains the axis of joint 1 and the point where axes 4 and 5 intersect. The orientation nearest to a general orientation is the one obtained by rotating the tool's pointing direction so that it lies in the plane, using a minimum amount of rotation. Without developing an explicit expression for this subspace, we will construct a method for projecting a general goal frame into it. Note that this entire discussion is for the case that the wrist frame and tool frame differ only by a translation along $\hat{Z}_w$.

In Fig. 4.9, we indicate the plane of the arm by its normal, $\hat{M}$, and the desired pointing direction of the tool by $\hat{Z}_T$. This pointing direction must be rotated by angle θ about some vector $\hat{K}$ in order to cause the new pointing direction, $\hat{Z}'_T$, to lie in the plane. It is clear that the $\hat{K}$ that minimizes θ lies in the plane and is orthogonal to both $\hat{Z}_T$ and $\hat{Z}'_T$.

For any given goal frame, $\hat{M}$ is defined as

$$\hat{M} = \frac{1}{\sqrt{p_x^2 + p_y^2}} \begin{bmatrix} -p_y \\ p_x \\ 0 \end{bmatrix}, \tag{4.84}$$

where p_x and p_y are the X and Y coordinates of the desired tool position. Then K is given by

$$K = \hat{M} \times \hat{Z}_T. \tag{4.85}$$

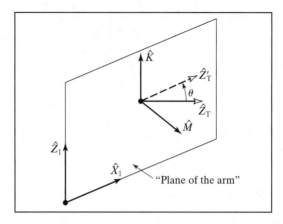

FIGURE 4.9: Rotating a goal frame into the Motoman's subspace.

The new $\hat{Z}'_T$ is

$$\hat{Z}'_T = \hat{K} \times \hat{M}. \tag{4.86}$$

The amount of rotation, θ, is given by

$$\cos\theta = \hat{Z}_T \cdot \hat{Z}'_T,$$
$$\sin\theta = (\hat{Z}_T \times \hat{Z}'_T) \cdot \hat{K}. \tag{4.87}$$

Using Rodriques's formula (see Exercise 2.20), we have

$$\hat{Y}'_T = c\theta\hat{Y}_T + s\theta(\hat{K} \times \hat{Y}_T) + (1 - c\theta)(\hat{K} \cdot \hat{Y}_T)\hat{K}. \tag{4.88}$$

Finally, we compute the remaining unknown column of the new rotation matrix of the tool as

$$\hat{X}'_T = \hat{Y}'_T \times \hat{Z}'_T. \tag{4.89}$$

Equations (4.84) through (4.89) describe a method of projecting a given general goal orientation into the subspace of the Motoman robot.

   Assuming that the given wrist frame, $^B_W T$, lies in the manipulator's subspace, we solve the kinematic equations as follows. In deriving the kinematic equations for the Motoman L-3, we formed the product of link transformations:

$$^0_5T = {}^0_1T\,{}^1_2T\,{}^2_3T\,{}^3_4T\,{}^4_5T. \tag{4.90}$$

If we let

$$^0_5T = \begin{bmatrix} r_{11} & r_{12} & r_{13} & p_x \\ r_{21} & r_{22} & r_{23} & p_y \\ r_{31} & r_{32} & r_{33} & p_z \\ 0 & 0 & 0 & 1 \end{bmatrix} \tag{4.91}$$

and premultiply both sides by $^0_1T^{-1}$, we have

$$^0_1T^{-1}\,{}^0_5T = {}^1_2T\,{}^2_3T\,{}^3_4T\,{}^4_5T, \tag{4.92}$$

where the left-hand side is

$$\begin{bmatrix} c_1 r_{11} + s_1 r_{21} & c_1 r_{12} + s_1 r_{22} & c_1 r_{13} + s_1 r_{23} & c_1 p_x + s_1 p_y \\ -r_{31} & -r_{32} & -r_{33} & -p_z \\ -s_1 r_{11} + c_1 r_{21} & -s_1 r_{12} + c_1 r_{22} & -s_1 r_{13} + c_1 r_{23} & -s_1 p_x + c_1 p_y \\ 0 & 0 & 0 & 1 \end{bmatrix} \qquad (4.93)$$

and the right-hand side is

$$\begin{bmatrix} * & * & s_{234} & * \\ * & * & -c_{234} & * \\ s_5 & c_5 & 0 & 0 \\ 0 & 0 & 0 & 1 \end{bmatrix}; \qquad (4.94)$$

in the latter, several of the elements have not been shown. Equating the (3,4) elements, we get

$$-s_1 p_x + c_1 p_y = 0, \qquad (4.95)$$

which gives us[5]

$$\theta_1 = \text{Atan2}(p_y, p_x). \qquad (4.96)$$

Equating the (3,1) and (3,2) elements, we get

$$s_5 = -s_1 r_{11} + c_1 r_{21},$$
$$c_5 = -s_1 r_{12} + c_1 r_{22}, \qquad (4.97)$$

from which we calculate θ_5 as

$$\theta_5 = \text{Atan2}(r_{21} c_1 - r_{11} s_1, r_{22} c_1 - r_{12} s_1). \qquad (4.98)$$

Equating the (2,3) and (1,3) elements, we get

$$c_{234} = r_{33},$$
$$s_{234} = c_1 r_{13} + s_1 r_{23}, \qquad (4.99)$$

which leads to

$$\theta_{234} = \text{Atan2}(r_{13} c_1 + r_{23} s_1, r_{33}). \qquad (4.100)$$

To solve for the individual angles θ_2, θ_3, and θ_4, we will take a geometric approach. Figure 4.10 shows the plane of the arm with point A at joint axis 2, point B at joint axis 3, and point C at joint axis 4.

From the law of cosines applied to triangle ABC, we have

$$\cos \theta_3 = \frac{p_x^2 + p_y^2 + p_z^2 - l_2^2 - l_3^2}{2 l_2 l_3}. \qquad (4.101)$$

Next, we have[6]

$$\theta_3 = \text{Atan2}\left(\sqrt{1 - \cos^2 \theta_3}, \cos \theta_3\right). \qquad (4.102)$$

---

[5] For this manipulator, a second solution would violate joint limits and so is not calculated.
[6] For this manipulator, a second solution would violate joint limits and so is not calculated.

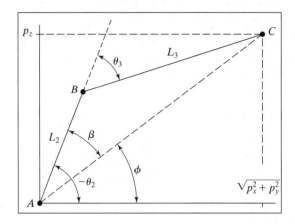

FIGURE 4.10: The plane of the Motoman manipulator.

From Fig. 4.10, we see that $\theta_2 = -\phi - \beta$, or

$$\theta_2 = -\text{Atan2}\left(p_z, \sqrt{p_x^2 + p_y^2}\right) - \text{Atan2}(l_3 \sin\theta_3, l_2 + l_3 \cos\theta_3). \tag{4.103}$$

Finally, we have

$$\theta_4 = \theta_{234} - \theta_2 - \theta_3. \tag{4.104}$$

Having solved for joint angles, we must perform the further computation to obtain the actuator values. Referring to Section 3.7, we solve equation (3.16) for the A_i:

$$A_1 = \frac{1}{k_1}(\theta_1 - \lambda_1),$$

$$A_2 = \frac{1}{k_2}\left(\sqrt{-2\alpha_2\beta_2 \cos\left(\theta_2 - \Omega_2 - \tan^{-1}\left(\frac{\phi_2}{\gamma_2}\right) + 270°\right) + \alpha_2^2 + \beta_2^2} - \lambda_2\right),$$

$$A_3 = \frac{1}{k_3}\left(\sqrt{-2\alpha_3\beta_3 \cos\left(\theta_2 + \theta_3 - \tan^{-1}\left(\frac{\phi_3}{\gamma_3}\right) + 90°\right) + \alpha_3^2 + \beta_3^2} - \lambda_3\right),$$

$$A_4 = \frac{1}{k_4}(180° + \lambda_4 - \theta_2 - \theta_3 - \theta_4),$$

$$A_5 = \frac{1}{k_5}(\lambda_5 - \theta_5). \tag{4.105}$$

The actuators have limited ranges of motion, so we must check that our computed solution is in range. This "in range" check is complicated by the fact that the mechanical arrangement makes actuators interact and affect each other's allowed range of motion. For the Motoman robot, actuators 2 and 3 interact in such a way that the following relationship must always be obeyed:

$$A_2 - 10,000 > A_3 > A_2 + 3000. \tag{4.106}$$

That is, the limits of actuator 3 are a function of the position of actuator 2. Similarly,

$$32,000 - A_4 < A_5 < 55,000. \qquad (4.107)$$

Now, one revolution of joint 5 corresponds to 25,600 actuator counts, so, when $A_4 > 2600$, there are two possible solutions for A_5. This is the only situation in which the Yasukawa Motoman L-3 has more than one solution.

4.8 THE STANDARD FRAMES

The ability to solve for joint angles is really the central element in many robot control systems. Again, consider the paradigm indicated in Fig. 4.11, which shows the standard frames.

The way these frames are used in a general robot system is as follows:

1. The user specifies to the system where the station frame is to be located. This might be at the corner of a work surface, as in Fig. 4.12, or even affixed to a moving conveyor belt. The station frame, {S}, is defined relative to the base frame, {B}.

2. The user specifies the description of the tool being used by the robot by giving the {T}-frame specification. Each tool the robot picks up could have a different {T} frame associated with it. Note that the same tool grasped in different ways requires different {T}-frame definitions. {T} is specified relative to {W}—that is, $^W_T T$.

3. The user specifies the goal point for a robot motion by giving the description of the goal frame, {G}, relative to the station frame. Often, the definitions of {T} and {S} remain fixed for several motions of the robot. In this case, once they are defined, the user simply gives a series of {G} specifications.

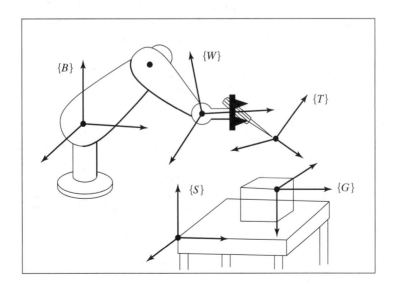

FIGURE 4.11: Location of the "standard" frames.

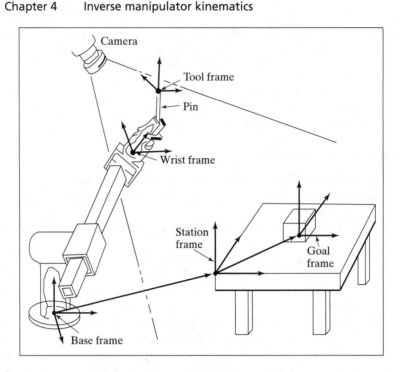

FIGURE 4.12: Example workstation.

In many systems, the tool frame definition ($^{W}_{T}T$) is constant (for example, it is defined with its origin at the center of the fingertips). Also, the station frame might be fixed or might easily be taught by the user with the robot itself. In such systems, the user need not be aware of the five standard frames—he or she simply thinks in terms of moving the tool to locations (goals) with respect to the work area specified by station frame.

4. The robot system calculates a series of joint angles to move the joints through in order that the tool frame will move from its initial location in a smooth manner until $\{T\} = \{G\}$ at the end of motion.

4.9 SOLVE-ING A MANIPULATOR

The SOLVE function implements Cartesian transformations and calls the inverse kinematics function. Thus, the inverse kinematics are generalized so that arbitrary tool-frame and station-frame definitions may be used with our basic inverse kinematics, which solves for the wrist frame relative to the base frame.

Given the goal-frame specification, $^{S}_{T}T$, SOLVE uses the tool and station definitions to calculate the location of $\{W\}$ relative to $\{B\}$, $^{B}_{W}T$:

$$^{B}_{W}T = {}^{B}_{S}T \, {}^{S}_{T}T \, {}^{W}_{T}T^{-1}. \tag{4.108}$$

Then, the inverse kinematics take $^{B}_{W}T$ as an input and calculate θ_1 through θ_n.

4.10   REPEATABILITY AND ACCURACY

Many industrial robots today move to goal points that have been taught. A **taught point** is one that the manipulator is moved to physically, and then the joint position sensors are read and the joint angles stored. When the robot is commanded to return to that point in space, each joint is moved to the stored value. In simple "teach and playback" manipulators such as these, the inverse kinematic problem never arises, because goal points are never specified in Cartesian coordinates. When a manufacturer specifies how precisely a manipulator can return to a taught point, he is specifying the **repeatability** of the manipulator.

Any time a goal position and orientation are specified in Cartesian terms, the inverse kinematics of the device must be computed in order to solve for the required joint angles. Systems that allow goals to be described in Cartesian terms are capable of moving the manipulator to points that were never taught—points in its workspace to which it has perhaps never gone before. We will call such points **computed points**. Such a capability is necessary for many manipulation tasks. For example, if a computer vision system is used to locate a part that the robot must grasp, the robot must be able to move to the Cartesian coordinates supplied by the vision sensor. The precision with which a computed point can be attained is called the **accuracy** of the manipulator.

The accuracy of a manipulator is bounded by the repeatability. Clearly, accuracy is affected by the precision of parameters appearing in the kinematic equations of the robot. Errors in knowledge of the Denavit–Hartenberg parameters will cause the inverse kinematic equations to calculate joint angle values that are in error. Hence, although the repeatability of most industrial manipulators is quite good, the accuracy is usually much worse and varies quite a bit from manipulator to manipulator. Calibration techniques can be devised that allow the accuracy of a manipulator to be improved through estimation of that particular manipulator's kinematic parameters [10].

4.11   COMPUTATIONAL CONSIDERATIONS

In many path-control schemes, which we will consider in Chapter 7, it is necessary to calculate the inverse kinematics of a manipulator at fairly high rates, for example, 30 Hz or faster. Therefore, computational efficiency is an issue. These speed requirements rule out the use of numerical-solution techniques that are iterative in nature; for this reason, we have not considered them.

Most of the general comments of Section 3.10, made for forward kinematics, also hold for the problem of inverse kinematics. For the inverse-kinematic case, a table-lookup Atan2 routine is often used to attain higher speeds.

Structure of the computation of multiple solutions is also important. It is generally fairly efficient to generate all of them in parallel, rather than pursuing one after another serially. Of course, in some applications, when all solutions are not required, substantial time is saved by computing only one.

When a geometric approach is used to develop an inverse-kinematic solution, it is sometimes possible to calculate multiple solutions by simple operations on the various angles solved for in obtaining the first solution. That is, the first solution is moderately expensive computationally, but the other solutions are found very quickly by summing and differencing angles, subtracting π, and so on.

BIBLIOGRAPHY

[1] B. Roth, J. Rastegar, and V. Scheinman, "On the Design of Computer Controlled Manipulators," *On the Theory and Practice of Robots and Manipulators*, Vol. 1, First CISM-IFToMM Symposium, September 1973, pp. 93–113.

[2] B. Roth, "Performance Evaluation of Manipulators from a Kinematic Viewpoint," *Performance Evaluation of Manipulators*, National Bureau of Standards, special publication, 1975.

[3] D. Pieper and B. Roth, "The Kinematics of Manipulators Under Computer Control," *Proceedings of the Second International Congress on Theory of Machines and Mechanisms*, Vol. 2, Zakopane, Poland, 1969, pp. 159–169.

[4] D. Pieper, "The Kinematics of Manipulators Under Computer Control," Unpublished Ph.D. Thesis, Stanford University, 1968.

[5] R.P. Paul, B. Shimano, and G. Mayer, "Kinematic Control Equations for Simple Manipulators," *IEEE Transactions on Systems, Man, and Cybernetics*, Vol. SMC-11, No. 6, 1981.

[6] L. Tsai and A. Morgan, "Solving the Kinematics of the Most General Six- and Five-degree-of-freedom Manipulators by Continuation Methods," Paper 84-DET-20, ASME Mechanisms Conference, Boston, October 7–10, 1984.

[7] C.S.G. Lee and M. Ziegler, "Geometric Approach in Solving Inverse Kinematics of PUMA Robots," *IEEE Transactions on Aerospace and Electronic Systems*, Vol. AES-20, No. 6, November 1984.

[8] W. Beyer, *CRC Standard Mathematical Tables*, 25th edition, CRC Press, Inc., Boca Raton, FL, 1980.

[9] R. Burington, *Handbook of Mathematical Tables and Formulas*, 5th edition, McGraw-Hill, New York, 1973.

[10] J. Hollerbach, "A Survey of Kinematic Calibration," in *The Robotics Review*, O. Khatib, J. Craig, and T. Lozano-Perez, Editors, MIT Press, Cambridge, MA, 1989.

[11] Y. Nakamura and H. Hanafusa, "Inverse Kinematic Solutions with Singularity Robustness for Robot Manipulator Control," *ASME Journal of Dynamic Systems, Measurement, and Control*, Vol. 108, 1986.

[12] D. Baker and C. Wampler, "On the Inverse Kinematics of Redundant Manipulators," *International Journal of Robotics Research*, Vol. 7, No. 2, 1988.

[13] L.W. Tsai, *Robot Analysis: The Mechanics of Serial and Parallel Manipulators*, Wiley, New York, 1999.

EXERCISES

4.1 [15] Sketch the fingertip workspace of the three-link manipulator of Chapter 3, Exercise 3.3 for the case $l_1 = 15.0$, $l_2 = 10.0$, and $l_3 = 3.0$.

4.2 [26] Derive the inverse kinematics of the three-link manipulator of Chapter 3, Exercise 3.3.

4.3 [12] Sketch the fingertip workspace of the 3-DOF manipulator of Chapter 3, Example 3.4.

4.4 [24] Derive the inverse kinematics of the 3-DOF manipulator of Chapter 3, Example 3.4.

4.5 [38] Write a Pascal (or C) subroutine that computes all possible solutions for the PUMA 560 manipulator that lie within the following joint limits:

$$-170.0 < \theta_1 < 170.0,$$
$$-225.0 < \theta_2 < 45.0,$$
$$-250.0 < \theta_3 < 75.0,$$
$$-135.0 < \theta_4 < 135.0,$$
$$-100.0 < \theta_5 < 100.0,$$
$$-180.0 < \theta_6 < 180.0.$$

Use the equations derived in Section 4.7 with these numerical values (in inches):

$$a_2 = 17.0,$$
$$a_3 = 0.8,$$
$$d_3 = 4.9,$$
$$d_4 = 17.0.$$

4.6 [15] Describe a simple algorithm for choosing the nearest solution from a set of possible solutions.

4.7 [10] Make a list of factors that might affect the repeatability of a manipulator. Make a second list of additional factors that affect the accuracy of a manipulator.

4.8 [12] Given a desired position and orientation of the hand of a three-link planar rotary-jointed manipulator, there are two possible solutions. If we add one more rotational joint (in such a way that the arm is still planar), how many solutions are there?

4.9 [26] Figure 4.13 shows a two-link planar arm with rotary joints. For this arm, the second link is half as long as the first—that is, $l_1 = 2l_2$. The joint range limits in

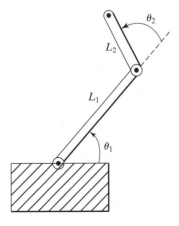

FIGURE 4.13: Two-link planar manipulator.

degrees are

$$0 < \theta_1 < 180,$$

$$-90 < \theta_2 < 180.$$

Sketch the approximate reachable workspace (an area) of the tip of link 2.

4.10 [23] Give an expression for the subspace of the manipulator of Chapter 3, Example 3.4.

4.11 [24] A 2-DOF positioning table is used to orient parts for arc-welding. The forward kinematics that locate the bed of the table (link 2) with respect to the base (link 0) are

$$
{}_2^0T = \begin{bmatrix}
c_1 c_2 & -c_1 s_2 & s_1 & l_2 s_1 + l_1 \\
s_2 & c_2 & 0 & 0 \\
-s_1 c_2 & s_1 s_2 & c_1 & l_2 c_1 + h_1 \\
0 & 0 & 0 & 1
\end{bmatrix}.
$$

Given any unit direction fixed in the frame of the bed (link 2), ${}^2\hat{V}$, give the inverse-kinematic solution for θ_1, θ_2 such that this vector is aligned with ${}^0\hat{Z}$ (i.e., upward). Are there multiple solutions? Is there a singular condition for which a unique solution cannot be obtained?

4.12 [22] In Fig. 4.14, two $3R$ mechanisms are pictured. In both cases, the three axes intersect at a point (and, over all configurations, this point remains fixed in space). The mechanism in Fig. 4.14(a) has link twists (α_i) of magnitude 90 degrees. The mechanism in Fig. 4.14(b) has one twist of ϕ in magnitude and the other of $180 - \phi$ in magnitude.

The mechanism in Fig. 4.14(a) can be seen to be in correspondence with Z–Y–Z Euler angles, and therefore we know that it suffices to orient link 3 (with arrow in figure) arbitrarily with respect to the fixed link 0. Because ϕ is not equal to 90 degrees, it turns out that the other mechanism cannot orient link 3 arbitrarily.

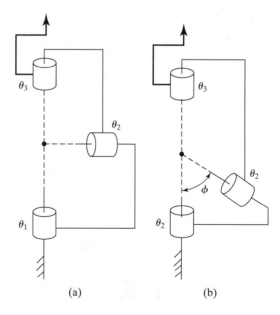

(a)

(b)

FIGURE 4.14: Two $3R$ mechanisms (Exercise 4.12).

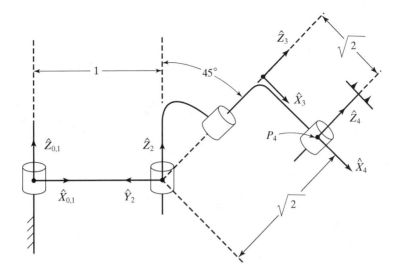

FIGURE 4.15: A $4R$ manipulator shown in the position $\Theta = [0, 90°, -90°, 0]^T$ (Exercise 4.16).

Describe the set of orientations that are *unattainable* with the second mechanism. Note that we assume that all joints can turn 360 degrees (i.e. no limits) and we assume that the links may pass through each other if need be (i.e., workspace not limited by self-collisions).

4.13 [13] Name two reasons for which closed-form analytic kinematic solutions are preferred over iterative solutions.

4.14 [14] There exist 6-DOF robots for which the kinematics are NOT closed-form solvable. Does there exist any 3-DOF robot for which the (position) kinematics are NOT closed-form solvable?

4.15 [38] Write a subroutine that solves quartic equations in closed form. (See [8, 9].)

4.16 [25] A $4R$ manipulator is shown schematically in Fig. 4.15. The nonzero link parameters are $a_1 = 1$, $\alpha_2 = 45°$, $d_3 = \sqrt{2}$, and $a_3 = \sqrt{2}$, and the mechanism is pictured in the configuration corresponding to $\Theta = [0, 90°, -90°, 0]^T$. Each joint has $\pm 180°$ as limits. Find all values of θ_3 such that

$$^{0}P_{4ORG} = [1.1, 1.5, 1.707]^T.$$

4.17 [25] A $4R$ manipulator is shown schematically in Fig. 4.16. The nonzero link parameters are $\alpha_1 = -90°$, $d_2 = 1$, $\alpha_2 = 45°$, $d_3 = 1$, and $a_3 = 1$, and the mechanism is pictured in the configuration corresponding to $\Theta = [0, 0, 90°, 0]^T$. Each joint has $\pm 180°$ as limits. Find all values of θ_3 such that

$$^{0}P_{4ORG} = [0.0, 1.0, 1.414]^T.$$

4.18 [15] Consider the RRP manipulator shown in Fig. 3.37. How many solutions do the (position) kinematic equations possess?

4.19 [15] Consider the RRR manipulator shown in Fig. 3.38. How many solutions do the (position) kinematic equations possess?

4.20 [15] Consider the RPP manipulator shown in Fig. 3.39. How many solutions do the (position) kinematic equations possess?

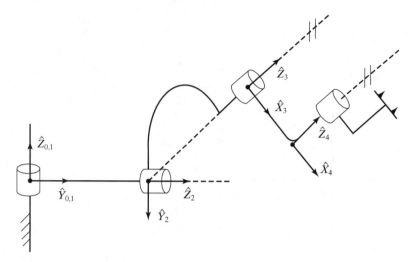

FIGURE 4.16: A $4R$ manipulator shown in the position $\Theta = [0, 0, 90°, 0]^T$ (Exercise 4.17).

4.21 [15] Consider the *PRR* manipulator shown in Fig. 3.40. How many solutions do the (position) kinematic equations possess?

4.22 [15] Consider the *PPP* manipulator shown in Fig. 3.41. How many solutions do the (position) kinematic equations possess?

4.23 [38] The following kinematic equations arise in a certain problem:

$$\sin \xi = a \sin \theta + b,$$
$$\sin \phi = c \cos \theta + d,$$
$$\psi = \xi + \phi.$$

Given a, b, c, d, and ψ, show that, in the general case, there are four solutions for θ. Give a special condition under which there are just two solutions for θ.

4.24 [20] Given the description of link frame $\{i\}$ in terms of link frame $\{i - 1\}$, find the four Denavit–Hartenberg parameters as functions of the elements of $_i^{i-1}T$.

PROGRAMMING EXERCISE (PART 4)

1. Write a subroutine to calculate the inverse kinematics for the three-link manipulator of Section 4.4. The routine should pass arguments in the form

```
Procedure INVKIN(VAR wrelb: frame; VAR current, near, far: vec3;
VAR sol: boolean);
```

where "wrelb," an input, is the wrist frame specified relative to the base frame; "current," an input, is the current position of the robot (given as a vector of joint angles); "near" is the nearest solution; "far" is the second solution; and "sol" is a flag that indicates whether solutions were found. (sol = FALSE if no solutions were found). The link lengths (meters) are

$$l_1 = l_2 = 0.5.$$

The joint ranges of motion are

$$-170° \leq \theta_i \leq 170°.$$

Test your routine by calling it back-to-back with KIN to demonstrate that they are indeed inverses of one another.

2. A tool is attached to link 3 of the manipulator. This tool is described by $^W_T T$, the tool frame relative to the wrist frame. Also, a user has described his work area, the station frame relative to the base of the robot, as $^B_S T$. Write the subroutine

```
Procedure SOLVE(VAR trels: frame; VAR current, near, far: vec3;
VAR sol: boolean);
```

where "trels" is the {T} frame specified relative to the {S} frame. Other parameters are exactly as in the INVKIN subroutine. The definitions of {T} and {S} should be globally defined variables or constants. SOLVE should use calls to TMULT, TINVERT, and INVKIN.

3. Write a main program that accepts a goal frame specified in terms of x, y, and ϕ. This goal specification is {T} relative to {S}, which is the way the user wants to specify goals.

The robot is using the same tool in the same working area as in Programming Exercise (Part 2), so {T} and {S} are defined as

$$^W_T T = [x\ y\ \theta] = [0.1\ 0.2\ 30.0],$$

$$^B_S T = [x\ y\ \theta] = [-0.1\ 0.3\ 0.0].$$

Calculate the joint angles for each of the following three goal frames:

$$[x_1\ y_1\ \phi_1] = [0.0\ 0.0\ -90.0],$$

$$[x_2\ y_2\ \phi_2] = [0.6\ -0.3\ 45.0],$$

$$[x_3\ y_3\ \phi_3] = [-0.4\ 0.3\ 120.0],$$

$$[x_4\ y_4\ \phi_4] = [0.8\ 1.4\ 30.0].$$

Assume that the robot will start with all angles equal to 0.0 and move to these three goals in sequence. The program should find the nearest solution with respect to the previous goal point. You should call SOLVE and WHERE back-to-back to make sure they are truly inverse functions.

MATLAB EXERCISE 4

This exercise focuses on the inverse-pose kinematics solution for the planar 3-DOF, 3R robot. (See Figures 3.6 and 3.7; the DH parameters are given in Figure 3.8.) The following fixed-length parameters are given: $L_1 = 4$, $L_2 = 3$, and $L_3 = 2$(m).

a) Analytically derive, by hand, the inverse-pose solution for this robot: Given $^0_H T$, calculate all possible multiple solutions for $\{\theta_1\ \theta_2\ \theta_3\}$. (Three methods are presented in the text—choose one of these.) Hint: To simplify the equations, first calculate $^0_3 T$ from $^0_H T$ and L_3.

b) Develop a MATLAB program to solve this planar 3R robot inverse-pose kinematics problem completely (i.e., to give all multiple solutions). Test your program, using the following input cases:

i) $_H^0 T = \begin{bmatrix} 1 & 0 & 0 & 9 \\ 0 & 1 & 0 & 0 \\ 0 & 0 & 1 & 0 \\ 0 & 0 & 0 & 1 \end{bmatrix}.$

ii) $_H^0 T = \begin{bmatrix} 0.5 & -0.866 & 0 & 7.5373 \\ 0.866 & 0.6 & 0 & 3.9266 \\ 0 & 0 & 1 & 0 \\ 0 & 0 & 0 & 1 \end{bmatrix}.$

iii) $_H^0 T = \begin{bmatrix} 0 & 1 & 0 & -3 \\ -1 & 0 & 0 & 2 \\ 0 & 0 & 1 & 0 \\ 0 & 0 & 0 & 1 \end{bmatrix}.$

iv) $_H^0 T = \begin{bmatrix} 0.866 & 0.5 & 0 & -3.1245 \\ -0.5 & 0.866 & 0 & 9.1674 \\ 0 & 0 & 1 & 0 \\ 0 & 0 & 0 & 1 \end{bmatrix}.$

For all cases, employ a circular check to validate your results: Plug each resulting set of joint angles (for each of the multiple solutions) back into the forward-pose kinematics MATLAB program to demonstrate that you get the originally commanded $_H^0 T$.

c) Check all results by means of the Corke MATLAB Robotics Toolbox. Try function *ikine()*.

CHAPTER 5

Jacobians: velocities and static forces

5.1 INTRODUCTION
5.2 NOTATION FOR TIME-VARYING POSITION AND ORIENTATION
5.3 LINEAR AND ROTATIONAL VELOCITY OF RIGID BODIES
5.4 MORE ON ANGULAR VELOCITY
5.5 MOTION OF THE LINKS OF A ROBOT
5.6 VELOCITY "PROPAGATION" FROM LINK TO LINK
5.7 JACOBIANS
5.8 SINGULARITIES
5.9 STATIC FORCES IN MANIPULATORS
5.10 JACOBIANS IN THE FORCE DOMAIN
5.11 CARTESIAN TRANSFORMATION OF VELOCITIES AND STATIC FORCES

5.1 INTRODUCTION

In this chapter, we expand our consideration of robot manipulators beyond static-positioning problems. We examine the notions of linear and angular velocity of a rigid body and use these concepts to analyze the motion of a manipulator. We also will consider forces acting on a rigid body, and then use these ideas to study the application of static forces with manipulators.

It turns out that the study of both velocities and static forces leads to a matrix entity called the **Jacobian**[1] of the manipulator, which will be introduced in this chapter.

The field of kinematics of mechanisms is not treated in great depth here. For the most part, the presentation is restricted to only those ideas which are fundamental to the particular problem of robotics. The interested reader is urged to study further from any of several texts on mechanics [1–3].

5.2 NOTATION FOR TIME-VARYING POSITION AND ORIENTATION

Before investigating the description of the motion of a rigid body, we briefly discuss some basics: the differentiation of vectors, the representation of angular velocity, and notation.

---

[1] Mathematicians call it the "Jacobian matrix," but roboticists usually shorten it to simply "Jacobian."

Differentiation of position vectors

As a basis for our consideration of velocities (and, in Chapter 6, accelerations), we need the following notation for the derivative of a vector:

$$^B V_Q = \frac{d}{dt} \, ^B Q = \lim_{\Delta t \to 0} \frac{^B Q(t + \Delta t) - \, ^B Q(t)}{\Delta t}. \tag{5.1}$$

The velocity of a position vector can be thought of as the linear velocity of the point in space represented by the position vector. From (5.1), we see that we are calculating the derivative of Q relative to frame $\{B\}$. For example, if Q is not changing in time relative to $\{B\}$, then the velocity calculated is zero—even if there is some other frame in which Q is varying. Thus, it is important to indicate the frame in which the vector is differentiated.

As with any vector, a velocity vector can be described in terms of any frame, and this frame of reference is noted with a leading superscript. Hence, the velocity vector calculated by (5.1), when expressed in terms of frame $\{A\}$, would be written

$$^A(^B V_Q) = \frac{^A d}{dt} \, ^B Q. \tag{5.2}$$

So we see that, in the general case, a velocity vector is associated with a point in space, but the numerical values describing the velocity of that point depend on two frames: one with respect to which the differentiation was done, and one in which the resulting velocity vector is expressed.

In (5.1), the calculated velocity is written in terms of the frame of differentiation, so the result could be indicated with a leading B superscript, but, for simplicity, when both superscripts are the same, we needn't indicate the outer one; that is, we write

$$^B(^B V_Q) = \, ^B V_Q. \tag{5.3}$$

Finally, we can always remove the outer, leading superscript by explicitly including the rotation matrix that accomplishes the change in reference frame (see Section 2.10); that is, we write

$$^A(^B V_Q) = \, ^A_B R \, ^B V_Q. \tag{5.4}$$

We will usually write expressions in the form of the right-hand side of (5.4) so that the symbols representing velocities will always mean the velocity in the frame of differentiation and will not have outer, leading superscripts.

Rather than considering a general point's velocity relative to an arbitrary frame, we will very often consider the velocity of the *origin of a frame* relative to some understood universe reference frame. For this special case, we define a shorthand notation,

$$\upsilon_C = \, ^U V_{CORG}, \tag{5.5}$$

where the point in question is the origin of frame $\{C\}$ and the reference frame is $\{U\}$. For example, we can use the notation υ_C to refer to the velocity of the origin of frame $\{C\}$; then $^A \upsilon_C$ is the velocity of the origin of frame $\{C\}$ expressed in terms of frame $\{A\}$ (though differentiation was done relative to $\{U\}$).

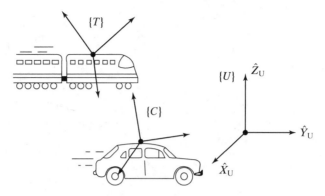

FIGURE 5.1: Example of some frames in linear motion.

EXAMPLE 5.1

Figure 5.1 shows a fixed universe frame, $\{U\}$, a frame attached to a train traveling at 100 mph, $\{T\}$, and a frame attached to a car traveling at 30 mph, $\{C\}$. Both vehicles are heading in the $\hat{X}$ direction of $\{U\}$. The rotation matrices, $^U_T R$ and $^U_C R$, are known and constant.

What is $\dfrac{^U d}{dt} {}^U P_{CORG}$?

$$\frac{^U d}{dt} {}^U P_{CORG} = {}^U V_{CORG} = v_C = 30\hat{X}.$$

What is $^C({}^U V_{TORG})$?

$$^C({}^U V_{TORG}) = {}^C v_T = {}^C_U R v_T = {}^C_U R(100\hat{X}) = {}^U_C R^{-1} 100\hat{X}.$$

What is $^C({}^T V_{CORG})$?

$$^C({}^T V_{CORG}) = {}^C_T R {}^T V_{CORG} = -{}^U_C R^{-1} {}^U_T R 70\hat{X}.$$

The angular velocity vector

We now introduce an **angular velocity vector**, using the symbol Ω. Whereas linear velocity describes an attribute of a point, angular velocity describes an attribute of a body. We always attach a frame to the bodies we consider, so we can also think of angular velocity as describing rotational motion of a frame.

In Fig. 5.2, $^A\Omega_B$ describes the rotation of frame $\{B\}$ relative to $\{A\}$. Physically, at any instant, the direction of $^A\Omega_B$ indicates the instantaneous axis of rotation of $\{B\}$ relative to $\{A\}$, and the magnitude of $^A\Omega_B$ indicates the speed of rotation. Again, like any vector, an angular velocity vector may be expressed in any coordinate system, and so another leading superscript may be added; for example, $^C(^A\Omega_B)$ is the angular velocity of frame $\{B\}$ relative to $\{A\}$ expressed in terms of frame $\{C\}$.

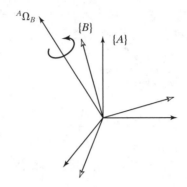

FIGURE 5.2: Frame $\{B\}$ is rotating with angular velocity $^A\Omega_B$ relative to frame $\{A\}$.

Again, we introduce a simplified notation for an important special case. This is simply the case in which there is an understood reference frame, so that it need not be mentioned in the notation:

$$\omega_C = {}^U\Omega_C. \tag{5.6}$$

Here, ω_C is the angular velocity of frame $\{C\}$ relative to some understood reference frame, $\{U\}$. For example, $^A\omega_C$ is the angular velocity of frame $\{C\}$ expressed in terms of $\{A\}$ (though the angular velocity is with respect to $\{U\}$).

5.3   LINEAR AND ROTATIONAL VELOCITY OF RIGID BODIES

In this section, we investigate the description of motion of a rigid body, at least as far as velocity is concerned. These ideas extend the notions of translations and orientations described in Chapter 2 to the time-varying case. In Chapter 6, we will further extend our study to considerations of acceleration.

As in Chapter 2, we attach a coordinate system to any body that we wish to describe. Then, motion of rigid bodies can be equivalently studied as the motion of frames relative to one another.

Linear velocity

Consider a frame $\{B\}$ attached to a rigid body. We wish to describe the motion of ^BQ relative to frame $\{A\}$, as in Fig. 5.3. We may consider $\{A\}$ to be fixed.

Frame $\{B\}$ is located relative to $\{A\}$, as described by a position vector, $^AP_{BORG}$, and a rotation matrix, ^A_BR. For the moment, we will assume that the orientation, ^A_BR, is not changing with time—that is, the motion of point Q relative to $\{A\}$ is due to $^AP_{BORG}$ or ^BQ changing in time.

Solving for the linear velocity of point Q in terms of $\{A\}$ is quite simple. Just express both components of the velocity in terms of $\{A\}$, and sum them:

$$^AV_Q = {}^AV_{BORG} + {}^A_BR\,{}^BV_Q. \tag{5.7}$$

Equation (5.7) is for only that case in which relative orientation of $\{B\}$ and $\{A\}$ remains constant.

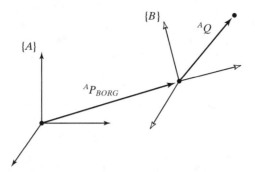

FIGURE 5.3: Frame $\{B\}$ is translating with velocity $^A V_{BORG}$ relative to frame $\{A\}$.

Rotational velocity

Now let us consider two frames with coincident origins and with zero linear relative velocity; their origins will remain coincident for all time. One or both could be attached to rigid bodies, but, for clarity, the rigid bodies are not shown in Fig. 5.4.

The orientation of frame $\{B\}$ with respect to frame $\{A\}$ is changing in time. As indicated in Fig. 5.4, rotational velocity of $\{B\}$ relative to $\{A\}$ is described by a vector called $^A\Omega_B$. We also have indicated a vector $^B Q$ that locates a point fixed in $\{B\}$. Now we consider the all-important question: How does a vector change with time as viewed from $\{A\}$ when it is fixed in $\{B\}$ and the systems are rotating?

Let us consider that the vector Q is constant as viewed from frame $\{B\}$; that is,

$$^B V_Q = 0. \tag{5.8}$$

Even though it is constant relative to $\{B\}$, it is clear that point Q will have a velocity as seen from $\{A\}$ that is due to the rotational velocity $^A\Omega_B$. To solve for the velocity of point Q, we will use an intuitive approach. Figure 5.5 shows two instants of time as vector Q rotates around $^A\Omega_B$. This is what an observer in $\{A\}$ would observe.

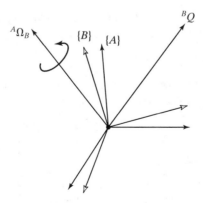

FIGURE 5.4: Vector $^B Q$, fixed in frame $\{B\}$, is rotating with respect to frame $\{A\}$ with angular velocity $^A\Omega_B$.

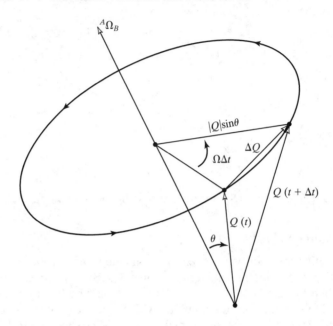

FIGURE 5.5: The velocity of a point due to an angular velocity.

By examining Fig. 5.5, we can figure out both the direction and the magnitude of the change in the vector as viewed from $\{A\}$. First, it is clear that the differential change in ^AQ must be perpendicular to both $^A\Omega_B$ and ^AQ. Second, we see from Fig. 5.5 that the magnitude of the differential change is

$$|\Delta Q| = (|^AQ| \sin\theta)(|^A\Omega_B|\Delta t). \tag{5.9}$$

These conditions on magnitude and direction immediately suggest the vector cross-product. Indeed, our conclusions about direction and magnitude are satisfied by the computational form

$$^AV_Q = {}^A\Omega_B \times {}^AQ. \tag{5.10}$$

In the general case, the vector Q could also be changing with respect to frame $\{B\}$, so, adding this component, we have

$$^AV_Q = {}^A({}^BV_Q) + {}^A\Omega_B \times {}^AQ. \tag{5.11}$$

Using a rotation matrix to remove the dual-superscript, and noting that the description of ^AQ at any instant is $^A_BR\,{}^BQ$, we end with

$$^AV_Q = {}^A_BR\,{}^BV_Q + {}^A\Omega_B \times {}^A_BR\,{}^BQ. \tag{5.12}$$

Simultaneous linear and rotational velocity

We can very simply expand (5.12) to the case where origins are not coincident by adding on the linear velocity of the origin to (5.12) to derive the general formula for velocity of a vector fixed in frame $\{B\}$ as seen from frame $\{A\}$:

$$^AV_Q = {}^AV_{BORG} + {}^A_BR\,{}^BV_Q + {}^A\Omega_B \times {}^A_BR\,{}^BQ \tag{5.13}$$

Equation (5.13) is the final result for the derivative of a vector in a moving frame as seen from a stationary frame.

5.4 MORE ON ANGULAR VELOCITY

In this section, we take a deeper look at angular velocity and, in particular, at the derivation of (5.10). Whereas the previous section took a geometric approach toward showing the validity of (5.10), here we take a mathematical approach. This section may be skipped by the first-time reader.

A property of the derivative of an orthonormal matrix

We can derive an interesting relationship between the derivative of an orthonormal matrix and a certain skew-symmetric matrix as follows. For any $n \times n$ orthonormal matrix, R, we have

$$RR^T = I_n \tag{5.14}$$

where I_n is the $n \times n$ identity matrix. Our interest, by the way, is in the case where $n = 3$ and R is a *proper* orthonormal matrix, or rotation matrix. Differentiating (5.14) yields

$$\dot{R}R^T + R\dot{R}^T = 0_n, \tag{5.15}$$

where 0_n is the $n \times n$ zero matrix. Eq. (5.15) may also be written

$$\dot{R}R^T + (\dot{R}R^T)^T = 0_n. \tag{5.16}$$

Defining

$$S = \dot{R}R^T, \tag{5.17}$$

we have, from (5.16), that

$$S + S^T = 0_n. \tag{5.18}$$

So, we see that S is a skew-symmetric matrix. Hence, a property relating the derivative of orthonormal matrices with skew-symmetric matrices exists and can be stated as

$$S = \dot{R}R^{-1}. \tag{5.19}$$

Velocity of a point due to rotating reference frame

Consider a fixed vector ^BP unchanging with respect to frame $\{B\}$. Its description in another frame $\{A\}$ is given as

$$^AP = {}^A_B R \, {}^BP. \tag{5.20}$$

If frame $\{B\}$ is rotating (i.e., the derivative ${}^A_B\dot{R}$ is nonzero), then ^AP will be changing even though ^BP is constant; that is,

$$^A\dot{P} = {}^A_B\dot{R} \, {}^BP, \tag{5.21}$$

or, using our notation for velocity,

$$^AV_P = {}^A_B\dot{R} \, {}^BP. \tag{5.22}$$

Now, rewrite (5.22) by substituting for ^{B}P, to obtain

$$^{A}V_{P} = {}^{A}_{B}\dot{R}\,{}^{A}_{B}R^{-1}\,{}^{A}P. \tag{5.23}$$

Making use of our result (5.19) for orthonormal matrices, we have

$$^{A}V_{P} = {}^{A}_{B}S\,{}^{A}P, \tag{5.24}$$

where we have adorned S with sub- and superscripts to indicate that it is the skew-symmetric matrix associated with the particular rotation matrix $^{A}_{B}R$. Because of its appearance in (5.24), and for other reasons to be seen shortly, the skew-symmetric matrix we have introduced is called the **angular-velocity matrix**.

Skew-symmetric matrices and the vector cross-product

If we assign the elements in a skew-symmetric matrix S as

$$S = \begin{bmatrix} 0 & -\Omega_{x} & \Omega_{y} \\ \Omega_{x} & 0 & -\Omega_{x} \\ -\Omega_{y} & \Omega_{x} & 0 \end{bmatrix}, \tag{5.25}$$

and define the 3×1 column vector

$$\Omega = \begin{bmatrix} \Omega_{x} \\ \Omega_{y} \\ \Omega_{z} \end{bmatrix}, \tag{5.26}$$

then it is easily verified that

$$SP = \Omega \times P, \tag{5.27}$$

where P is any vector, and $\times$ is the vector cross-product.

The 3×1 vector Ω, which corresponds to the 3×3 angular-velocity matrix, is called the **angular-velocity vector** and was already introduced in Section 5.2.

Hence, our relation (5.24) can be written

$$^{A}V_{P} = {}^{A}\Omega_{B} \times {}^{A}P, \tag{5.28}$$

where we have shown the notation for Ω indicating that it is the angular-velocity vector specifying the motion of frame $\{B\}$ with respect to frame $\{A\}$.

Gaining physical insight concerning the angular-velocity vector

Having concluded that there exists some vector Ω such that (5.28) is true, we now wish to gain some insight as to its physical meaning. Derive Ω by direct differentiation of a rotation matrix; that is,

$$\dot{R} = \lim_{\Delta t \to 0} \frac{R(t + \Delta t) - R(t)}{\Delta t}. \tag{5.29}$$

Now, write $R(t + \Delta t)$ as the composition of two matrices, namely,

$$R(t + \Delta t) = R_{K}(\Delta\theta)R(t), \tag{5.30}$$

where, over the interval Δt, a small rotation of $\Delta\theta$ has occurred about axis $\hat{K}$. Using (5.30), write (5.29) as

$$\dot{R} = \lim_{\Delta t \to 0} \left(\frac{R_K(\Delta\theta) - I_3}{\Delta t} R(t) \right); \tag{5.31}$$

that is,

$$\dot{R} = \left(\lim_{\Delta t \to 0} \frac{R_K(\Delta\theta) - I_3}{\Delta t} \right) R(t). \tag{5.32}$$

Now, from small angle substitutions in (2.80), we have

$$R_K(\Delta\theta) = \begin{bmatrix} 1 & -k_z\Delta\theta & k_y\Delta\theta \\ k_z\Delta\theta & 1 & -k_x\Delta\theta \\ -k_y\Delta\theta & k_x\Delta\theta & 1 \end{bmatrix}. \tag{5.33}$$

So, (5.32) may be written

$$\dot{R} = \left(\lim_{\Delta t \to 0} \frac{\begin{bmatrix} 0 & -k_z\Delta\theta & k_y\Delta\theta \\ k_z\Delta\theta & 0 & -k_x\Delta\theta \\ -k_y\Delta\theta & k_x\Delta\theta & 0 \end{bmatrix}}{\Delta t} \right) R(t). \tag{5.34}$$

Finally, dividing the matrix through by Δt and then taking the limit, we have

$$\dot{R} = \begin{bmatrix} 0 & -k_z\dot{\theta} & k_y\dot{\theta} \\ k_z\dot{\theta} & 0 & -k_x\dot{\theta} \\ -k_y\dot{\theta} & k_x\dot{\theta} & 0 \end{bmatrix} R(t). \tag{5.35}$$

Hence, we see that

$$\dot{R}R^{-1} = \begin{bmatrix} 0 & -\Omega_z & \Omega_y \\ \Omega_z & 0 & -\Omega_x \\ -\Omega_y & \Omega_x & 0 \end{bmatrix}, \tag{5.36}$$

where

$$\Omega = \begin{bmatrix} \Omega_x \\ \Omega_y \\ \Omega_z \end{bmatrix} = \begin{bmatrix} k_x\dot{\theta} \\ k_y\dot{\theta} \\ k_z\dot{\theta} \end{bmatrix} = \dot{\theta}\hat{K}. \tag{5.37}$$

The physical meaning of the angular-velocity vector Ω is that, at any instant, the change in orientation of a rotating frame can be viewed as a rotation about some axis $\hat{K}$. This **instantaneous axis of rotation**, taken as a unit vector and then scaled by the speed of rotation about that axis ($\dot{\theta}$), yields the angular-velocity vector.

Other representations of angular velocity

Other representations of angular velocity are possible; for example, imagine that the angular velocity of a rotating body is available as rates of the set of Z–Y–Z Euler angles:

$$\dot{\Theta}_{Z'Y'Z'} = \begin{bmatrix} \dot{\alpha} \\ \dot{\beta} \\ \dot{\gamma} \end{bmatrix}. \tag{5.38}$$

Given this style of description, or any other using one of the 24 **angle sets**, we would like to derive the equivalent angular-velocity vector.

We have seen that

$$\dot{R}R^T = \begin{bmatrix} 0 & -\Omega_z & \Omega_y \\ \Omega_z & 0 & -\Omega_x \\ -\Omega_y & \Omega_x & 0 \end{bmatrix}. \tag{5.39}$$

From this matrix equation, one can extract three independent equations, namely,

$$\Omega_x = \dot{r}_{31}r_{21} + \dot{r}_{32}r_{22} + \dot{r}_{33}r_{23},$$

$$\Omega_y = \dot{r}_{11}r_{31} + \dot{r}_{12}r_{32} + \dot{r}_{13}r_{33}, \tag{5.40}$$

$$\Omega_z = \dot{r}_{21}r_{11} + \dot{r}_{22}r_{12} + \dot{r}_{23}r_{13}.$$

From (5.40) and a symbolic description of R in terms of an angle set, one can derive the expressions that relate the angle-set velocities to the equivalent angular-velocity vector. The resulting expressions can be cast in matrix form—for example, for Z–Y–Z Euler angles,

$$\Omega = E_{Z'Y'Z'}(\Theta_{Z'Y'Z'})\dot{\Theta}_{Z'Y'Z'}. \tag{5.41}$$

That is, $E(\cdot)$ is a Jacobian relating an angle-set velocity vector to the angular-velocity vector and is a function of the instantaneous values of the angle set. The form of $E(\cdot)$ depends on the particular angle set it is developed for; hence, a subscript is added to indicate which.

EXAMPLE 5.2

Construct the E matrix that relates Z–Y–Z Euler angles to the angular-velocity vector; that is, find $E_{Z'Y'Z'}$ in (5.41).

Using (2.72) and (5.40) and doing the required symbolic differentiations yields

$$E_{Z'Y'Z'} = \begin{bmatrix} 0 & -s\alpha & c\alpha s\beta \\ 0 & c\alpha & s\alpha s\beta \\ 1 & 0 & c\beta \end{bmatrix}. \tag{5.42}$$

5.5 MOTION OF THE LINKS OF A ROBOT

In considering the motions of robot links, we will always use link frame {0} as our reference frame. Hence, v_i is the linear velocity of the origin of link frame {i}, and ω_i is the angular velocity of link frame {i}.

At any instant, each link of a robot in motion has some linear and angular velocity. Figure 5.6 indicates these vectors for link i. In this case, it is indicated that they are written in frame {i}.

5.6 VELOCITY "PROPAGATION" FROM LINK TO LINK

We now consider the problem of calculating the linear and angular velocities of the links of a robot. A manipulator is a chain of bodies, each one capable of motion

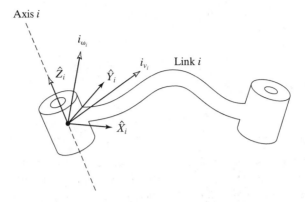

FIGURE 5.6: The velocity of link i is given by vectors v_i and ω_i, which may be written in any frame, even frame $\{i\}$.

relative to its neighbors. Because of this structure, we can compute the velocity of each link in order, starting from the base. The velocity of link $i + 1$ will be that of link i, plus whatever new velocity components were added by joint $i + 1$.[2]

As indicated in Fig. 5.6, let us now think of each link of the mechanism as a rigid body with linear and angular velocity vectors describing its motion. Further, we will express these velocities with respect to the link frame itself rather than with respect to the base coordinate system. Figure 5.7 shows links i and $i + 1$, along with their velocity vectors defined in the link frames.

Rotational velocities can be added when both ω vectors are written with respect to the same frame. Therefore, the angular velocity of link $i + 1$ is the same

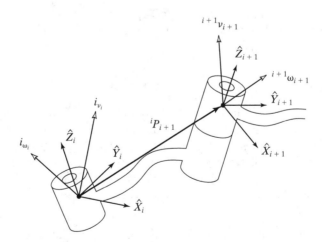

FIGURE 5.7: Velocity vectors of neighboring links.

---

[2]Remember that linear velocity is associated with a point, but angular velocity is associated with a body. Hence, the term "velocity of a link" here means the linear velocity of the origin of the link frame and the rotational velocity of the link.

as that of link i plus a new component caused by rotational velocity at joint $i + 1$. This can be written in terms of frame $\{i\}$ as

$$^i\omega_{i+1} = {}^i\omega_i + {}^i_{i+1}R\,\dot\theta_{i+1}\,{}^{i+1}\hat Z_{i+1}. \tag{5.43}$$

Note that

$$\dot\theta_{i+1}\,{}^{i+1}\hat Z_{i+1} = {}^{i+1}\begin{bmatrix} 0 \\ 0 \\ \dot\theta_{i+1} \end{bmatrix}. \tag{5.44}$$

We have made use of the rotation matrix relating frames $\{i\}$ and $\{i + 1\}$ in order to represent the added rotational component due to motion at the joint in frame $\{i\}$. The rotation matrix rotates the axis of rotation of joint $i + 1$ into its description in frame $\{i\}$, so that the two components of angular velocity can be added.

By premultiplying both sides of (5.43) by $^{i+1}_i R$ we can find the description of the angular velocity of link $i + 1$ with respect to frame $\{i + 1\}$:

$$^{i+1}\omega_{i+1} = {}^{i+1}_i R\,{}^i\omega_i + \dot\theta_{i+1}\,{}^{i+1}\hat Z_{i+1}. \tag{5.45}$$

The linear velocity of the origin of frame $\{i + 1\}$ is the same as that of the origin of frame $\{i\}$ plus a new component caused by rotational velocity of link i. This is exactly the situation described by (5.13), with one term vanishing because $^i P_{i+1}$ is constant in frame $\{i\}$. Therefore, we have

$$^i v_{i+1} = {}^i v_i + {}^i\omega_i \times {}^i P_{i+1}. \tag{5.46}$$

Premultiplying both sides by $^{i+1}_i R$, we compute

$$^{i+1}v_{i+1} = {}^{i+1}_i R({}^i v_i + {}^i\omega_i \times {}^i P_{i+1}). \tag{5.47}$$

Equations (5.45) and (5.47) are perhaps the most important results of this chapter. The corresponding relationships for the case that joint $i + 1$ is prismatic are

$$^{i+1}\omega_{i+1} = {}^{i+1}_i R\,{}^i\omega_i,$$

$$^{i+1}v_{i+1} = {}^{i+1}_i R({}^i v_i + {}^i\omega_i \times {}^i P_{i+1}) + \dot d_{i+1}\,{}^{i+1}\hat Z_{i+1}. \tag{5.48}$$

Applying these equations successively from link to link, we can compute $^N\omega_N$ and $^N v_N$, the rotational and linear velocities of the last link. Note that the resulting velocities are expressed in terms of frame $\{N\}$. This turns out to be useful, as we will see later. If the velocities are desired in terms of the base coordinate system, they can be rotated into base coordinates by multiplication with $^0_N R$.

EXAMPLE 5.3

A two-link manipulator with rotational joints is shown in Fig. 5.8. Calculate the velocity of the tip of the arm as a function of joint rates. Give the answer in two forms—in terms of frame $\{3\}$ and also in terms of frame $\{0\}$.

Frame $\{3\}$ has been attached at the end of the manipulator, as shown in Fig. 5.9, and we wish to find the velocity of the origin of this frame expressed in frame $\{3\}$. As a second part of the problem, we will express these velocities in frame $\{0\}$ as

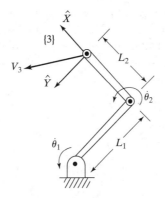

FIGURE 5.8: A two-link manipulator.

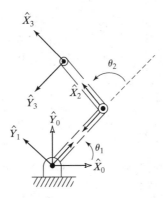

FIGURE 5.9: Frame assignments for the two-link manipulator.

well. We will start by attaching frames to the links as we have done before (shown in Fig. 5.9).

We will use (5.45) and (5.47) to compute the velocity of the origin of each frame, starting from the base frame {0}, which has zero velocity. Because (5.45) and (5.47) will make use of the link transformations, we compute them:

$$
{}^0_1T = \begin{bmatrix} c_1 & -s_1 & 0 & 0 \\ s_1 & c_1 & 0 & 0 \\ 0 & 0 & 1 & 0 \\ 0 & 0 & 0 & 1 \end{bmatrix},
$$

$$
{}^1_2T = \begin{bmatrix} c_2 & -s_2 & 0 & l_1 \\ s_2 & c_2 & 0 & 0 \\ 0 & 0 & 1 & 0 \\ 0 & 0 & 0 & 1 \end{bmatrix}, \qquad (5.49)
$$

$$
{}^2_3T = \begin{bmatrix} 1 & 0 & 0 & l_2 \\ 0 & 1 & 0 & 0 \\ 0 & 0 & 1 & 0 \\ 0 & 0 & 0 & 1 \end{bmatrix}.
$$

Note that these correspond to the manipulator of Example 3.3 with joint 3 permanently fixed at zero degrees. The final transformation between frames {2} and {3} need not be cast as a standard link transformation (though it might be helpful to do so). Then, using (5.45) and (5.47) sequentially from link to link, we calculate

$$
{}^1\omega_1 = \begin{bmatrix} 0 \\ 0 \\ \dot{\theta}_1 \end{bmatrix}, \tag{5.50}
$$

$$
{}^1\upsilon_1 = \begin{bmatrix} 0 \\ 0 \\ 0 \end{bmatrix}, \tag{5.51}
$$

$$
{}^2\omega_2 = \begin{bmatrix} 0 \\ 0 \\ \dot{\theta}_1 + \dot{\theta}_2 \end{bmatrix}, \tag{5.52}
$$

$$
{}^2\upsilon_2 = \begin{bmatrix} c_2 & s_2 & 0 \\ -s_2 & c_2 & 0 \\ 0 & 0 & 1 \end{bmatrix} \begin{bmatrix} 0 \\ l_1\dot{\theta}_1 \\ 0 \end{bmatrix} = \begin{bmatrix} l_1 s_2 \dot{\theta}_1 \\ l_1 c_2 \dot{\theta}_1 \\ 0 \end{bmatrix}, \tag{5.53}
$$

$$
{}^3\omega_3 = {}^2\omega_2, \tag{5.54}
$$

$$
{}^3\upsilon_3 = \begin{bmatrix} l_1 s_2 \dot{\theta}_1 \\ l_1 c_2 \dot{\theta}_1 + l_2(\dot{\theta}_1 + \dot{\theta}_2) \\ 0 \end{bmatrix}. \tag{5.55}
$$

Equation (5.55) is the answer. Also, the rotational velocity of frame {3} is found in (5.54).

   To find these velocities with respect to the nonmoving base frame, we rotate them with the rotation matrix ${}^0_3R$, which is

$$
{}^0_3R = {}^0_1R\ {}^1_2R\ {}^2_3R = \begin{bmatrix} c_{12} & -s_{12} & 0 \\ s_{12} & c_{12} & 0 \\ 0 & 0 & 1 \end{bmatrix}. \tag{5.56}
$$

This rotation yields

$$
{}^0\upsilon_3 = \begin{bmatrix} -l_1 s_1 \dot{\theta}_1 - l_2 s_{12}(\dot{\theta}_1 + \dot{\theta}_2) \\ l_1 c_1 \dot{\theta}_1 + l_2 c_{12}(\dot{\theta}_1 + \dot{\theta}_2) \\ 0 \end{bmatrix}. \tag{5.57}
$$

   It is important to point out the two distinct uses for (5.45) and (5.47). First, they can be used as a means of deriving analytical expressions, as in Example 5.3 above. Here, we manipulate the symbolic equations until we arrive at a form such as (5.55), which will be evaluated with a computer in some application. Second, they can be used directly to compute (5.45) and (5.47) as they are written. They can easily be written as a subroutine, which is then applied iteratively to compute link velocities. As such, they could be used for any manipulator, without the need of deriving the equations for a particular manipulator. However, the computation then yields a numeric result with the structure of the equations hidden. We are often interested in the structure of an analytic result such as (5.55). Also, if we bother to do the work (that is, (5.50) through (5.57)), we generally will find that there are fewer computations left for the computer to perform in the final application.

5.7   JACOBIANS

The Jacobian is a multidimensional form of the derivative. Suppose, for example, that we have six functions, each of which is a function of six independent variables:

$$y_1 = f_1(x_1, x_2, x_3, x_4, x_5, x_6),$$
$$y_2 = f_2(x_1, x_2, x_3, x_4, x_5, x_6),$$
$$\vdots$$
$$y_6 = f_6(x_1, x_2, x_3, x_4, x_5, x_6).$$

(5.58)

We could also use vector notation to write these equations:

$$Y = F(X).$$

(5.59)

Now, if we wish to calculate the differentials of y_i as a function of differentials of x_j, we simply use the chain rule to calculate, and we get

$$\delta y_1 = \frac{\partial f_1}{\partial x_1} \delta x_1 + \frac{\partial f_1}{\partial x_2} \delta x_2 + \cdots + \frac{\partial f_1}{\partial x_6} \delta x_6,$$

$$\delta y_2 = \frac{\partial f_2}{\partial x_1} \delta x_1 + \frac{\partial f_2}{\partial x_2} \delta x_2 + \cdots + \frac{\partial f_2}{\partial x_6} \delta x_6,$$

$$\vdots$$

$$\delta y_6 = \frac{\partial f_6}{\partial x_1} \delta x_1 + \frac{\partial f_6}{\partial x_2} \delta x_2 + \cdots + \frac{\partial f_6}{\partial x_6} \delta x_6,$$

(5.60)

which again might be written more simply in vector notation:

$$\delta Y = \frac{\partial F}{\partial X} \delta X.$$

(5.61)

The 6×6 matrix of partial derivatives in (5.61) is what we call the Jacobian, J. Note that, if the functions $f_1(X)$ through $f_6(X)$ are nonlinear, then the partial derivatives are a function of the x_i, so, we can use the notation

$$\delta Y = J(X)\delta X.$$

(5.62)

By dividing both sides by the differential time element, we can think of the Jacobian as mapping velocities in X to those in Y:

$$\dot{Y} = J(X)\dot{X}.$$

(5.63)

At any particular instant, X has a certain value, and $J(X)$ is a linear transformation. At each new time instant, X has changed, and therefore, so has the linear transformation. Jacobians are time-varying linear transformations.

In the field of robotics, we generally use Jacobians that relate joint velocities to Cartesian velocities of the tip of the arm—for example,

$$^0v = {}^0J(\Theta)\dot{\Theta}, \tag{5.64}$$

where Θ is the vector of joint angles of the manipulator and v is a vector of Cartesian velocities. In (5.64), we have added a leading superscript to our Jacobian notation to indicate in which frame the resulting Cartesian velocity is expressed. Sometimes this superscript is omitted when the frame is obvious or when it is unimportant to the development. Note that, for any given configuration of the manipulator, joint rates are related to velocity of the tip in a linear fashion, yet this is only an instantaneous relationship—in the next instant, the Jacobian has changed slightly. For the general case of a six-jointed robot, the Jacobian is 6×6, $\dot{\Theta}$ is 6×1, and ^0v is 6×1. This 6×1 Cartesian velocity vector is the 3×1 linear velocity vector and the 3×1 rotational velocity vector stacked together:

$$^0v = \begin{bmatrix} ^0v \\ ^0\omega \end{bmatrix}. \tag{5.65}$$

Jacobians of any dimension (including nonsquare) can be defined. The number of rows equals the number of degrees of freedom in the Cartesian space being considered. The number of columns in a Jacobian is equal to the number of joints of the manipulator. In dealing with a planar arm, for example, there is no reason for the Jacobian to have more than three rows, although, for redundant planar manipulators, there could be arbitrarily many columns (one for each joint).

In the case of a two-link arm, we can write a 2×2 Jacobian that relates joint rates to end-effector velocity. From the result of Example 5.3, we can easily determine the Jacobian of our two-link arm. The Jacobian written in frame {3} is seen (from (5.55)) to be

$$^3J(\Theta) = \begin{bmatrix} l_1 s_2 & 0 \\ l_1 c_2 + l_2 & l_2 \end{bmatrix}, \tag{5.66}$$

and the Jacobian written in frame {0} is (from (5.57))

$$^0J(\Theta) = \begin{bmatrix} -l_1 s_1 - l_2 s_{12} & -l_2 s_{12} \\ l_1 c_1 + l_2 c_{12} & l_2 c_{12} \end{bmatrix}. \tag{5.67}$$

Note that, in both cases, we have chosen to write a square matrix that relates joint rates to end-effector velocity. We could also consider a 3×2 Jacobian that would include the angular velocity of the end-effector.

Considering (5.58) through (5.62), which define the Jacobian, we see that the Jacobian might also be found by directly differentiating the kinematic equations of the mechanism. This is straightforward for linear velocity, but there is no 3×1 orientation vector whose derivative is ω. Hence, we have introduced a method to derive the Jacobian by using successive application of (5.45) and (5.47). There are several other methods that can be used (see, for example, [4]), one of which will be introduced shortly in Section 5.8. One reason for deriving Jacobians via the method presented is that it helps prepare us for material in Chapter 6, in which we will find that similar techniques apply to calculating the dynamic equations of motion of a manipulator.

Changing a Jacobian's frame of reference

Given a Jacobian written in frame $\{B\}$, that is,

$$\begin{bmatrix} {}^B\upsilon \\ {}^B\omega \end{bmatrix} = {}^B v = {}^B J(\Theta)\dot{\Theta}, \tag{5.68}$$

we might be interested in giving an expression for the Jacobian in another frame, $\{A\}$. First, note that a 6×1 Cartesian velocity vector given in $\{B\}$ is described relative to $\{A\}$ by the transformation

$$\begin{bmatrix} {}^A\upsilon \\ {}^A\omega \end{bmatrix} = \begin{bmatrix} {}^A_B R & 0 \\ \hline 0 & {}^A_B R \end{bmatrix} \begin{bmatrix} {}^B\upsilon \\ {}^B\omega \end{bmatrix}. \tag{5.69}$$

Hence, we can write

$$\begin{bmatrix} {}^A\upsilon \\ {}^A\omega \end{bmatrix} = \begin{bmatrix} {}^A_B R & 0 \\ \hline 0 & {}^A_B R \end{bmatrix} {}^B J(\Theta)\dot{\Theta}. \tag{5.70}$$

Now it is clear that changing the frame of reference of a Jacobian is accomplished by means of the following relationship:

$$^A J(\Theta) = \begin{bmatrix} {}^A_B R & 0 \\ \hline 0 & {}^A_B R \end{bmatrix} {}^B J(\Theta). \tag{5.71}$$

5.8 SINGULARITIES

Given that we have a linear transformation relating joint velocity to Cartesian velocity, a reasonable question to ask is: Is this matrix invertible? That is, is it nonsingular? If the matrix is nonsingular, then we can invert it to calculate joint rates from given Cartesian velocities:

$$\dot{\Theta} = J^{-1}(\Theta)v. \tag{5.72}$$

This is an important relationship. For example, say that we wish the hand of the robot to move with a certain velocity vector in Cartesian space. Using (5.72), we could calculate the necessary joint rates at each instant along the path. The real question of invertibility is: Is the Jacobian invertible for all values of Θ? If not, where is it not invertible?

Most manipulators have values of Θ where the Jacobian becomes singular. Such locations are called **singularities of the mechanism** or **singularities** for short. All manipulators have singularities at the boundary of their workspace, and most have loci of singularities inside their workspace. An in-depth study of the classification of singularities is beyond the scope of this book—for more information, see [5]. For our purposes, and without giving rigorous definitions, we will class singularities into two categories:

1. **Workspace-boundary singularities** occur when the manipulator is fully stretched out or folded back on itself in such a way that the end-effector is at or very near the boundary of the workspace.

2. **Workspace-interior singularities** occur away from the workspace boundary; they generally are caused by a lining up of two or more joint axes.

When a manipulator is in a singular configuration, it has lost one or more degrees of freedom (as viewed from Cartesian space). This means that there is some direction (or subspace) in Cartesian space along which it is impossible to move the hand of the robot, no matter what joint rates are selected. It is obvious that this happens at the workspace boundary of robots.

EXAMPLE 5.4

Where are the singularities of the simple two-link arm of Example 5.3? What is the physical explanation of the singularities? Are they workspace-boundary singularities or workspace-interior singularities?

To find the singular points of a mechanism, we must examine the determinant of its Jacobian. Where the determinant is equal to zero, the Jacobian has lost full rank and is singular:

$$DET[J(\Theta)] = \begin{bmatrix} l_1 s_2 & 0 \\ l_1 c_2 + l_2 & l_2 \end{bmatrix} = l_1 l_2 s_2 = 0. \tag{5.73}$$

Clearly, a singularity of the mechanism exists when θ_2 is 0 or 180 degrees. Physically, when $\theta_2 = 0$, the arm is stretched straight out. In this configuration, motion of the end-effector is possible along only one Cartesian direction (the one perpendicular to the arm). Therefore, the mechanism has lost one degree of freedom. Likewise, when $\theta_2 = 180$, the arm is folded completely back on itself, and motion of the hand again is possible only in one Cartesian direction instead of two. We will class these singularities as workspace-boundary singularities, because they exist at the edge of the manipulator's workspace. Note that the Jacobian written with respect to frame {0}, or any other frame, would have yielded the same result.

The danger in applying (5.72) in a robot control system is that, at a singular point, the inverse Jacobian blows up! This results in joint rates approaching infinity as the singularity is approached.

EXAMPLE 5.5

Consider the two-link robot from Example 5.3 as it is moving its end-effector along the $\hat{X}$ axis at 1.0 m/s, as in Fig. 5.10. Show that joint rates are reasonable when far from a singularity, but that, as a singularity is approached at $\theta_2 = 0$, joint rates tend to infinity.

We start by calculating the inverse of the Jacobian written in {0}:

$$^0 J^{-1}(\Theta) = \frac{1}{l_1 l_2 s_2} \begin{bmatrix} l_2 c_{12} & l_2 s_{12} \\ -l_1 c_1 - l_2 c_{12} & -l_1 s_1 - l_2 s_{12} \end{bmatrix}. \tag{5.74}$$

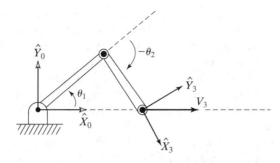

FIGURE 5.10: A two-link manipulator moving its tip at a constant linear velocity.

Then, using Eq. (5.74) for a velocity of 1 m/s in the $\hat{X}$ direction, we can calculate joint rates as a function of manipulator configuration:

$$\dot{\theta}_1 = \frac{c_{12}}{l_1 s_2}, \tag{5.75}$$

$$\dot{\theta}_2 = -\frac{c_1}{l_2 s_2} - \frac{c_{12}}{l_1 s_2}.$$

Clearly, as the arm stretches out toward $\theta_2 = 0$, both joint rates go to infinity.

EXAMPLE 5.6

For the PUMA 560 manipulator, give two examples of singularities that can occur.

There is singularity when θ_3 is near -90.0 degrees. Calculation of the exact value of θ_3 is left as an exercise. (See Exercise 5.14.) In this situation, links 2 and 3 are "stretched out," just like the singular location of the two-link manipulator in Example 5.3. This is classed as a workspace-boundary singularity.

Whenever $\theta_5 = 0.0$ degrees, the manipulator is in a singular configuration. In this configuration, joint axes 4 and 6 line up—both of their actions would result in the same end-effector motion, so it is as if a degree of freedom has been lost. Because this can occur interior to the workspace envelope, we will class it as a workspace-interior singularity.

5.9 STATIC FORCES IN MANIPULATORS

The chainlike nature of a manipulator leads us quite naturally to consider how forces and moments "propagate" from one link to the next. Typically, the robot is pushing on something in the environment with the chain's free end (the end-effector) or is perhaps supporting a load at the hand. We wish to solve for the joint torques that must be acting to keep the system in static equilibrium.

In considering static forces in a manipulator, we first lock all the joints so that the manipulator becomes a structure. We then consider each link in this structure and write a force-moment balance relationship in terms of the link frames. Finally, we compute what static torque must be acting about the joint axis in order for the manipulator to be in static equilibrium. In this way, we solve for the set of joint torques needed to support a static load acting at the end-effector.

In this section, we will not be considering the force on the links due to gravity (that will be left until chapter 6). The static forces and torques we are considering at the joints are those caused by a static force or torque (or both) acting on the last link—for example, as when the manipulator has its end-effector in contact with the environment.

We define special symbols for the force and torque exerted by a neighbor link:

f_i = force exerted on link i by link $i - 1$,

n_i = torque exerted on link i by link $i - 1$.

We will use our usual convention for assigning frames to links. Figure 5.11 shows the static forces and moments (excluding the gravity force) acting on link i. Summing the forces and setting them equal to zero, we have

$$^i f_i - {}^i f_{i+1} = 0. \tag{5.76}$$

Summing torques about the origin of frame $\{i\}$, we have

$$^i n_i - {}^i n_{i+1} - {}^i P_{i+1} \times {}^i f_{i+1} = 0. \tag{5.77}$$

If we start with a description of the force and moment applied by the hand, we can calculate the force and moment applied by each link, working from the last link down to the base (link 0). To do this, we formulate the force-moment expressions (5.76) and (5.77) such that they specify iterations from higher numbered links to lower numbered links. The result can be written as

$$^i f_i = {}^i f_{i+1}, \tag{5.78}$$

$$^i n_i = {}^i n_{i+1} + {}^i P_{i+1} \times {}^i f_{i+1}. \tag{5.79}$$

In order to write these equations in terms of only forces and moments defined within their own link frames, we transform with the rotation matrix describing frame

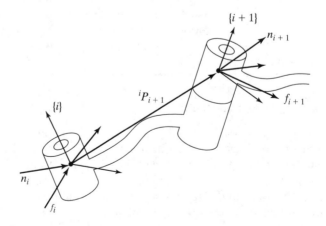

FIGURE 5.11: Static force-moment balance for a single link.

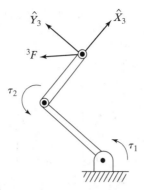

FIGURE 5.12: A two-link manipulator applying a force at its tip.

$\{i + 1\}$ relative to frame $\{i\}$. This leads to our most important result for static force "propagation" from link to link:

$$^i f_i = {}^i_{i+1}R \, {}^{i+1}f_{i+1},$$

(5.80)

$$^i n_i = {}^i_{i+1}R \, {}^{i+1}n_{i+1} + {}^i P_{i+1} \times {}^i f_i.$$

(5.81)

Finally, this important question arises: What torques are needed at the joints in order to balance the reaction forces and moments acting on the links? All components of the force and moment vectors are resisted by the structure of the mechanism itself, except for the torque about the joint axis. Therefore, to find the joint torque required to maintain the static equilibrium, the dot product of the joint-axis vector with the moment vector acting on the link is computed:

$$\tau_i = {}^i n_i^T \, {}^i \hat{Z}_i.$$

(5.82)

In the case that joint i is prismatic, we compute the joint actuator force as

$$\tau_i = {}^i f_i^T \, {}^i \hat{Z}_i.$$

(5.83)

Note that we are using the symbol τ even for a linear joint force.

As a matter of convention, we generally define the positive direction of joint torque as the direction which would tend to move the joint in the direction of increasing joint angle.

Equations (5.80) through (5.83) give us a means to compute the joint torques needed to apply any force or moment with the end-effector of a manipulator in the static case.

EXAMPLE 5.7

The two-link manipulator of Example 5.3 is applying a force vector ^3F with its end-effector. (Consider this force to be acting at the origin of $\{3\}$.) Find the required joint torques as a function of configuration and of the applied force. (See Fig. 5.12.)

We apply Eqs. (5.80) through (5.82), starting from the last link and going toward the base of the robot:

$$^2f_2 = \begin{bmatrix} f_x \\ f_y \\ 0 \end{bmatrix}, \tag{5.84}$$

$$^2n_2 = l_2\hat{X}_2 \times \begin{bmatrix} f_x \\ f_y \\ 0 \end{bmatrix} = \begin{bmatrix} 0 \\ 0 \\ l_2 f_y \end{bmatrix}, \tag{5.85}$$

$$^1f_1 = \begin{bmatrix} c_2 & -s_2 & 0 \\ s_2 & c_2 & 0 \\ 0 & 0 & 1 \end{bmatrix} \begin{bmatrix} f_x \\ f_y \\ 0 \end{bmatrix} = \begin{bmatrix} c_2 f_x - s_2 f_y \\ s_2 f_x + c_2 f_y \\ 0 \end{bmatrix}, \tag{5.86}$$

$$^1n_1 = \begin{bmatrix} 0 \\ 0 \\ l_2 f_y \end{bmatrix} + l_1\hat{X}_1 \times {}^1f_1 = \begin{bmatrix} 0 \\ 0 \\ l_1 s_2 f_x + l_1 c_2 f_y + l_2 f_y \end{bmatrix}. \tag{5.87}$$

Therefore, we have

$$\tau_1 = l_1 s_2 f_x + (l_2 + l_1 c_2) f_y, \tag{5.88}$$

$$\tau_2 = l_2 f_y. \tag{5.89}$$

This relationship can be written as a matrix operator:

$$\tau = \begin{bmatrix} l_1 s_2 & l_2 + l_1 c_2 \\ 0 & l_2 \end{bmatrix} \begin{bmatrix} f_x \\ f_y \end{bmatrix}. \tag{5.90}$$

It is not a coincidence that this matrix is the transpose of the Jacobian that we found in (5.66)!

5.10   JACOBIANS IN THE FORCE DOMAIN

We have found joint torques that will exactly balance forces at the hand in the static situation. When forces act on a mechanism, work (in the technical sense) is done if the mechanism moves through a displacement. Work is defined as a force acting through a distance and is a scalar with units of energy. The principle of **virtual work** allows us to make certain statements about the static case by allowing the amount of this displacement to go to an infinitesimal. Work has the units of energy, so it must be the same measured in any set of generalized coordinates. Specifically, we can equate the work done in Cartesian terms with the work done in joint-space terms. In the multidimensional case, work is the dot product of a vector force or torque and a vector displacement. Thus, we have

$$\mathcal{F} \cdot \delta\chi = \tau \cdot \delta\Theta, \tag{5.91}$$

where $\mathcal{F}$ is a 6×1 Cartesian force-moment vector acting at the end-effector, $\delta\chi$ is a 6×1 infinitesimal Cartesian displacement of the end-effector, τ is a 6×1 vector of torques at the joints, and $\delta\Theta$ is a 6×1 vector of infinitesimal joint displacements. Expression (5.91) can also be written as

$$\mathcal{F}^T \delta\chi = \tau^T \delta\Theta. \tag{5.92}$$

The definition of the Jacobian is

$$\delta\chi = J\delta\Theta, \tag{5.93}$$

so we may write

$$\mathcal{F}^T J\delta\theta = \tau^T \delta\Theta, \tag{5.94}$$

which must hold for all $\delta\Theta$; hence, we have

$$\mathcal{F}^T J = \tau^T. \tag{5.95}$$

Transposing both sides yields this result:

$$\tau = J^T \mathcal{F}. \tag{5.96}$$

Equation (5.96) verifies in general what we saw in the particular case of the two-link manipulator in Example 5.6: The Jacobian transpose maps Cartesian forces acting at the hand into equivalent joint torques. When the Jacobian is written with respect to frame {0}, then force vectors written in {0} can be transformed, as is made clear by the following notation:

$$\tau = {}^0J^T \, {}^0\mathcal{F}. \tag{5.97}$$

When the Jacobian loses full rank, there are certain directions in which the end-effector cannot exert static forces even if desired. That is, in (5.97), if the Jacobian is singular, $\mathcal{F}$ could be increased or decreased in certain directions (those defining the null-space of the Jacobian [6]) without effect on the value calculated for τ. This also means that, near singular configurations, mechanical advantage tends toward infinity, such that, with small joint torques, large forces could be generated at the end-effector.[3] Thus, singularities manifest themselves in the force domain as well as in the position domain.

Note that (5.97) is a very interesting relationship, in that it allows us to convert a Cartesian quantity into a joint-space quantity without calculating any inverse kinematic functions. We will make use of this when we consider the problem of control in later chapters.

5.11   CARTESIAN TRANSFORMATION OF VELOCITIES AND STATIC FORCES

We might wish to think in terms of 6×1 representations of general velocity of a body:

$$\upsilon = \begin{bmatrix} \upsilon \\ \omega \end{bmatrix}. \tag{5.98}$$

Likewise, we could consider 6×1 representations of general force vectors, such as

$$\mathcal{F} = \begin{bmatrix} F \\ N \end{bmatrix}, \tag{5.99}$$

---

[3]Consider a two-link planar manipulator nearly outstretched with the end-effector in contact with a reaction surface. In this configuration, arbitrarily large forces could be exerted by "small" joint torques.

where F is a 3×1 force vector and N is a 3×1 moment vector. It is then natural to think of 6×6 transformations that map these quantities from one frame to another. This is exactly what we have already done in considering the propagation of velocities and forces from link to link. Here, we write (5.45) and (5.47) in matrix-operator form to transform general velocity vectors in frame $\{A\}$ to their description in frame $\{B\}$.

The two frames involved here are rigidly connected, so $\dot{\theta}_{i+1}$, appearing in (5.45), is set to zero in deriving the relationship

$$\begin{bmatrix} {}^B\upsilon_B \\ {}^B\omega_B \end{bmatrix} = \begin{bmatrix} {}^B_A R & -{}^B_A R \, {}^A P_{BORG} \times \\ 0 & {}^B_A R \end{bmatrix} \begin{bmatrix} {}^A\upsilon_A \\ {}^A\omega_A \end{bmatrix}, \tag{5.100}$$

where the cross product is understood to be the matrix operator

$$P \times = \begin{bmatrix} 0 & -p_x & p_y \\ p_x & 0 & -p_x \\ -p_y & p_x & 0 \end{bmatrix}. \tag{5.101}$$

Now, (5.100) relates velocities in one frame to those in another, so the 6×6 operator will be called a **velocity transformation**; we will use the symbol T_υ. In this case, it is a velocity transformation that maps velocities in $\{A\}$ into velocities in $\{B\}$, so we use the following notation to express (5.100) compactly:

$$ {}^B\upsilon_B = {}^B_A T_\upsilon \, {}^A\upsilon_A. \tag{5.102}$$

We can invert (5.100) in order to compute the description of velocity in terms of $\{A\}$, given the quantities in $\{B\}$:

$$\begin{bmatrix} {}^A\upsilon_A \\ {}^A\omega_A \end{bmatrix} = \begin{bmatrix} {}^A_B R & {}^A P_{BORG} \times {}^A_B R \\ 0 & {}^A_B R \end{bmatrix} \begin{bmatrix} {}^B\upsilon_B \\ {}^B\omega_B \end{bmatrix}, \tag{5.103}$$

or

$$ {}^A\upsilon_A = {}^A_B T_\upsilon \, {}^B\upsilon_B. \tag{5.104}$$

Note that these mappings of velocities from frame to frame depend on ${}^A_B T$ (or its inverse) and so must be interpreted as instantaneous results, unless the relationship between the two frames is static. Similarly, from (5.80) and (5.81), we write the 6×6 matrix that transforms general force vectors written in terms of $\{B\}$ into their description in frame $\{A\}$, namely,

$$\begin{bmatrix} {}^A F_A \\ {}^A N_A \end{bmatrix} = \begin{bmatrix} {}^A_B R & 0 \\ {}^A P_{BORG} \times {}^A_B R & {}^A_B R \end{bmatrix} \begin{bmatrix} {}^B F_B \\ {}^B N_B \end{bmatrix}, \tag{5.105}$$

which may be written compactly as

$$ {}^A\mathcal{F}_A = {}^A_B T_f \, {}^B\mathcal{F}_B, \tag{5.106}$$

where T_f is used to denote a **force-moment transformation**.

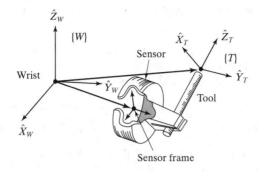

FIGURE 5.13: Frames of interest with a force sensor.

Velocity and force transformations are similar to Jacobians in that they relate velocities and forces in different coordinate systems. As with Jacobians, we have

$$_B^A T_f = {_B^A T_v^T},\qquad(5.107)$$

as can be verified by examining (5.105) and (5.103).

EXAMPLE 5.8

Figure 5.13 shows an end-effector holding a tool. Located at the point where the end-effector attaches to the manipulator is a force-sensing wrist. This is a device that can measure the forces and torques applied to it.

Consider the output of this sensor to be a 6×1 vector, $^S\mathcal{F}$, composed of three forces and three torques expressed in the sensor frame, $\{S\}$. Our real interest is in knowing the forces and torques applied at the tip of the tool, $^T\mathcal{F}$. Find the 6×6 transformation that transforms the force-moment vector from $\{S\}$ to the tool frame, $\{T\}$. The transform relating $\{T\}$ to $\{S\}$, $_T^S T$, is known. (Note that $\{S\}$ here is the sensor frame, not the station frame.)

This is simply an application of (5.106). First, from $_T^S T$, we calculate the inverse, $_S^T T$, which is composed of $_S^T R$ and $^T P_{SORG}$. Then we apply (5.106) to obtain

$$^T\mathcal{F}_T = {_S^T T_f}\, ^S\mathcal{F}_S,\qquad(5.108)$$

where

$$_S^T T_f = \begin{bmatrix} _S^T R & 0 \\ ^T P_{SORG} \times {_S^T R} & _S^T R \end{bmatrix}.\qquad(5.109)$$

BIBLIOGRAPHY

[1] K. Hunt, *Kinematic Geometry of Mechanisms*, Oxford University Press, New York, 1978.

[2] K.R. Symon, *Mechanics*, 3rd edition, Addison-Wesley, Reading, MA, 1971.

[3] I. Shames, *Engineering Mechanics*, 2nd edition, Prentice-Hall, Englewood Cliffs, NJ, 1967.

[4] D. Orin and W. Schrader, "Efficient Jacobian Determination for Robot Manipulators," in *Robotics Research: The First International Symposium*, M. Brady and R.P. Paul, Editors, MIT Press, Cambridge, MA, 1984.

[5] B. Gorla and M. Renaud, *Robots Manipulateurs*, Cepadues-Editions, Toulouse, 1984.

[6] B. Noble, *Applied Linear Algebra*, Prentice-Hall, Englewood Cliffs, NJ, 1969.

[7] J.K. Salisbury and J. Craig, "Articulated Hands: Kinematic and Force Control Issues," *International Journal of Robotics Research*, Vol. 1, No. 1, Spring 1982.

[8] C. Wampler, "Wrist Singularities: Theory and Practice," in *The Robotics Review 2*, O. Khatib, J. Craig, and T. Lozano-Perez, Editors, MIT Press, Cambridge, MA, 1992.

[9] D.E. Whitney, "Resolved Motion Rate Control of Manipulators and Human Prostheses," *IEEE Transactions on Man-Machine Systems*, 1969.

EXERCISES

5.1 [10] Repeat Example 5.3, but using the Jacobian written in frame {0}. Are the results the same as those of Example 5.3?

5.2 [25] Find the Jacobian of the manipulator with three degrees of freedom from Exercise 3 of Chapter 3. Write it in terms of a frame {4} located at the tip of the hand and having the same orientation as frame {3}.

5.3 [35] Find the Jacobian of the manipulator with three degrees of freedom from Exercise 3 of Chapter 3. Write it in terms of a frame {4} located at the tip of the hand and having the same orientation as frame {3}. Derive the Jacobian in three different ways: velocity propagation from base to tip, static force propagation from tip to base, and by direct differentiation of the kinematic equations.

5.4 [8] Prove that singularities in the force domain exist at the same configurations as singularities in the position domain.

5.5 [39] Calculate the Jacobian of the PUMA 560 in frame {6}.

5.6 [47] Is it true that any mechanism with three revolute joints and nonzero link lengths must have a locus of singular points interior to its workspace?

5.7 [7] Sketch a figure of a mechanism with three degrees of freedom whose linear velocity Jacobian is the 3×3 identity matrix over all configurations of the manipulator. Describe the kinematics in a sentence or two.

5.8 [18] General mechanisms sometimes have certain configurations, called "isotropic points," where the columns of the Jacobian become orthogonal and of equal magnitude [7]. For the two-link manipulator of Example 5.3, find out if any isotropic points exist. Hint: Is there a requirement on l_1 and l_2?

5.9 [50] Find the conditions necessary for isotropic points to exist in a general manipulator with six degrees of freedom. (See Exercise 5.8.)

5.10 [7] For the two-link manipulator of Example 5.2, give the transformation that would map joint torques into a 2×1 force vector, ^3F, at the hand.

5.11 [14] Given

$$^A_B T = \begin{bmatrix} 0.866 & -0.500 & 0.000 & 10.0 \\ 0.500 & 0.866 & 0.000 & 0.0 \\ 0.000 & 0.000 & 1.000 & 5.0 \\ 0 & 0 & 0 & 1 \end{bmatrix},$$

if the velocity vector at the origin of $\{A\}$ is

$$
^A v = \begin{bmatrix} 0.0 \\ 2.0 \\ -3.0 \\ 1.414 \\ 1.414 \\ 0.0 \end{bmatrix},
$$

find the 6×1 velocity vector with reference point the origin of $\{B\}$.

5.12 [15] For the three-link manipulator of Exercise 3.3, give a set of joint angles for which the manipulator is at a workspace-boundary singularity and another set of angles for which the manipulator is at a workspace-interior singularity.

5.13 [9] A certain two-link manipulator has the following Jacobian:

$$
^0 J(\Theta) = \begin{bmatrix} -l_1 s_1 - l_2 s_{12} & -l_2 s_{12} \\ l_1 c_1 + l_2 c_{12} & l_2 c_{12} \end{bmatrix}.
$$

Ignoring gravity, what are the joint torques required in order that the manipulator will apply a static force vector $^0 F = 10 \hat{X}_0$?

5.14 [18] If the link parameter a_3 of the PUMA 560 were zero, a workspace-boundary singularity would occur when $\theta_3 = -90.0°$. Give an expression for the value of θ_3 where the singularity occurs, and show that, if a_3 were zero, the result would be $\theta_3 = -90.0°$. *Hint:* In this configuration, a straight line passes through joint axes 2 and 3 and the point where axes 4, 5, and 6 intersect.

5.15 [24] Give the 3×3 Jacobian that calculates linear velocity of the tool tip from the three joint rates for the manipulator of Example 3.4 in Chapter 3. Give the Jacobian in frame $\{0\}$.

5.16 [20] A $3R$ manipulator has kinematics that correspond exactly to the set of Z–Y–Z Euler angles (i.e., the forward kinematics are given by (2.72) with $\alpha = \theta_1$, $\beta = \theta_2$, and $\gamma = \theta_3$). Give the Jacobian relating joint velocities to the angular velocity of the final link.

5.17 [31] Imagine that, for a general 6-DOF robot, we have available $^0 \hat{Z}_i$ and $^0 P_{iorg}$ for all i—that is, we know the values for the unit Z vectors of each link frame in terms of the base frame and we know the locations of the origins of all link frames in terms of the base frame. Let us also say that we are interested in the velocity of the tool point (fixed relative to link n) and that we know $^0 P_{tool}$ also. Now, for a revolute joint, the velocity of the tool tip due to the velocity of joint i is given by

$$
^0 v_i = \dot{\theta}_i \; ^0 \hat{Z}_i \times (^0 P_{tool} - {}^0 P_{iorg}) \tag{5.110}
$$

and the angular velocity of link n due to the velocity of this joint is given by

$$
^0 \omega_i = \dot{\theta}_i \; ^0 \hat{Z}_i. \tag{5.111}
$$

The total linear and angular velocity of the tool is given by the sum of the $^0 v_i$ and $^0 \omega_i$ respectively. Give equations analogous to (5.110) and (5.111) for the case of joint i prismatic, and write the 6×6 Jacobian matrix of an arbitrary 6-DOF manipulator in terms of the $\hat{Z}_i$, P_{iorg}, and P_{tool}.

5.18 [18] The kinematics of a $3R$ robot are given by

$$
^0_3 T = \begin{bmatrix} c_1 c_{23} & -c_1 s_{23} & s_1 & l_1 c_1 + l_2 c_1 c_2 \\ s_1 c_{23} & -s_1 s_{23} & -c_1 & l_1 s_1 + l_2 s_1 c_2 \\ s_{23} & c_{23} & 0 & l_2 s_2 \\ 0 & 0 & 0 & 1 \end{bmatrix}.
$$

Find $^0J(\Theta)$, which, when multiplied by the joint velocity vector, gives the linear velocity of the origin of frame {3} relative to frame {0}.

5.19 [15] The position of the origin of link 2 for an *RP* manipulator is given by

$$^0P_{2ORG} = \begin{bmatrix} a_1c_1 - d_2s_1 \\ a_1s_1 + d_2c_1 \\ 0 \end{bmatrix}.$$

Give the 2×2 Jacobian that relates the two joint rates to the linear velocity of the origin of frame {2}. Give a value of Θ where the device is at a singularity.

5.20 [20] Explain what might be meant by the statement: "An n-DOF manipulator at a singularity can be treated as a redundant manipulator in a space of dimensionality $n - 1$."

PROGRAMMING EXERCISE (PART 5)

1. Two frames, {A} and {B}, are not moving relative to one another—that is, ^A_BT is constant. In the planar case, we define the velocity of frame {A} as

$$^Av_A = \begin{bmatrix} ^A\dot{x}_A \\ ^A\dot{y}_A \\ ^A\dot{\theta}_A \end{bmatrix}.$$

Write a routine that, given ^A_BT and ^Av_A, computes ^Bv_B. *Hint*: This is the planar analog of (5.100). Use a procedure heading something like (or equivalent C):

```
Procedure Veltrans (VAR brela: frame; VAR vrela, vrelb: vec3);
```

where "vrela" is the velocity relative to frame {A}, or ^Av_A, and "vrelb" is the output of the routine (the velocity relative to frame {B}), or ^Bv_B.

2. Determine the 3×3 Jacobian of the three-link planar manipulator (from Example 3.3). In order to derive the Jacobian, you should use velocity-propagation analysis (as in Example 5.2) or static-force analysis (as in Example 5.6). Hand in your work showing how you derived the Jacobian.

Write a routine to compute the Jacobian in frame {3}—that is, $^3J(\Theta)$—as a function of the joint angles. Note that frame {3} is the standard link frame with origin on the axis of joint 3. Use a procedure heading something like (or equivalent C):

```
Procedure Jacobian (VAR theta: vec3; Var Jac: mat33);
```

The manipulator data are $l_2 = l_2 = 0.5$ meters.

3. A tool frame and a station frame are defined as follows by the user for a certain task (units are meters and degrees):

$$^W_TT = [x\ y\ \theta] = [0.1\ 0.2\ 30.0],$$

$$^B_ST = [x\ y\ \theta] = [0.0\ 0.0\ 0.0].$$

At a certain instant, the tool tip is at the position

$$^S_TT = [x\ y\ \theta] = [0.6\ -0.3\ 45.0].$$

At the same instant, the joint rates (in deg/sec) are measured to be

$$\dot{\Theta} = [\dot{\theta}_1 \; \dot{\theta}_2 \; \dot{\theta}_3] = [20.0 \; -10.0 \; 12.0].$$

Calculate the linear and angular velocity of the tool tip relative to its own frame, that is, $^T v_T$. If there is more than one possible answer, calculate all possible answers.

MATLAB EXERCISE 5

This exercise focuses on the Jacobian matrix and determinant, simulated resolved-rate control, and inverse statics for the planar 3-DOF, 3R robot. (See Figures 3.6 and 3.7; the DH parameters are given in Figure 3.8.)

The resolved-rate control method [9] is based on the manipulator velocity equation $^k \dot{X} = {}^k J \dot{\Theta}$, where $^k J$ is the Jacobian matrix, $\dot{\Theta}$ is the vector of relative joint rates, $^k \dot{X}$ is the vector of commanded Cartesian velocities (both translational and rotational), and k is the frame of expression for the Jacobian matrix and Cartesian velocities. This figure shows a block diagram for simulating the resolved-rate control algorithm:

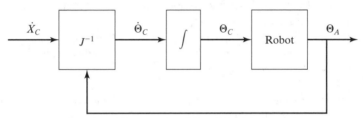

Resolved-Rate-Algorithm Block Diagram

As is seen in the figure, the resolved-rate algorithm calculates the required commanded joint rates $\dot{\Theta}_C$ to provide the commanded Cartesian velocities $\dot{X}_C$; this diagram must be calculated at every simulated time step. The Jacobian matrix changes with configuration Θ_A. For simulation purposes, assume that the commanded joint angles Θ_C are always identical to the actual joint angles achieved, Θ_A (a result rarely true in the real world). For the planar 3-DOF, 3R robot assigned, the velocity equations $^k \dot{X} = {}^k J \dot{\Theta}$ for $k = 0$ are

$$^0 \left\{ \begin{array}{c} \dot{x} \\ \dot{y} \\ \omega_z \end{array} \right\} = {}^0 \begin{bmatrix} -L_1 s_1 - L_2 s_{12} - L_3 s_{123} & -L_2 s_{12} - L_3 s_{123} & -L_3 s_{123} \\ L_1 c_1 + L_2 c_{12} + L_3 c_{123} & L_2 c_{12} + L_3 c_{123} & L_3 c_{123} \\ 1 & 1 & 1 \end{bmatrix} \left\{ \begin{array}{c} \dot{\theta}_1 \\ \dot{\theta}_2 \\ \dot{\theta}_3 \end{array} \right\},$$

where $s_{123} = \sin(\theta_1 + \theta_2 + \theta_3)$, $c_{123} = \cos(\theta_1 + \theta_2 + \theta_3)$, and so on. Note that $^0 \dot{X}$ gives the Cartesian velocities of the origin of the hand frame (at the center of the grippers in Figure 3.6) with respect to the origin of the base frame {0}, expressed in {0} coordinates.

Now, most industrial robots cannot command $\dot{\Theta}_C$ directly, so we must first integrate these commanded relative joint rates to commanded joint angles Θ_C, which can be commanded to the robot at every time step. In practice, the simplest possible integration scheme works well, assuming a small control time step Δt: $\Theta_{new} = \Theta_{old} + \dot{\Theta} \Delta t$. In your MATLAB resolved-rate simulation, assume that the commanded Θ_{new} can be achieved perfectly by the virtual robot. (Chapters 6 and 9 present dynamics and control material for which we do not have to make this simplifying assumption.) Be sure to update the

Jacobian matrix with the new configuration Θ_{new} before completing the resolved-rate calculations for the next time step.

Develop a MATLAB program to calculate the Jacobian matrix and to simulate resolved-rate control for the planar 3R robot. Given the robot lengths $L_1 = 4$, $L_2 = 3$, and $L_3 = 2$ (m); the initial joint angles $\Theta = \{\theta_1 \ \theta_2 \ \theta_3\}^T = \{10° \ 20° \ 30°\}^T$, and the constant commanded Cartesian rates $^0\{\dot{X}\} = \{\dot{x} \ \dot{y} \ \omega_z\}^T = \{0.2 \ -0.3 \ -0.2\}^T$ (m/s, m/s, rad/s), simulate for exactly 5 sec, using time steps of exactly $dt = 0.1$ sec. In the same program loop, calculate the inverse-statics problem—that is, calculate the joint torques $T = \{\tau_1 \ \tau_2 \ \tau_3\}^T$ (Nm), given the constant commanded Cartesian wrench $^0\{W\} = \{f_x \ f_y \ m_z\}^T = \{1 \ 2 \ 3\}^T$ (N, N, Nm). Also, in the same loop, animate the robot to the screen during each time step, so that you can watch the simulated motion to verify that it is correct.

a) For the specific numbers assigned, present five plots (each set on a separate graph, please):

1. the three active joint rates $\dot{\Theta} = \{\dot{\theta}_1 \ \dot{\theta}_2 \ \dot{\theta}_3\}^T$ vs. time;

2. the three active joint angles $\Theta = \{\theta_1 \ \theta_2 \ \theta_3\}^T$ vs. time;

3. the three Cartesian components of $^0_H T$, $X = \{x \ y \ \phi\}^T$ (rad is fine for ϕ so that it will fit) vs. time;

4. the Jacobian matrix determinant $|J|$ vs. time—comment on nearness to singularities during the simulated resolved-rate motion;

5. the three active joint torques $T = \{\tau_1 \ \tau_2 \ \tau_3\}^T$ vs. time.

Carefully label (by hand is fine!) each component on each plot; also, label the axes with names and units.

b) Check your Jacobian matrix results for the initial and final joint-angle sets by means of the Corke MATLAB Robotics Toolbox. Try function *jacob0()*. **Caution:** The toolbox Jacobian functions are for motion of {3} with respect to {0}, not for {H} with respect to {0} as in the problem assignment. The preceding function gives the Jacobian result in {0} coordinates; *jacobn()* would give results in {3} coordinates.

C H A P T E R   6

Manipulator dynamics

6.1 INTRODUCTION

Our study of manipulators so far has focused on kinematic considerations only. We have studied static positions, static forces, and velocities; but we have never considered *the forces required to cause motion*. In this chapter, we consider the equations of motion for a manipulator—the way in which motion of the manipulator arises from torques applied by the actuators or from external forces applied to the manipulator.

Dynamics of mechanisms is a field in which many books have been written. Indeed, one can spend years studying the field. Obviously, we cannot cover the material in the completeness it deserves. However, certain formulations of the dynamics problem seem particularly well suited to application to manipulators. In particular, methods which make use of the serial-chain nature of manipulators are natural candidates for our study.

There are two problems related to the dynamics of a manipulator that we wish to solve. In the first problem, we are given a trajectory point, Θ, $\dot{\Theta}$, and $\ddot{\Theta}$, and we wish to find the required vector of joint torques, τ. This formulation of dynamics is useful for the problem of controlling the manipulator (Chapter 10). The second problem is to calculate how the mechanism will move under application of a set of joint torques. That is, given a torque vector, τ, calculate the resulting motion of the manipulator, Θ, $\dot{\Theta}$, and $\ddot{\Theta}$. This is useful for simulating the manipulator.

6.2 ACCELERATION OF A RIGID BODY

We now extend our analysis of rigid-body motion to the case of accelerations. At any instant, the linear and angular velocity vectors have derivatives that are called the linear and angular accelerations, respectively. That is,

$$
{}^B\dot{V}_Q = \frac{d}{dt}\,{}^BV_Q = \lim_{\Delta t \to o} \frac{{}^BV_Q(t + \Delta t) - {}^BV_Q(t)}{\Delta t}, \tag{6.1}
$$

and

$$
{}^A\dot{\Omega}_B = \frac{d}{dt}\,{}^A\Omega_B = \lim_{\Delta t \to o} \frac{{}^A\Omega_B(t + \Delta t) - {}^A\Omega_B(t)}{\Delta t}. \tag{6.2}
$$

As with velocities, when the reference frame of the differentiation is understood to be some universal reference frame, $\{U\}$, we will use the notation

$$
\dot{v}_A = {}^U\dot{V}_{AORG} \tag{6.3}
$$

and

$$
\dot{\omega}_A = {}^U\dot{\Omega}_A. \tag{6.4}
$$

Linear acceleration

We start by restating (5.12), an important result from Chapter 5, which describes the velocity of a vector ${}^BQ$ as seen from frame $\{A\}$ when the origins are coincident:

$$
{}^AV_Q = {}^A_BR\,{}^BV_Q + {}^A\Omega_B \times {}^A_BR\,{}^BQ. \tag{6.5}
$$

The left-hand side of this equation describes how ${}^AQ$ is changing in time. So, because origins are coincident, we could rewrite (6.5) as

$$
\frac{d}{dt}({}^A_BR\,{}^BQ) = {}^A_BR\,{}^BV_Q + {}^A\Omega_B \times {}^A_BR\,{}^BQ. \tag{6.6}
$$

This form of the equation will be useful when deriving the corresponding acceleration equation.

By differentiating (6.5), we can derive expressions for the acceleration of ${}^BQ$ as viewed from $\{A\}$ when the origins of $\{A\}$ and $\{B\}$ coincide:

$$
{}^A\dot{V}_Q = \frac{d}{dt}({}^A_BR\,{}^BV_Q) + {}^A\dot{\Omega}_B \times {}^A_BR\,{}^BQ + {}^A\Omega_B \times \frac{d}{dt}({}^A_BR\,{}^BQ). \tag{6.7}
$$

Now we apply (6.6) twice—once to the first term, and once to the last term. The right-hand side of equation (6.7) becomes

$$
\begin{aligned}
{}^A_BR\,{}^B\dot{V}_Q &+ {}^A\Omega_B \times {}^A_BR\,{}^BV_Q + {}^A\dot{\Omega}_B \times {}^A_BR\,{}^BQ \\
&+ {}^A\Omega_B \times ({}^A_BR\,{}^BV_Q + {}^A\Omega_B \times {}^A_BR\,{}^BQ).
\end{aligned} \tag{6.8}
$$

Combining two terms, we get

$$
{}^A_BR\,{}^B\dot{V}_Q + 2\,{}^A\Omega_B \times {}^A_BR\,{}^BV_Q + {}^A\dot{\Omega}_B \times {}^A_BR\,{}^BQ + {}^A\Omega_B \times ({}^A\Omega_B \times {}^A_BR\,{}^BQ). \tag{6.9}
$$

Finally, to generalize to the case in which the origins are not coincident, we add one term which gives the linear acceleration of the origin of $\{B\}$, resulting in the final general formula:

$$^A\dot{V}_{BORG} + {}^A_BR\,{}^B\dot{V}_Q + 2{}^A\Omega_B \times {}^A_BR\,{}^BV_Q + {}^A\dot{\Omega}_B \times {}^A_BR\,{}^BQ \tag{6.10}$$

$$+ {}^A\Omega_B \times ({}^A\Omega_B \times {}^A_BR\,{}^BQ).$$

A particular case that is worth pointing out is when ^BQ is constant, or

$$^BV_Q = {}^B\dot{V}_Q = 0. \tag{6.11}$$

In this case, (6.10) simplifies to

$$^A\dot{V}_Q = {}^A\dot{V}_{BORG} + {}^A\Omega_B \times ({}^A\Omega_B \times {}^A_BR\,{}^BQ) + {}^A\dot{\Omega}_B \times {}^A_BR\,{}^BQ. \tag{6.12}$$

We will use this result in calculating the linear acceleration of the links of a manipulator with rotational joints. When a prismatic joint is present, the more general form of (6.10) will be used.

Angular acceleration

Consider the case in which $\{B\}$ is rotating relative to $\{A\}$ with $^A\Omega_B$ and $\{C\}$ is rotating relative to $\{B\}$ with $^B\Omega_C$. To calculate $^A\Omega_C$, we sum the vectors in frame $\{A\}$:

$$^A\Omega_C = {}^A\Omega_B + {}^A_BR\,{}^B\Omega_C. \tag{6.13}$$

By differentiating, we obtain

$$^A\dot{\Omega}_C = {}^A\dot{\Omega}_B + \frac{d}{dt}({}^A_BR\,{}^B\Omega_C). \tag{6.14}$$

Now, applying (6.6) to the last term of (6.14), we get

$$^A\dot{\Omega}_C = {}^A\dot{\Omega}_B + {}^A_BR\,{}^B\dot{\Omega}_C + {}^A\Omega_B \times {}^A_BR\,{}^B\Omega_C. \tag{6.15}$$

We will use this result to calculate the angular acceleration of the links of a manipulator.

6.3   MASS DISTRIBUTION

In systems with a single degree of freedom, we often talk about the mass of a rigid body. In the case of rotational motion about a single axis, the notion of the *moment of inertia* is a familiar one. For a rigid body that is free to move in three dimensions, there are infinitely many possible rotation axes. In the case of rotation about an arbitrary axis, we need a complete way of characterizing the mass distribution of a rigid body. Here, we introduce the **inertia tensor**, which, for our purposes, can be thought of as a generalization of the scalar moment of inertia of an object.

We shall now define a set of quantities that give information about the distribution of mass of a rigid body relative to a reference frame. Figure 6.1 shows a rigid body with an attached frame. Inertia tensors can be defined relative to any frame, but we will always consider the case of an inertia tensor defined for a frame

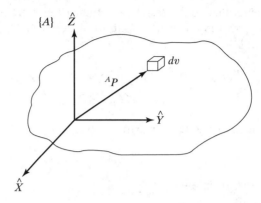

FIGURE 6.1: The inertia tensor of an object describes the object's mass distribution. Here, the vector $^A P$ locates the differential volume element, dv.

attached to the rigid body. Where it is important, we will indicate, with a leading superscript, the frame of reference of a given inertia tensor. The inertia tensor relative to frame $\{A\}$ is expressed in the matrix form as the 3×3 matrix

$$^A I = \begin{bmatrix} I_{xx} & -I_{xy} & -I_{xz} \\ -I_{xy} & I_{yy} & -I_{yz} \\ -I_{xz} & -I_{yz} & I_{zz} \end{bmatrix}, \tag{6.16}$$

where the scalar elements are given by

$$I_{xx} = \iiint_V (y^2 + z^2)\rho dv,$$

$$I_{yy} = \iiint_V (x^2 + z^2)\rho dv,$$

$$I_{zz} = \iiint_V (x^2 + y^2)\rho dv, \tag{6.17}$$

$$I_{xy} = \iiint_V xy\rho dv,$$

$$I_{xz} = \iiint_V xz\rho dv,$$

$$I_{yz} = \iiint_V yz\rho dv,$$

in which the rigid body is composed of differential volume elements, dv, containing material of density ρ. Each volume element is located with a vector, $^A P = [xyz]^T$, as shown in Fig. 6.1.

The elements I_{xx}, I_{yy}, and I_{zz} are called the **mass moments of inertia**. Note that, in each case, we are integrating the mass elements, ρdv, times the squares of the perpendicular distances from the corresponding axis. The elements with mixed indices are called the **mass products of inertia**. This set of six independent quantities will, for a given body, depend on the position and orientation of the frame in which they are defined. If we are free to choose the orientation of the reference frame, it is possible to cause the products of inertia to be zero. The axes of the reference frame when so aligned are called the **principal axes** and the corresponding mass moments are the **principal moments** of inertia.

EXAMPLE 6.1

Find the inertia tensor for the rectangular body of uniform density ρ with respect to the coordinate system shown in Fig. 6.2.

First, we compute I_{xx}. Using volume element $dv = dx \, dy \, dz$, we get

$$
\begin{aligned}
I_{xx} &= \int_0^h \int_0^l \int_0^\omega (y^2 + z^2)\rho \, dx \, dy \, dz \\
&= \int_0^h \int_0^l (y^2 + z^2)\omega\rho \, dy \, dz \qquad\qquad (6.18) \\
&= \int_0^h \left(\frac{l^3}{3} + z^2 l \right) \omega\rho \, dz \\
&= \left(\frac{hl^3\omega}{3} + \frac{h^3 l\omega}{3} \right) \rho \\
&= \frac{m}{3}(l^2 + h^2),
\end{aligned}
$$

where m is the total mass of the body. Permuting the terms, we can get I_{yy} and I_{zz} by inspection:

$$
I_{yy} = \frac{m}{3}(\omega^2 + h^2) \qquad\qquad (6.19)
$$

and

$$
I_{zz} = \frac{m}{3}(l^2 + \omega^2). \qquad\qquad (6.20)
$$

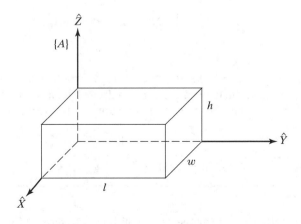

FIGURE 6.2: A body of uniform density.

We next compute I_{xy}:

$$I_{xy} = \int_0^h \int_0^l \int_0^\omega xy\rho \, dx \, dy \, dz$$

$$= \int_0^h \int_0^l \frac{\omega^2}{2} y\rho \, dy \, dz \qquad (6.21)$$

$$= \int_0^h \frac{\omega^2 l^2}{4} \rho \, dz$$

$$= \frac{m}{4}\omega l.$$

Permuting the terms, we get

$$I_{xz} = \frac{m}{4}h\omega \qquad (6.22)$$

and

$$I_{yz} = \frac{m}{4}hl. \qquad (6.23)$$

Hence, the inertia tensor for this object is

$$^A I = \begin{bmatrix} \frac{m}{3}(l^2 + h^2) & -\frac{m}{4}\omega l & -\frac{m}{4}h\omega \\ -\frac{m}{4}\omega l & \frac{m}{3}(\omega^2 + h^2) & -\frac{m}{4}hl \\ -\frac{m}{4}h\omega & -\frac{m}{4}hl & \frac{m}{3}(l^2 + \omega^2) \end{bmatrix}. \qquad (6.24)$$

As noted, the inertia tensor is a function of the location and orientation of the reference frame. A well-known result, the **parallel-axis theorem**, is one way of computing how the inertia tensor changes under *translations* of the reference coordinate system. The parallel-axis theorem relates the inertia tensor in a frame with origin at the center of mass to the inertia tensor with respect to another reference frame. Where $\{C\}$ is located at the center of mass of the body, and $\{A\}$ is an arbitrarily translated frame, the theorem can be stated [1] as

$$^A I_{zz} = {}^C I_{zz} + m(x_c^2 + y_c^2),$$

$$^A I_{xy} = {}^C I_{xy} - mx_c y_c, \qquad (6.25)$$

where $P_c = [x_c, y_c, z_c]^T$ locates the center of mass relative to $\{A\}$. The remaining moments and products of inertia are computed from permutations of $x, y,$ and z in (6.25). The theorem may be stated in vector–matrix form as

$$^A I = {}^C I + m[P_c^T P_c I_3 - P_c P_c^T], \qquad (6.26)$$

where I_3 is the 3×3 identity matrix.

EXAMPLE 6.2

Find the inertia tensor for the same solid body described for Example 6.1 when it is described in a coordinate system with origin at the body's center of mass.

We can apply the parallel-axis theorem, (6.25), where

$$
\begin{bmatrix} x_c \\ y_c \\ z_c \end{bmatrix} = \frac{1}{2} \begin{bmatrix} \omega \\ l \\ h \end{bmatrix}.
$$

Next, we find

$$
{}^C I_{zz} = \frac{m}{12}(\omega^2 + l^2),
$$

$$
{}^C I_{xy} = 0. \tag{6.27}
$$

The other elements are found by symmetry. The resulting inertia tensor written in the frame at the center of mass is

$$
{}^C I = \begin{bmatrix} \frac{m}{12}(h^2 + l^2) & 0 & 0 \\ 0 & \frac{m}{12}(\omega^2 + h^2) & 0 \\ 0 & 0 & \frac{m}{12}(l^2 + \omega^2) \end{bmatrix}. \tag{6.28}
$$

The result is diagonal, so frame {C} must represent the principal axes of this body.

Some additional facts about inertia tensors are as follows:

1. If two axes of the reference frame form a plane of symmetry for the mass distribution of the body, the products of inertia having as an index the coordinate that is normal to the plane of symmetry will be zero.
2. Moments of inertia must always be positive. Products of inertia may have either sign.
3. The sum of the three moments of inertia is invariant under orientation changes in the reference frame.
4. The eigenvalues of an inertia tensor are the principal moments for the body. The associated eigenvectors are the principal axes.

Most manipulators have links whose geometry and composition are somewhat complex, so that the application of (6.17) is difficult in practice. A pragmatic option is actually to measure rather than to calculate the moment of inertia of each link by using a measuring device (e.g., an *inertia pendulum*).

6.4 NEWTON'S EQUATION, EULER'S EQUATION

We will consider each link of a manipulator as a rigid body. If we know the location of the center of mass and the inertia tensor of the link, then its mass distribution is completely characterized. In order to move the links, we must accelerate and decelerate them. The forces required for such motion are a function of the acceleration desired and of the mass distribution of the links. Newton's equation, along with its rotational analog, Euler's equation, describes how forces, inertias, and accelerations relate.

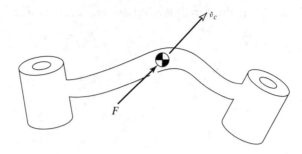

FIGURE 6.3: A force F acting at the center of mass of a body causes the body to accelerate at $\dot{v}_C$.

Newton's equation

Figure 6.3 shows a rigid body whose center of mass is accelerating with acceleration $\dot{v}_C$. In such a situation, the force, F, acting at the center of mass and causing this acceleration is given by Newton's equation

$$F = m\dot{v}_C,\tag{6.29}$$

where m is the total mass of the body.

Euler's equation

Figure 6.4 shows a rigid body rotating with angular velocity ω and with angular acceleration $\dot{\omega}$. In such a situation, the moment N, which must be acting on the body to cause this motion, is given by Euler's equation

$$N = {}^C I \dot{\omega} + \omega \times {}^C I \omega,\tag{6.30}$$

where ${}^C I$ is the inertia tensor of the body written in a frame, $\{C\}$, whose origin is located at the center of mass.

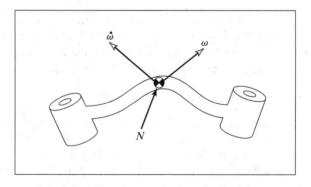

FIGURE 6.4: A moment N is acting on a body, and the body is rotating with velocity ω and accelerating at $\dot{\omega}$.

6.5 ITERATIVE NEWTON–EULER DYNAMIC FORMULATION

We now consider the problem of computing the torques that correspond to a given trajectory of a manipulator. We assume we know the position, velocity, and acceleration of the joints, $(\Theta, \dot{\Theta}, \ddot{\Theta})$. With this knowledge, and with knowledge of the kinematics and the mass-distribution information of the robot, we can calculate the joint torques required to cause this motion. The algorithm presented is based upon the method published by Luh, Walker, and Paul in [2].

Outward iterations to compute velocities and accelerations

In order to compute inertial forces acting on the links, it is necessary to compute the rotational velocity and linear and rotational acceleration of the center of mass of each link of the manipulator at any given instant. These computations will be done in an iterative way, starting with link 1 and moving successively, link by link, *outward* to link n.

The "propagation" of rotational velocity from link to link was discussed in Chapter 5 and is given (for joint $i + 1$ rotational) by

$$^{i+1}\omega_{i+1} = {}^{i+1}_{i}R\,{}^{i}\omega_{i} + \dot{\theta}_{i+1}\,{}^{i+1}\hat{Z}_{i+1}. \tag{6.31}$$

From (6.15), we obtain the equation for transforming angular acceleration from one link to the next:

$$^{i+1}\dot{\omega}_{i+1} = {}^{i+1}_{i}R\,{}^{i}\dot{\omega}_{i} + {}^{i+1}_{i}R\,{}^{i}\omega_{i} \times \dot{\theta}_{i+1}\,{}^{i+1}\hat{Z}_{i+1} + \ddot{\theta}_{i+1}\,{}^{i+1}\hat{Z}_{i+1}. \tag{6.32}$$

When joint $i + 1$ is prismatic, this simplifies to

$$^{i+1}\dot{\omega}_{i+1} = {}^{i+1}_{i}R\,{}^{i}\omega_{i}. \tag{6.33}$$

The linear acceleration of each link-frame origin is obtained by the application of (6.12):

$$^{i+1}\dot{v}_{i+1} = {}^{i+1}_{i}R[{}^{i}\omega_{i} \times {}^{i}P_{i+1} + {}^{i}\omega_{i} \times ({}^{i}\omega_{i} \times {}^{i}P_{i+1}) + {}^{i}\dot{v}_{i}], \tag{6.34}$$

For prismatic joint $i + 1$, (6.34) becomes (from (6.10))

$$^{i+1}\dot{v}_{i+1} = {}^{i+1}_{i}R({}^{i}\dot{\omega}_{i} \times {}^{i}P_{i+1} + {}^{i}\omega_{i} \times ({}^{i}\omega_{i} \times {}^{i}P_{i+1}) + {}^{i}\dot{v}_{i})$$

$$+ 2\,{}^{i+1}\omega_{i+1} \times \dot{d}_{i+1}\,{}^{i+1}\hat{Z}_{i+1} + \ddot{d}_{i+1}\,{}^{i+1}\hat{Z}_{i+1}. \tag{6.35}$$

We also will need the linear acceleration of the center of mass of each link, which also can be found by applying (6.12):

$$^{i}\dot{v}_{C_i} = {}^{i}\dot{\omega}_{i} \times {}^{i}P_{C_i} + {}^{i}\omega_{i} \times ({}^{i}\omega_{i} + {}^{i}P_{C_i}) + {}^{i}\dot{v}_{i}, \tag{6.36}$$

Here, we imagine a frame, $\{C_i\}$, attached to each link, having its origin located at the center of mass of the link and having the same orientation as the link frame, $\{i\}$. Equation (6.36) doesn't involve joint motion at all and so is valid for joint $i + 1$, regardless of whether it is revolute or prismatic.

Note that the application of the equations to link 1 is especially simple, because $^{0}\omega_{0} = {}^{0}\dot{\omega}_{0} = 0$.

The force and torque acting on a link

Having computed the linear and angular accelerations of the mass center of each link, we can apply the Newton–Euler equations (Section 6.4) to compute the inertial force and torque acting at the center of mass of each link. Thus we have

$$F_i = m\dot{v}_{C_i},$$

$$N_i = {}^{C_i}I\dot{\omega}_i + \omega_i \times {}^{C_i}I\omega_i, \tag{6.37}$$

where $\{C_i\}$ has its origin at the center of mass of the link and has the same orientation as the link frame, $\{i\}$.

Inward iterations to compute forces and torques

Having computed the forces and torques acting on each link, we now need to calculate the joint torques that will result in these net forces and torques being applied to each link.

  We can do this by writing a force-balance and moment-balance equation based on a free-body diagram of a typical link. (See Fig. 6.5.) Each link has forces and torques exerted on it by its neighbors and in addition experiences an inertial force and torque. In Chapter 5, we defined special symbols for the force and torque exerted by a neighbor link, which we repeat here:

  f_i = force exerted on link i by link $i - 1$,
  n_i = torque exerted on link i by link $i - 1$.

  By summing the forces acting on link i, we arrive at the force-balance relationship:

$$ {}^iF_i = {}^if_i - {}^i_{i+1}R^{i+1}f_{i+1}. \tag{6.38}$$

  By summing torques about the center of mass and setting them equal to zero, we arrive at the torque-balance equation:

$$ {}^iN_i = {}^in_i - {}^in_{i+1} + (-{}^iP_{C_i}) \times {}^if_i - ({}^iP_{i+1} - {}^iP_{C_i}) \times {}^if_{i+1}. \tag{6.39}$$

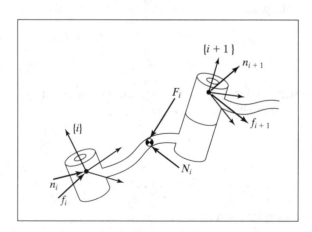

FIGURE 6.5: The force balance, including inertial forces, for a single manipulator link.

Using the result from the force-balance relation (6.38) and adding a few rotation matrices, we can write (6.39) as

$$^iN_i = {}^in_i - {}^i_{i+1}R\,^{i+1}n_{i+1} - {}^iP_{C_i} \times {}^iF_i - {}^iP_{i+1} \times {}^i_{i+1}R\,^{i+1}f_{i+1}. \tag{6.40}$$

Finally, we can rearrange the force and torque equations so that they appear as iterative relationships from higher numbered neighbor to lower numbered neighbor:

$$^if_i = {}^i_{i+1}R\,^{i+1}f_{i+1} + {}^iF_i, \tag{6.41}$$

$$^in_i = {}^iN_i + {}^i_{i+1}R\,^{i+1}n_{i+1} + {}^iP_{C_i} \times {}^iF_i + {}^iP_{i+1} \times {}^i_{i+1}R\,^{i+1}f_{i+1}. \tag{6.42}$$

These equations are evaluated link by link, starting from link n and working inward toward the base of the robot. These *inward force iterations* are analogous to the static force iterations introduced in Chapter 5, except that inertial forces and torques are now considered at each link.

As in the static case, the required joint torques are found by taking the $\hat{Z}$ component of the torque applied by one link on its neighbor:

$$\tau_i = {}^in_i^T\,{}^i\hat{Z}_i. \tag{6.43}$$

For joint i prismatic, we use

$$\tau_i = {}^if_i^T\,{}^i\hat{Z}_i, \tag{6.44}$$

where we have used the symbol τ for a linear actuator force.

Note that, for a robot moving in free space, $^{N+1}f_{N+1}$ and $^{N+1}n_{N+1}$ are set equal to zero, and so the first application of the equations for link n is very simple. If the robot is in contact with the environment, the forces and torques due to this contact can be included in the force balance by having nonzero $^{N+1}f_{N+1}$ and $^{N+1}n_{N+1}$.

The iterative Newton–Euler dynamics algorithm

The complete algorithm for computing joint torques from the motion of the joints is composed of two parts. First, link velocities and accelerations are iteratively computed from link 1 out to link n and the Newton–Euler equations are applied to each link. Second, forces and torques of interaction and joint actuator torques are computed recursively from link n back to link 1. The equations are summarized next for the case of all joints rotational:

Outward iterations: $i : 0 \rightarrow 5$

$$^{i+1}\omega_{i+1} = {}^{i+1}_i R \, {}^i\omega_i + \dot{\theta}_{i+1} \, {}^{i+1}\hat{Z}_{i+1}, \tag{6.45}$$

$$^{i+1}\dot{\omega}_{i+1} = {}^{i+1}_i R \, {}^i\dot{\omega}_i + {}^{i+1}_i R \, {}^i\omega_i \times \dot{\theta}_{i+1} \, {}^{i+1}\hat{Z}_{i+1} + \ddot{\theta}_{i+1} \, {}^{i+1}\hat{Z}_{i+1}, \tag{6.46}$$

$$^{i+1}\dot{v}_{i+1} = {}^{i+1}_i R({}^i\dot{\omega}_i \times {}^iP_{i+1} + {}^i\omega_i \times ({}^i\omega_i \times {}^iP_{i+1}) + {}^i\dot{v}_i), \tag{6.47}$$

$$^{i+1}\dot{v}_{C_{i+1}} = {}^{i+1}\dot{\omega}_{i+1} \times {}^{i+1}P_{C_{i+1}}$$
$$+ {}^{i+1}\omega_{i+1} \times ({}^{i+1}\omega_{i+1} \times {}^{i+1}P_{C_{i+1}}) + {}^{i+1}\dot{v}_{i+1}, \tag{6.48}$$

$$^{i+1}F_{i+1} = m_{i+1} \, {}^{i+1}\dot{v}_{C_{i+1}}, \tag{6.49}$$

$$^{i+1}N_{i+1} = {}^{C_{i+1}}I_{i+1} \, {}^{i+1}\dot{\omega}_{i+1} + {}^{i+1}\omega_{i+1} \times {}^{C_{i+1}}I_{i+1} \, {}^{i+1}\omega_{i+1}. \tag{6.50}$$

Inward iterations: $i : 6 \rightarrow 1$

$$^if_i = {}^i_{i+1}R \, {}^{i+1}f_{i+1} + {}^iF_i, \tag{6.51}$$

$$^in_i = {}^iN_i + {}^i_{i+1}R \, {}^{i+1}n_{i+1} + {}^iP_{C_i} \times {}^iF_i$$
$$+ {}^iP_{i+1} \times {}^i_{i+1}R \, {}^{i+1}f_{i+1}, \tag{6.52}$$

$$\tau_i = {}^in_i^T \, {}^i\hat{Z}_i. \tag{6.53}$$

Inclusion of gravity forces in the dynamics algorithm

The effect of gravity loading on the links can be included quite simply by setting $^0\dot{v}_0 = G$, where G has the magnitude of the gravity vector but points in the opposite direction. This is equivalent to saying that the base of the robot is accelerating upward with 1 g acceleration. This fictitious upward acceleration causes exactly the same effect on the links as gravity would. So, with no extra computational expense, the gravity effect is calculated.

6.6   ITERATIVE VS. CLOSED FORM

Equations (6.46) through (6.53) give a computational scheme whereby, given the joint positions, velocities, and accelerations, we can compute the required joint torques. As with our development of equations to compute the Jacobian in Chapter 5, these relations can be used in two ways: as a numerical computational algorithm, or as an algorithm used analytically to develop symbolic equations.

Use of the equations as a numerical computational algorithm is attractive because the equations apply to any robot. Once the inertia tensors, link masses, P_{C_i} vectors, and $^{i+1}_i R$ matrices are specified for a particular manipulator, the equations can be applied directly to compute the joint torques corresponding to any motion.

However, we often are interested in obtaining better insight into the structure of the equations. For example, what is the form of the gravity terms? How does the magnitude of the gravity effects compare with the magnitude of the inertial effects? To investigate these and other questions, it is often useful to write closed-form dynamic equations. These equations can be derived by applying the recursive

Newton–Euler equations symbolically to Θ, $\dot{\Theta}$, and $\ddot{\Theta}$. This is analogous to what we did in Chapter 5 to derive the symbolic form of the Jacobian.

6.7   AN EXAMPLE OF CLOSED-FORM DYNAMIC EQUATIONS

Here we compute the closed-form dynamic equations for the two-link planar manipulator shown in Fig. 6.6. For simplicity, we assume that the mass distribution is extremely simple: All mass exists as a point mass at the distal end of each link. These masses are m_1 and m_2.

First, we determine the values of the various quantities that will appear in the recursive Newton–Euler equations. The vectors that locate the center of mass for each link are

$$^1P_{C_1} = l_1\hat{X}_1,$$

$$^2P_{C_2} = l_2\hat{X}_2.$$

Because of the point-mass assumption, the inertia tensor written at the center of mass for each link is the zero matrix:

$$^{C_1}I_1 = 0,$$

$$^{C_2}I_2 = 0.$$

There are no forces acting on the end-effector, so we have

$$f_3 = 0,$$

$$n_3 = 0.$$

The base of the robot is not rotating; hence, we have

$$\omega_0 = 0,$$

$$\dot{\omega}_0 = 0.$$

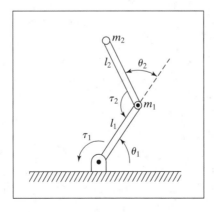

FIGURE 6.6: Two-link planar manipulator with point masses at distal ends of links.

To include gravity forces, we will use

$$^0\dot{v}_0 = g\hat{Y}_0.$$

The rotation between successive link frames is given by

$$
{}^i_{i+1}R = \begin{bmatrix} c_{i+1} & -s_{i+1} & 0.0 \\ s_{i+1} & c_{i+1} & 0.0 \\ 0.0 & 0.0 & 1.0 \end{bmatrix},
$$

$$
{}^{i+1}_{i}R = \begin{bmatrix} c_{i+1} & s_{i+1} & 0.0 \\ -s_{i+1} & c_{i+1} & 0.0 \\ 0.0 & 0.0 & 1.0 \end{bmatrix}.
$$

We now apply equations (6.46) through (6.53).
   The outward iterations for link 1 are as follows:

$$
{}^1\omega_1 = \dot{\theta}_1\, {}^1\hat{Z}_1 = \begin{bmatrix} 0 \\ 0 \\ \dot{\theta}_1 \end{bmatrix},
$$

$$
{}^1\dot{\omega}_1 = \ddot{\theta}_1\, {}^1\hat{Z}_1 = \begin{bmatrix} 0 \\ 0 \\ \ddot{\theta}_1 \end{bmatrix},
$$

$$
{}^1\dot{v}_1 = \begin{bmatrix} c_1 & s_1 & 0 \\ -s_1 & c_1 & 0 \\ 0 & 0 & 1 \end{bmatrix} \begin{bmatrix} 0 \\ g \\ 0 \end{bmatrix} = \begin{bmatrix} gs_1 \\ gc_1 \\ 0 \end{bmatrix},
$$

$$
{}^1\dot{v}_{C_1} = \begin{bmatrix} 0 \\ l_1\ddot{\theta}_1 \\ 0 \end{bmatrix} + \begin{bmatrix} -l_1\dot{\theta}_1^2 \\ 0 \\ 0 \end{bmatrix} + \begin{bmatrix} gs_1 \\ gc_1 \\ 0 \end{bmatrix} = \begin{bmatrix} -l_1\dot{\theta}_1^2 + gs_1 \\ l_1\ddot{\theta}_1 + gc_1 \\ 0 \end{bmatrix},
$$

$$
{}^1F_1 = \begin{bmatrix} -m_1 l_1\dot{\theta}_1^2 + m_1 gs_1 \\ m_1 l_1\ddot{\theta}_1 + m_1 gc_1 \\ 0 \end{bmatrix},
$$

$$
{}^1N_1 = \begin{bmatrix} 0 \\ 0 \\ 0 \end{bmatrix}. \tag{6.54}
$$

The outward iterations for link 2 are as follows:

$$
{}^2\omega_2 = \begin{bmatrix} 0 \\ 0 \\ \dot{\theta}_1 + \dot{\theta}_2 \end{bmatrix},
$$

$$
{}^2\dot{\omega}_2 = \begin{bmatrix} 0 \\ 0 \\ \ddot{\theta}_1 + \ddot{\theta}_2 \end{bmatrix},
$$

$$^2\dot{v}_2 = \begin{bmatrix} c_2 & s_2 & 0 \\ -s_2 & c_2 & 0 \\ 0 & 0 & 1 \end{bmatrix} \begin{bmatrix} -l_1\dot{\theta}_1^2 + gs_1 \\ l_1\ddot{\theta}_1 + gc_1 \\ 0 \end{bmatrix} = \begin{bmatrix} l_1\ddot{\theta}_1 s_2 - l_1\dot{\theta}_1^2 c_2 + gs_{12} \\ l_1\ddot{\theta}_1 c_2 + l_1\dot{\theta}_1^2 s_2 + gc_{12} \\ 0 \end{bmatrix},$$

$$^2\dot{v}_{C_2} = \begin{bmatrix} 0 \\ l_2(\ddot{\theta}_1 + \ddot{\theta}_2) \\ 0 \end{bmatrix} + \begin{bmatrix} -l_2(\dot{\theta}_1 + \dot{\theta}_2)^2 \\ 0 \\ 0 \end{bmatrix}$$

$$+ \begin{bmatrix} l_1\ddot{\theta}_1 s_2 - l_1\dot{\theta}_1^2 c_2 + gs_{12} \\ l_1\ddot{\theta}_1 c_2 + l_1\dot{\theta}_1^2 s_2 + gc_{12} \\ 0 \end{bmatrix}, \tag{6.55}$$

$$^2F_2 = \begin{bmatrix} m_2 l_1\ddot{\theta}_1 s_2 - m_2 l_1\dot{\theta}_1^2 c_2 + m_2 gs_{12} - m_2 l_2(\dot{\theta}_1 + \dot{\theta}_2)^2 \\ m_2 l_1\ddot{\theta}_1 c_2 + m_2 l_1\dot{\theta}_1^2 s_2 + m_2 gc_{12} + m_2 l_2(\ddot{\theta}_1 + \ddot{\theta}_2) \\ 0 \end{bmatrix},$$

$$^2N_2 = \begin{bmatrix} 0 \\ 0 \\ 0 \end{bmatrix}.$$

The inward iterations for link 2 are as follows:

$$^2f_2 = {}^2F_2,$$

$$^2n_2 = \begin{bmatrix} 0 \\ 0 \\ m_2 l_1 l_2 c_2\ddot{\theta}_1 + m_2 l_1 l_2 s_2\dot{\theta}_1^2 + m_2 l_2 gc_{12} + m_2 l_2^2(\ddot{\theta}_1 + \ddot{\theta}_2) \end{bmatrix}. \tag{6.56}$$

The inward iterations for link 1 are as follows:

$$^1f_1 = \begin{bmatrix} c_2 & -s_2 & 0 \\ s_2 & c_2 & 0 \\ 0 & 0 & 1 \end{bmatrix} \begin{bmatrix} m_2 l_1 s_2\ddot{\theta}_1 - m_2 l_1 c_2\dot{\theta}_1^2 + m_2 gs_{12} - m_2 l_2(\dot{\theta}_1 + \dot{\theta}_2)^2 \\ m_2 l_1 c_2\ddot{\theta}_1 + m_2 l_1 s_2\dot{\theta}_1^2 + m_2 gc_{12} + m_2 l_2(\ddot{\theta}_1 + \ddot{\theta}_2) \\ 0 \end{bmatrix}$$

$$+ \begin{bmatrix} -m_1 l_1\dot{\theta}_1^2 + m_1 gs_1 \\ m_1 l_1\ddot{\theta}_1 + m_1 gc_1 \\ 0 \end{bmatrix},$$

$$^1n_1 = \begin{bmatrix} 0 \\ 0 \\ m_2 l_1 l_2 c_2\ddot{\theta}_1 + m_2 l_1 l_2 s_2\dot{\theta}_1^2 + m_2 l_2 gc_{12} + m_2 l_2^2(\ddot{\theta}_1 + \ddot{\theta}_2) \end{bmatrix}$$

$$+ \begin{bmatrix} 0 \\ 0 \\ m_1 l_1^2\ddot{\theta}_1 + m_1 l_1 gc_1 \end{bmatrix}$$

$$+ \begin{bmatrix} 0 \\ 0 \\ m_2 l_1^2\ddot{\theta}_1 - m_2 l_1 l_2 s_2(\dot{\theta}_1 + \dot{\theta}_2)^2 + m_2 l_1 gs_2 s_{12} \\ + m_2 l_1 l_2 c_2(\ddot{\theta}_1 + \ddot{\theta}_2) + m_2 l_1 gc_2 c_{12} \end{bmatrix}. \tag{6.57}$$

Extracting the $\hat{Z}$ components of the $^i n_i$, we find the joint torques:

$$\tau_1 = m_2 l_2^2 (\ddot{\theta}_1 + \ddot{\theta}_2) + m_2 l_1 l_2 c_2 (2\ddot{\theta}_1 + \ddot{\theta}_2) + (m_1 + m_2) l_1^2 \ddot{\theta}_1 - m_2 l_1 l_2 s_2 \dot{\theta}_2^2$$
$$- 2m_2 l_1 l_2 s_2 \dot{\theta}_1 \dot{\theta}_2 + m_2 l_2 g c_{12} + (m_1 + m_2) l_1 g c_1,$$
$$\tau_2 = m_2 l_1 l_2 c_2 \ddot{\theta}_1 + m_2 l_1 l_2 s_2 \dot{\theta}_1^2 + m_2 l_2 g c_{12} + m_2 l_2^2 (\ddot{\theta}_1 + \ddot{\theta}_2). \qquad (6.58)$$

Equations (6.58) give expressions for the torque at the actuators as a function of joint position, velocity, and acceleration. Note that these rather complex functions arose from one of the simplest manipulators imaginable. Obviously, the closed-form equations for a manipulator with six degrees of freedom will be quite complex.

6.8 THE STRUCTURE OF A MANIPULATOR'S DYNAMIC EQUATIONS

It is often convenient to express the dynamic equations of a manipulator in a single equation that hides some of the details, but shows some of the structure of the equations.

The state-space equation

When the Newton–Euler equations are evaluated symbolically for any manipulator, they yield a dynamic equation that can be written in the form

$$\tau = M(\Theta)\ddot{\Theta} + V(\Theta, \dot{\Theta}) + G(\Theta), \qquad (6.59)$$

where $M(\Theta)$ is the $n \times n$ **mass matrix** of the manipulator, $V(\Theta, \dot{\Theta})$ is an $n \times 1$ vector of centrifugal and Coriolis terms, and $G(\Theta)$ is an $n \times 1$ vector of gravity terms. We use the term **state-space equation** because the term $V(\Theta, \dot{\Theta})$, appearing in (6.59), has both position and velocity dependence [3].

Each element of $M(\Theta)$ and $G(\Theta)$ is a complex function that depends on Θ, the position of all the joints of the manipulator. Each element of $V(\Theta, \dot{\Theta})$ is a complex function of both Θ and $\dot{\Theta}$.

We may separate the various types of terms appearing in the dynamic equations and form the mass matrix of the manipulator, the centrifugal and Coriolis vector, and the gravity vector.

EXAMPLE 6.3

Give $M(\Theta)$, $V(\Theta, \dot{\Theta})$, and $G(\Theta)$ for the manipulator of Section 6.7.

Equation (6.59) defines the manipulator mass matrix, $M(\Theta)$; it is composed of all those terms which multiply $\ddot{\Theta}$ and is a function of Θ. Therefore, we have

$$M(\Theta) = \begin{bmatrix} l_2^2 m_2 + 2l_1 l_2 m_2 c_2 + l_1^2 (m_1 + m_2) & l_2^2 m_2 + l_1 l_2 m_2 c_2 \\ l_2^2 m_2 + l_1 l_2 m_2 c_2 & l_2^2 m_2 \end{bmatrix}. \qquad (6.60)$$

Any manipulator mass matrix is symmetric and positive definite, and is, therefore, always invertible.

The velocity term, $V(\Theta, \dot{\Theta})$, contains all those terms that have any dependence on joint velocity. Thus, we obtain

$$V(\Theta, \dot{\Theta}) = \begin{bmatrix} -m_2 l_1 l_2 s_2 \dot{\theta}_2^2 - 2m_2 l_1 l_2 s_2 \dot{\theta}_1 \dot{\theta}_2 \\ m_2 l_1 l_2 s_2 \dot{\theta}_1^2 \end{bmatrix}. \qquad (6.61)$$

A term like $-m_2 l_1 l_2 s_2 \dot{\theta}_2^2$ is caused by a **centrifugal force**, and is recognized as such because it depends on the square of a joint velocity. A term such as $-2 m_2 l_1 l_2 s_2 \dot{\theta}_1 \dot{\theta}_2$ is caused by a **Coriolis force** and will always contain the product of two different joint velocities.

The gravity term, $G(\Theta)$, contains all those terms in which the gravitational constant, g, appears. Therefore, we have

$$G(\Theta) = \left[\begin{array}{c} m_2 l_2 g c_{12} + (m_1 + m_2) l_1 g c_1 \\ m_2 l_2 g c_{12} \end{array} \right]. \tag{6.62}$$

Note that the gravity term depends only on Θ and not on its derivatives.

The configuration-space equation

By writing the velocity-dependent term, $V(\Theta, \dot{\Theta})$, in a different form, we can write the dynamic equations as

$$\tau = M(\Theta)\ddot{\Theta} + B(\Theta)[\dot{\Theta}\dot{\Theta}] + C(\Theta)[\dot{\Theta}^2] + G(\Theta), \tag{6.63}$$

where $B(\Theta)$ is a matrix of dimensions $n \times n(n-1)/2$ of Coriolis coefficients, $[\dot{\Theta}\dot{\Theta}]$ is an $n(n-1)/2 \times 1$ vector of joint velocity products given by

$$[\dot{\Theta}\dot{\Theta}] = [\dot{\theta}_1 \dot{\theta}_2 \ \dot{\theta}_1 \dot{\theta}_3 \ \ldots \ \dot{\theta}_{n-1} \dot{\theta}_n]^T, \tag{6.64}$$

$C(\Theta)$ is an $n \times n$ matrix of centrifugal coefficients, and $[\dot{\Theta}^2]$ is an $n \times 1$ vector given by

$$[\dot{\theta}_1^2 \ \dot{\theta}_2^2 \ldots \dot{\theta}_n^2]^T. \tag{6.65}$$

We will call (6.63) the **configuration-space equation**, because the matrices are functions only of manipulator position [3].

In this form of the dynamic equations, the complexity of the computation is seen to be in the form of computing various parameters which are a function of only the manipulator position, Θ. This is important in applications (such as computer control of a manipulator) in which the dynamic equations must be updated as the manipulator moves. (Equation (6.63) gives a form in which parameters are a function of joint position only and can be updated at a rate related to how fast the manipulator is changing configuration.) We will consider this form again with regard to the problem of manipulator control in Chapter 10.

EXAMPLE 6.4

Give $B(\Theta)$ and $C(\Theta)$ (from (6.63)) for the manipulator of Section 6.7.

For this simple two-link manipulator, we have

$$[\dot{\Theta}\dot{\Theta}] = [\dot{\theta}_1 \dot{\theta}_2],$$

$$[\dot{\Theta}^2] = \left[\begin{array}{c} \dot{\theta}_1^2 \\ \dot{\theta}_2^2 \end{array} \right]. \tag{6.66}$$

So we see that

$$B(\Theta) = \begin{bmatrix} -2m_2 l_1 l_2 s_2 \\ 0 \end{bmatrix} \tag{6.67}$$

and

$$C(\Theta) = \begin{bmatrix} 0 & -m_2 l_1 l_2 s_2 \\ m_2 l_1 l_2 s_2 & 0 \end{bmatrix}. \tag{6.68}$$

6.9 LAGRANGIAN FORMULATION OF MANIPULATOR DYNAMICS

The Newton–Euler approach is based on the elementary dynamic formulas (6.29) and (6.30) and on an analysis of forces and moments of constraint acting between the links. As an alternative to the Newton–Euler method, in this section we briefly introduce the **Lagrangian dynamic formulation**. Whereas the Newton–Euler formulation might be said to be a "force balance" approach to dynamics, the Lagrangian formulation is an "energy-based" approach to dynamics. Of course, for the same manipulator, both will give the same equations of motion. Our statement of Lagrangian dynamics will be brief and somewhat specialized to the case of a serial-chain mechanical manipulator with rigid links. For a more complete and general reference, see [4].

We start by developing an expression for the kinetic energy of a manipulator. The kinetic energy of the ith link, k_i, can be expressed as

$$k_i = \tfrac{1}{2} m_i v_{C_i}^T v_{C_i} + \tfrac{1}{2} {}^i\omega_i^T {}^{C_i}I_i {}^i\omega_i, \tag{6.69}$$

where the first term is kinetic energy due to linear velocity of the link's center of mass and the second term is kinetic energy due to angular velocity of the link. The total kinetic energy of the manipulator is the sum of the kinetic energy in the individual links—that is,

$$k = \sum_{i=1}^{n} k_i. \tag{6.70}$$

The v_{C_i} and ${}^i\omega_i$ in (6.69) are functions of Θ and $\dot{\Theta}$, so we see that the kinetic energy of a manipulator can be described by a scalar formula as a function of joint position and velocity, $k(\Theta, \dot{\Theta})$. In fact, the kinetic energy of a manipulator is given by

$$k(\Theta, \dot{\Theta}) = \tfrac{1}{2} \dot{\Theta}^T M(\Theta) \dot{\Theta}, \tag{6.71}$$

where $M(\Theta)$ is the $n \times n$ manipulator mass matrix already introduced in Section 6.8. An expression of the form of (6.71) is known as a **quadratic form** [5], since when expanded out, the resulting scalar equation is composed solely of terms whose dependence on the $\dot{\theta}_i$ is quadratic. Further, because the total kinetic energy must always be positive, the manipulator mass matrix must be a so-called **positive definite** matrix. Positive definite matrices are those having the property that their quadratic form is always a positive scalar. Equation (6.71) can be seen to be analogous to the familiar expression for the kinetic energy of a point mass:

$$k = \tfrac{1}{2} m v^2. \tag{6.72}$$

The fact that a manipulator mass matrix must be positive definite is analogous to the fact that a scalar mass is always a positive number.

The potential energy of the ith link, u_i, can be expressed as

$$u_i = -m_i \, {}^0g^T \, {}^0P_{C_i} + u_{ref_i}, \qquad (6.73)$$

where ${}^0g$ is the 3×1 gravity vector, ${}^0P_{C_i}$ is the vector locating the center of mass of the ith link, and u_{ref_i} is a constant chosen so that the minimum value of u_i is zero.[1] The total potential energy stored in the manipulator is the sum of the potential energy in the individual links—that is,

$$u = \sum_{i=1}^{n} u_i. \qquad (6.74)$$

Because the ${}^0P_{C_i}$ in (6.73) are functions of Θ, we see that the potential energy of a manipulator can be described by a scalar formula as a function of joint position, $u(\Theta)$.

The Lagrangian dynamic formulation provides a means of deriving the equations of motion from a scalar function called the **Lagrangian**, which is defined as the difference between the kinetic and potential energy of a mechanical system. In our notation, the Lagrangian of a manipulator is

$$\mathcal{L}(\Theta, \dot{\Theta}) = k(\Theta, \dot{\Theta}) - u(\Theta). \qquad (6.75)$$

The equations of motion for the manipulator are then given by

$$\frac{d}{dt} \frac{\partial \mathcal{L}}{\partial \dot{\Theta}} - \frac{\partial \mathcal{L}}{\partial \Theta} = \tau, \qquad (6.76)$$

where τ is the $n \times 1$ vector of actuator torques. In the case of a manipulator, this equation becomes

$$\frac{d}{dt} \frac{\partial k}{\partial \dot{\Theta}} - \frac{\partial k}{\partial \Theta} + \frac{\partial u}{\partial \Theta} = \tau, \qquad (6.77)$$

where the arguments of $k(\cdot)$ and $u(\cdot)$ have been dropped for brevity.

EXAMPLE 6.5

The links of an RP manipulator, shown in Fig. 6.7, have inertia tensors

$$
{}^{C_1}I_1 = \begin{bmatrix} I_{xx1} & 0 & 0 \\ 0 & I_{yy1} & 0 \\ 0 & 0 & I_{zz1} \end{bmatrix},
$$

$$
{}^{C_2}I_2 = \begin{bmatrix} I_{xx2} & 0 & 0 \\ 0 & I_{yy2} & 0 \\ 0 & 0 & I_{zz2} \end{bmatrix}, \qquad (6.78)
$$

---

[1] Actually, only the partial derivative of the potential energy with respect to Θ will appear in the dynamics, so this constant is arbitrary. This corresponds to defining the potential energy relative to an arbitrary zero reference height.

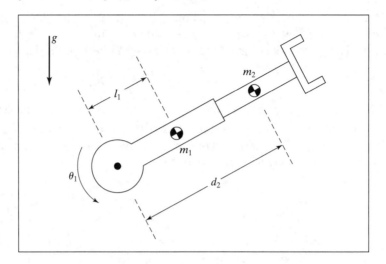

FIGURE 6.7: The RP manipulator of Example 6.5.

and total mass m_1 and m_2. As shown in Fig. 6.7, the center of mass of link 1 is located at a distance l_1 from the joint-1 axis, and the center of mass of link 2 is at the variable distance d_2 from the joint-1 axis. Use Lagrangian dynamics to determine the equation of motion for this manipulator.

Using (6.69), we write the kinetic energy of link 1 as

$$k_1 = \tfrac{1}{2} m_1 l_1^2 \dot{\theta}_1^2 + \tfrac{1}{2} I_{zz1} \dot{\theta}_1^2 \tag{6.79}$$

and the kinetic energy of link 2 as

$$k_2 = \tfrac{1}{2} m_2 (d_2^2 \dot{\theta}_1^2 + \dot{d}_2^2) + \tfrac{1}{2} I_{zz2} \dot{\theta}_1^2. \tag{6.80}$$

Hence, the total kinetic energy is given by

$$k(\Theta, \dot{\Theta}) = \tfrac{1}{2} (m_1 l_1^2 + I_{zz1} + I_{zz2} + m_2 d_2^2) \dot{\theta}_1^2 + \tfrac{1}{2} m_2 \dot{d}_2^2. \tag{6.81}$$

Using (6.73), we write the potential energy of link 1 as

$$u_1 = m_1 l_1 g \sin(\theta_1) + m_1 l_1 g \tag{6.82}$$

and the potential energy of link 2 as

$$u_2 = m_2 g d_2 \sin(\theta_1) + m_2 g d_{2max}, \tag{6.83}$$

where d_{2max} is the maximum extension of joint 2. Hence, the total potential energy is given by

$$u(\Theta) = g(m_1 l_1 + m_2 d_2) \sin(\theta_1) + m_1 l_1 g + m_2 g d_{2max}. \tag{6.84}$$

Next, we take partial derivatives as needed for (6.77):

$$\frac{\partial k}{\partial \dot{\Theta}} = \begin{bmatrix} (m_1 l_1^2 + I_{zz1} + I_{zz2} + m_2 d_2^2)\dot{\theta}_1 \\ m_2 d_2 \end{bmatrix}, \tag{6.85}$$

$$\frac{\partial k}{\partial \Theta} = \begin{bmatrix} 0 \\ m_2 d_2 \dot{\theta}_1^2 \end{bmatrix}, \tag{6.86}$$

$$\frac{\partial u}{\partial \Theta} = \begin{bmatrix} g(m_1 l_1 + m_2 d_2)\cos(\theta_1) \\ gm_2 \sin(\theta_1) \end{bmatrix}. \tag{6.87}$$

Finally, substituting into (6.77), we have

$$\tau_1 = (m_1 l_1^2 + I_{zz1} + I_{zz2} + m_2 d_2^2)\ddot{\theta}_1 + 2m_2 d_2 \dot{\theta}_1 \dot{d}_2$$
$$+ (m_1 l_1 + m_2 d_2)g\cos(\theta_1), \tag{6.88}$$
$$\tau_2 = m_2 \ddot{d}_2 - m_2 d_2 \dot{\theta}_1^2 + m_2 g \sin(\theta_1).$$

From (6.89), we can see that

$$M(\Theta) = \begin{bmatrix} (m_1 l_1^2 + I_{zz1} + I_{zz2} + m_2 d_2^2) & 0 \\ 0 & m_2 \end{bmatrix},$$

$$V(\Theta, \dot{\Theta}) = \begin{bmatrix} 2m_2 d_2 \dot{\theta}_1 \dot{d}_2 \\ -m_2 d_2 \dot{\theta}_1^2 \end{bmatrix}, \tag{6.89}$$

$$G(\Theta) = \begin{bmatrix} (m_1 l_1 + m_2 d_2)g\cos(\theta_1) \\ m_2 g \sin(\theta_1) \end{bmatrix}.$$

6.10 FORMULATING MANIPULATOR DYNAMICS IN CARTESIAN SPACE

Our dynamic equations have been developed in terms of the position and time derivatives of the manipulator joint angles, or in **joint space,** with the general form

$$\tau = M(\Theta)\ddot{\Theta} + V(\Theta, \dot{\Theta}) + G(\Theta). \tag{6.90}$$

We developed this equation in joint space because we could use the serial-link nature of the mechanism to advantage in deriving the equations. In this section, we discuss the formulation of the dynamic equations that relate acceleration of the end-effector expressed in Cartesian space to Cartesian forces and moments acting at the end-effector.

The Cartesian state-space equation

As explained in Chapters 10 and 11, it might be desirable to express the dynamics of a manipulator with respect to Cartesian variables in the general form [6]

$$\mathcal{F} = M_x(\Theta)\ddot{\chi} + V_x(\Theta, \dot{\Theta}) + G_x(\Theta), \tag{6.91}$$

where $\mathcal{F}$ is a force–torque vector acting on the end-effector of the robot, and χ is an appropriate Cartesian vector representing position and orientation of the end-effector [7]. Analogous to the joint-space quantities, $M_x(\Theta)$ is the **Cartesian mass matrix**, $V_x(\Theta, \dot{\Theta})$ is a vector of velocity terms in Cartesian space, and $G_x(\Theta)$ is a vector of gravity terms in Cartesian space. Note that the fictitious forces acting on the end-effector, $\mathcal{F}$, could in fact be applied by the actuators at the joints by using the relationship

$$\tau = J^T(\Theta)\mathcal{F}, \tag{6.92}$$

where the Jacobian, $J(\Theta)$, is written in the same frame as $\mathcal{F}$ and $\ddot{\chi}$, usually the tool frame, $\{T\}$.

We can derive the relationship between the terms of (6.90) and those of (6.91) in the following way. First, we premultiply (6.90) by the inverse of the Jacobian transpose to obtain

$$J^{-T}\tau = J^{-T}M(\Theta)\ddot{\Theta} + J^{-T}V(\Theta, \dot{\Theta}) + J^{-T}G(\Theta), \tag{6.93}$$

or

$$\mathcal{F} = J^{-T}M(\Theta)\ddot{\Theta} + J^{-T}V(\Theta, \dot{\Theta}) + J^{-T}G(\Theta). \tag{6.94}$$

Next, we develop a relationship between joint space and Cartesian acceleration, starting with the definition of the Jacobian,

$$\dot{\chi} = J\dot{\Theta}, \tag{6.95}$$

and differentiating to obtain

$$\ddot{\chi} = \dot{J}\dot{\Theta} + J\ddot{\Theta}. \tag{6.96}$$

Solving (6.96) for joint-space acceleration leads to

$$\ddot{\Theta} = J^{-1}\ddot{\chi} - J^{-1}\dot{J}\dot{\Theta}. \tag{6.97}$$

Substituting (6.97) into (6.94), we have

$$\mathcal{F} = J^{-T}M(\Theta)J^{-1}\ddot{\chi} - J^{-T}M(\Theta)J^{-1}\dot{J}\dot{\Theta} + J^{-T}V(\Theta, \dot{\Theta}) + J^{-T}G(\Theta), \tag{6.98}$$

from which we derive the expressions for the terms in the Cartesian dynamics as

$$M_x(\Theta) = J^{-T}(\Theta)M(\Theta)J^{-1}(\Theta),$$
$$V_x(\Theta, \dot{\Theta}) = J^{-T}(\Theta)(V(\Theta, \dot{\Theta}) - M(\Theta)J^{-1}(\Theta)\dot{J}(\Theta)\dot{\Theta}), \tag{6.99}$$
$$G_x(\Theta) = J^{-T}(\Theta)G(\Theta).$$

Note that the Jacobian appearing in equations (6.100) is written in the same frames as $\mathcal{F}$ and χ in (6.91); the choice of this frame is arbitrary.[2] Note that, when the manipulator approaches a singularity, certain quantities in the Cartesian dynamics become infinite.

---

[2]Certain choices could facilitate computation.

EXAMPLE 6.6

Derive the Cartesian-space form of the dynamics for the two-link planar arm of Section 6.7. Write the dynamics in terms of a frame attached to the end of the second link.

For this manipulator, we have already obtained the dynamics (in Section 6.7) and the Jacobian (equation (5.66)), which we restate here:

$$J(\Theta) = \begin{bmatrix} l_1 s_2 & 0 \\ l_1 c_2 + l_2 & l_2 \end{bmatrix}. \tag{6.100}$$

First, compute the inverse Jacobian:

$$J^{-1}(\Theta) = \frac{1}{l_1 l_2 s_2} \begin{bmatrix} l_2 & 0 \\ -l_1 c_2 - l_2 & l_1 s_2 \end{bmatrix}. \tag{6.101}$$

Next, obtain the time derivative of the Jacobian:

$$\dot{J}(\Theta) = \begin{bmatrix} l_1 c_2 \dot{\theta}_2 & 0 \\ -l_1 s_2 \dot{\theta}_2 & 0 \end{bmatrix}. \tag{6.102}$$

Using (6.100) and the results of Section 6.7, we get

$$M_x(\Theta) = \begin{bmatrix} m_2 + \frac{m_1}{s_2^2} & 0 \\ 0 & m_2 \end{bmatrix},$$

$$V_x(\Theta, \dot{\Theta}) = \begin{bmatrix} -(m_2 l_1 c_2 + m_2 l_2)\dot{\theta}_1^2 - m_2 l_2 \dot{\theta}_2^2 - (2m_2 l_2 + m_2 l_1 c_2 + m_1 l_1 \frac{c_2}{s_2^2})\dot{\theta}_1 \dot{\theta}_2 \\ m_2 l_1 s_2 \dot{\theta}_1^2 + l_1 m_2 s_2 \dot{\theta}_1 \dot{\theta}_2 \end{bmatrix},$$

$$G_x(\Theta) = \begin{bmatrix} m_1 g \frac{c_1}{s_2} + m_2 g s_{12} \\ m_2 g c_{12} \end{bmatrix}. \tag{6.103}$$

When $s_2 = 0$, the manipulator is in a singular position, and some of the dynamic terms go to infinity. For example, when $\theta_2 = 0$ (arm stretched straight out), the effective Cartesian mass of the end-effector becomes infinite in the $\hat{X}_2$ direction of the link-2 tip frame, as expected. In general, at a singular configuration there is a certain direction, the *singular direction* in which motion is impossible, but general motion in the subspace "orthogonal" to this direction is possible [8].

The Cartesian configuration space torque equation

Combining (6.91) and (6.92), we can write equivalent joint torques with the dynamics expressed in Cartesian space:

$$\tau = J^T(\Theta)(M_x(\Theta)\ddot{\chi} + V_x(\Theta, \dot{\Theta}) + G_x(\Theta)). \tag{6.104}$$

We will find it useful to write this equation in the form

$$\tau = J^T(\Theta)M_x(\Theta)\ddot{\chi} + B_x(\Theta)[\dot{\Theta}\dot{\Theta}] + C_x(\Theta)[\dot{\Theta}^2] + G(\Theta), \tag{6.105}$$

where $B_x(\Theta)$ is a matrix of dimension $n \times n(n-1)/2$ of Coriolis coefficients, $[\dot{\Theta}\dot{\Theta}]$ is an $n(n-1)/2 \times 1$ vector of joint velocity products given by

$$[\dot{\Theta}\dot{\Theta}] = [\dot{\theta}_1\dot{\theta}_2 \ \dot{\theta}_1 \ \dot{\theta}_3 \ldots \dot{\theta}_{n-1}\dot{\theta}_n]^T, \tag{6.106}$$

$C_x(\Theta)$ is an $n \times n$ matrix of centrifugal coefficients, and $[\dot{\Theta}^2]$ is an $n \times 1$ vector given by

$$[\dot{\theta}_1^2 \ \dot{\theta}_2^2 \ \ldots \dot{\theta}_n^2]^T. \tag{6.107}$$

Note that, in (6.105), $G(\Theta)$ is the same as in the joint-space equation, but in general, $B_x(\Theta) \neq B(\Theta)$ and $C_x(\Theta) \neq C(\Theta)$.

EXAMPLE 6.7

Find $B_x(\Theta)$ and $C_x(\Theta)$ (from (6.105)) for the manipulator of Section 6.7.
   If we form the product $J^T(\Theta)V_x(\Theta, \dot{\Theta})$, we find that

$$B_x(\Theta) = \begin{bmatrix} m_1 l_1^2 \frac{c_2}{s_2} - m_2 l_1 l_2 s_2 \\ m_2 l_1 l_2 s_2 \end{bmatrix} \tag{6.108}$$

and

$$C_x(\Theta) = \begin{bmatrix} 0 & -m_2 l_1 l_2 s_2 \\ m_2 l_1 l_2 s_2 & 0 \end{bmatrix}. \tag{6.109}$$

6.11 INCLUSION OF NONRIGID BODY EFFECTS

It is important to realize that the dynamic equations we have derived *do not encompass all the effects acting on a manipulator.* They include only those forces which arise from rigid body mechanics. The most important source of forces that are *not* included is friction. All mechanisms are, of course, affected by frictional forces. In present-day manipulators, in which significant gearing is typical, the forces due to friction can actually be quite large—perhaps equaling 25% of the torque required to move the manipulator in typical situations.

   In order to make dynamic equations reflect the reality of the physical device, it is important to model (at least approximately) these forces of friction. A very simple model for friction is **viscous friction**, in which the torque due to friction is proportional to the velocity of joint motion. Thus, we have

$$\tau_{friction} = v\dot{\theta}, \tag{6.110}$$

where v is a viscous-friction constant. Another possible simple model for friction, **Coulomb friction**, is sometimes used. Coulomb friction is constant except for a sign dependence on the joint velocity and is given by

$$\tau_{friction} = c \ sgn(\dot{\theta}), \tag{6.111}$$

where c is a Coulomb-friction constant. The value of c is often taken at one value when $\dot{\theta} = 0$, the static coefficient, but at a lower value, the dynamic coefficient, when $\dot{\theta} \neq 0$. Whether a joint of a particular manipulator exhibits viscous or Coulomb friction is a complicated issue of lubrication and other effects. A reasonable model is to include both, because both effects are likely:

$$\tau_{friction} = c\, sgn(\dot{\theta}) + v\dot{\theta}. \tag{6.112}$$

It turns out that, in many manipulator joints, friction also displays a dependence on the joint position. A major cause of this effect might be gears that are not perfectly round—their eccentricity would cause friction to change according to joint position. So a fairly complex friction model would have the form

$$\tau_{friction} = f(\theta, \dot{\theta}). \tag{6.113}$$

These friction models are then added to the other dynamic terms derived from the rigid-body model, yielding the more complete model

$$\tau = M(\Theta)\ddot{\Theta} + V(\Theta, \dot{\Theta}) + G(\Theta) + F(\Theta, \dot{\Theta}). \tag{6.114}$$

There are also other effects, which are neglected in this model. For example, the assumption of rigid body links means that we have failed to include bending effects (which give rise to resonances) in our equations of motion. However, these effects are extremely difficult to model and are beyond the scope of this book. (See [9, 10].)

6.12 DYNAMIC SIMULATION

To simulate the motion of a manipulator, we must make use of a model of the dynamics such as the one we have just developed. Given the dynamics written in closed form as in (6.59), simulation requires solving the dynamic equation for acceleration:

$$\ddot{\Theta} = M^{-1}(\Theta)[\tau - V(\Theta, \dot{\Theta}) - G(\Theta) - F(\Theta, \dot{\Theta})]. \tag{6.115}$$

We can then apply any of several known **numerical integration** techniques to integrate the acceleration to compute future positions and velocities.

Given initial conditions on the motion of the manipulator, usually in the form

$$\Theta(0) = \Theta_0,$$

$$\dot{\Theta}(0) = 0, \tag{6.116}$$

we integrate (6.115) forward in time numerically by steps of size Δt. There are many methods of performing numerical integration [11]. Here, we introduce the simplest integration scheme, called *Euler integration*: Starting with $t = 0$, iteratively compute

$$\dot{\Theta}(t + \Delta t) = \dot{\Theta}(t) + \ddot{\Theta}(t)\Delta t,$$

$$\Theta(t + \Delta t) = \Theta(t) + \dot{\Theta}(t)\Delta t + \tfrac{1}{2}\ddot{\Theta}(t)\Delta t^2, \tag{6.117}$$

where, for each iteration, (6.115) is computed to calculate $\ddot{\Theta}$. In this way, the position, velocity, and acceleration of the manipulator caused by a certain input torque function can be computed numerically.

Euler integration is conceptually simple, but other, more sophisticated integration techniques are recommended for accurate and efficient simulation [11]. How to select the size of Δt is an issue that is often discussed. It should be sufficiently small that breaking continuous time into these small increments is a reasonable approximation. It should be sufficiently large that an excessive amount of computer time is not required to compute a simulation.

6.13 COMPUTATIONAL CONSIDERATIONS

Because the dynamic equations of motion for typical manipulators are so complex, it is important to consider computational issues. In this section, we restrict our attention to joint-space dynamics. Some issues of computational efficiency of Cartesian dynamics are discussed in [7, 8].

A historical note concerning efficiency

Counting the number of multiplications and additions for the equations (6.46)–(6.53) when taking into consideration the simple first outward computation and simple last inward computation, we get

$$126n - 99 \text{ multiplications,}$$

$$106n - 92 \text{ additions,}$$

where n is the number of links (here, at least two). Although still somewhat complex, the formulation is tremendously efficient in comparison with some previously suggested formulations of manipulator dynamics. The first formulation of the dynamics for a manipulator [12, 13] was done via a fairly straightforward Lagrangian approach whose required computations came out to be approximately [14]

$$32n^4 + 86n^3 + 171n^2 + 53n - 128 \text{ multiplications,}$$

$$25n^4 + 66n^3 + 129n^2 + 42n - 96 \text{ additions.}$$

For a typical case, $n = 6$, the iterative Newton–Euler scheme is about 100 times more efficient! The two approaches must of course yield equivalent equations, and numeric calculations would yield exactly the same results, but the structure of the equations is quite different. This is not to say that a Lagrangian approach cannot be made to produce efficient equations. Rather, this comparison indicates that, in formulating a computational scheme for this problem, care must be taken as regards efficiency. The relative efficiency of the method we have presented stems from posing the computations as iterations from link to link and in the particulars of how the various quantities are represented [15].

Renaud [16] and Liegois et al. [17] made early contributions concerning the formulation of the mass-distribution descriptions of the links. While studying the modeling of human limbs, Stepanenko and Vukobratovic [18] began investigating a "Newton–Euler" approach to dynamics instead of the somewhat more traditional Lagrangian approach. This work was revised for efficiency by Orin et al. [19] in an application to the legs of walking robots. Orin's group improved the efficiency somewhat by writing the forces and moments in the local link-reference frames instead of in the inertial frame. They also noticed the sequential nature of calculations from one link to the next and speculated that an efficient recursive formulation might exist. Armstrong [20] and Luh, Walker, and Paul

[2] paid close attention to details of efficiency and published an algorithm that is $O(n)$ in complexity. This was accomplished by setting up the calculations in an iterative (or recursive) nature and by expressing the velocities and accelerations of the links in the local link frames. Hollerbach [14] and Silver [15] further explored various computational algorithms. Hollerbach and Sahar [21] showed that, for certain specialized geometries, the complexity of the algorithm would reduce further.

Efficiency of closed form vs. that of iterative form

The iterative scheme introduced in this chapter is quite efficient as a general means of computing the dynamics of any manipulator, but closed-form equations derived for a particular manipulator will usually be even more efficient. Consider the two-link planar manipulator of Section 6.7. Plugging $n = 2$ into the formulas given in Section 6.13, we find that our iterative scheme would require 153 multiplications and 120 additions to compute the dynamics of a general two-link. However, our particular two-link arm happens to be quite simple: It is planar, and the masses are treated as point masses. So, if we consider the closed-form equations that we worked out in Section 6.7, we see that computation of the dynamics in this form requires about 30 multiplications and 13 additions. This is an extreme case, because the particular manipulator is very simple, but it illustrates the point that symbolic closed-form equations are likely to be the most efficient formulation of dynamics. Several authors have published articles showing that, for any given manipulator, customized closed-form dynamics are more efficient than even the best of the general schemes [22–27].

Hence, if manipulators are designed to be *simple* in the kinematic and dynamic sense, they will have dynamic equations that are simple. We might define a **kinematically simple** manipulator to be a manipulator that has many (or all) of its link twists equal to $0°$, $90°$, or $-90°$ and many of its link lengths and offsets equal to zero. We might define a **dynamically simple** manipulator as one for which each link-inertia tensor is diagonal in frame $\{C_i\}$.

The drawback of formulating closed-form equations is simply that it currently requires a fair amount of human effort. However, symbolic manipulation programs that can derive the closed-form equations of motion of a device and automatically factor out common terms and perform trigonometric substitutions have been developed [25, 28–30].

Efficient dynamics for simulation

When dynamics are to be computed for the purpose of performing a numerical simulation of a manipulator, we are interested in solving for the joint accelerations, given the manipulator's current position and velocity and the input torques. An efficient computational scheme must therefore address both the computation of the dynamic equations studied in this chapter and efficient schemes for solving equations (for joint accelerations) and performing numerical integration. Several efficient methods for dynamic simulation of manipulators are reported in [31].

Memorization schemes

In any computational scheme, a trade-off can be made between computations and memory usage. In the problem of computing the dynamic equation of a manipulator (6.59), we have implicitly assumed that, when a value of τ is needed, it is computed as quickly as possible from Θ, $\dot{\Theta}$, and $\ddot{\Theta}$ at run time. If we wish, we can trade off this computational burden at the cost of a tremendously large memory by precomputing (6.59) for all possible Θ, $\dot{\Theta}$, and $\ddot{\Theta}$ values (suitably quantized). Then, when dynamic information is needed, the answer is found by table lookup.

The size of the memory required is large. Assume that each joint angle range is quantized to ten discrete values; likewise, assume that velocities and accelerations are quantized to ten ranges each. For a six-jointed manipulator, the number of cells in the (Θ, $\dot{\Theta}$, $\ddot{\Theta}$) quantized space is $(10 \times 10 \times 10)^6$. In each of these cells, there are six torque values. Assuming each torque value requires one computer word, this memory size is 6×10^{18} words! Also, note that the table needs to be recomputed for a change in the mass of the load—or else another dimension needs to be added to account for all possible loads.

There are many intermediate solutions that trade off memory for computation in various ways. For example, if the matrices appearing in equation (6.63) were precomputed, the table would have only one dimension (in Θ) rather than three. After the functions of Θ are looked up, a modest amount of computation (given by (6.63)) is done. For more details and for other possible parameterizations of this problem, see [3] and [6].

BIBLIOGRAPHY

[1] I. Shames, *Engineering Mechanics*, 2nd edition, Prentice-Hall, Englewood Cliffs, NJ, 1967.

[2] J.Y.S. Luh, M.W. Walker, and R.P. Paul, "On-Line Computational Scheme for Mechanical Manipulators," *Transactions of the ASME Journal of Dynamic Systems, Measurement, and Control*, 1980.

[3] M. Raibert, "Mechanical Arm Control Using a State Space Memory," SME paper MS77-750, 1977.

[4] K.R. Symon, *Mechanics*, 3rd edition, Addison-Wesley, Reading, MA, 1971.

[5] B. Noble, *Applied Linear Algebra*, Prentice-Hall, Englewood Cliffs, NJ, 1969.

[6] O. Khatib, "Commande Dynamique dans L'Espace Operationnel des Robots Manipulateurs en Presence d'Obstacles," These de Docteur-Ingenieur. Ecole Nationale Superieure de l'Aeronautique et de L'Espace (ENSAE), Toulouse.

[7] O. Khatib, "Dynamic Control of Manipulators in Operational Space," Sixth IFTOMM Congress on Theory of Machines and Mechanisms, New Delhi, December 15–20, 1983.

[8] O. Khatib, "The Operational Space Formulation in Robot Manipulator Control," 15th ISIR, Tokyo, September 11–13, 1985.

[9] E. Schmitz, "Experiments on the End-Point Position Control of a Very Flexible One-Link Manipulator," Unpublished Ph.D. Thesis, Department of Aeronautics and Astronautics, Stanford University, SUDAAR No. 547, June 1985.

[10] W. Book, "Recursive Lagrangian Dynamics of Flexible Manipulator Arms," *International Journal of Robotics Research*, Vol. 3, No. 3, 1984.

[11] S. Conte and C. DeBoor, *Elementary Numerical Analysis: An Algorithmic Approach*, 2nd edition, McGraw-Hill, New York, 1972.

[12] J. Uicker, "On the Dynamic Analysis of Spatial Linkages Using 4×4 Matrices," Unpublished Ph.D dissertation, Northwestern University, Evanston, IL, 1965.

[13] J. Uicker, "Dynamic Behaviour of Spatial Linkages," *ASME Mechanisms*, Vol. 5, No. 68, pp. 1–15.

[14] J.M. Hollerbach, "A Recursive Lagrangian Formulation of Manipulator Dynamics and a Comparative Study of Dynamics Formulation Complexity," in *Robot Motion*, M. Brady et al., Editors, MIT Press, Cambridge, MA, 1983.

[15] W. Silver, "On the Equivalence of Lagrangian and Newton–Euler Dynamics for Manipulators," *International Journal of Robotics Research*, Vol. 1, No. 2, pp. 60–70.

[16] M. Renaud, "Contribution à l'Etude de la Modélisation et de la Commande des Systèmes Mécaniques Articulés," Thèse de Docteur-Ingénieur, Université Paul Sabatier, Toulouse, December 1975.

[17] A. Liegois, W. Khalil, J.M. Dumas, and M. Renaud, "Mathematical Models of Interconnected Mechanical Systems," Symposium on the Theory and Practice of Robots and Manipulators, Poland, 1976.

[18] Y. Stepanenko and M. Vukobratovic, "Dynamics of Articulated Open-Chain Active Mechanisms," *Math-Biosciences* Vol. 28, 1976, pp. 137–170.

[19] D.E. Orin et al, "Kinematic and Kinetic Analysis of Open-Chain Linkages Utilizing Newton–Euler Methods," *Math-Biosciences* Vol. 43, 1979, pp. 107–130.

[20] W.W. Armstrong, "Recursive Solution to the Equations of Motion of an N-Link Manipulator," *Proceedings of the 5th World Congress on the Theory of Machines and Mechanisms*, Montreal, July 1979.

[21] J.M. Hollerbach and G. Sahar, "Wrist-Partitioned Inverse Accelerations and Manipulator Dynamics," MIT AI Memo No. 717, April 1983.

[22] T.K. Kanade, P.K. Khosla, and N. Tanaka, "Real-Time Control of the CMU Direct Drive Arm II Using Customized Inverse Dynamics," *Proceedings of the 23rd IEEE Conference on Decision and Control*, Las Vegas, NV, December 1984.

[23] A. Izaguirre and R.P. Paul, "Computation of the Inertial and Gravitational Coefficients of the Dynamic Equations for a Robot Manipulator with a Load," *Proceedings of the 1985 International Conference on Robotics and Automation*, pp. 1024–1032, St. Louis, March 1985.

[24] B. Armstrong, O. Khatib, and J. Burdick, "The Explicit Dynamic Model and Inertial Parameters of the PUMA 560 Arm," *Proceedings of the 1986 IEEE International Conference on Robotics and Automation*, San Francisco, April 1986, pp. 510–518.

[25] J.W. Burdick, "An Algorithm for Generation of Efficient Manipulator Dynamic Equations," *Proceedings of the 1986 IEEE International Conference on Robotics and Automation*, San Francisco, April 7–11, 1986, pp. 212–218.

[26] T.R. Kane and D.A. Levinson, "The Use of Kane's Dynamical Equations in Robotics," *The International Journal of Robotics Research*, Vol. 2, No. 3, Fall 1983, pp. 3–20.

[27] M. Renaud, "An Efficient Iterative Analytical Procedure for Obtaining a Robot Manipulator Dynamic Model," First International Symposium of Robotics Research, NH, August 1983.

[28] W. Schiehlen, "Computer Generation of Equations of Motion," in *Computer Aided Analysis and Optimization of Mechanical System Dynamics*, E.J. Haug, Editor, Springer-Verlag, Berlin & New York, 1984.

[29] G. Cesareo, F. Nicolo, and S. Nicosia, "DYMIR: A Code for Generating Dynamic Model of Robots," in *Advanced Software in Robotics*, Elsevier Science Publishers, North-Holland, 1984.

[30] J. Murray, and C. Neuman, "ARM: An Algebraic Robot Dynamic Modelling Program," IEEE International Conference on Robotics, Atlanta, March 1984.

[31] M. Walker and D. Orin, "Efficient Dynamic Computer Simulation of Robotic Mechanisms," *ASME Journal of Dynamic Systems, Measurement, and Control*, Vol. 104, 1982.

EXERCISES

6.1 [12] Find the inertia tensor of a right cylinder of homogeneous density with respect to a frame with origin at the center of mass of the body.

6.2 [32] Construct the dynamic equations for the two-link manipulator in Section 6.7 when each link is modeled as a rectangular solid of homogeneous density. Each link has dimensions l_i, ω_i, and h_i and total mass m_i.

6.3 [43] Construct the dynamic equations for the three-link manipulator of Chapter 3, Exercise 3.3. Consider each link to be a rectangular solid of homogeneous density with dimensions l_i, ω_i, and h_i and total mass m_i.

6.4 [13] Write the set of equations that correspond to (6.46)–(6.53) for the case where the mechanism could have sliding joints.

6.5 [30] Construct the dynamic equations for the two-link nonplanar manipulator shown in Fig. 6.8. Assume that all the mass of the links can be considered as a point mass located at the distal (outermost) end of the link. The mass values are m_1 and m_2, and the link lengths are l_1 and l_2. This manipulator is like the first two links of the arm in Exercise 3.3. Assume further that viscous friction is acting at each joint, with coefficients v_1 and v_2.

6.6 [32] Derive the Cartesian space form of the dynamics for the two-link planar manipulator of Section 6.7 in terms of the base frame. *Hint*: See Example 6.5, but use the Jacobian written in the base frame.

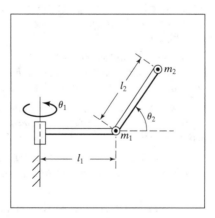

FIGURE 6.8: Two-link nonplanar manipulator with point masses at distal ends of links.

6.7 [18] How many memory locations would be required to store the dynamic equations of a general three-link manipulator in a table? Quantize each joint's position, velocity, and acceleration into 16 ranges. Make any assumptions needed.

6.8 [32] Derive the dynamic equations for the two-link manipulator shown in Fig. 4.6. Link 1 has an inertia tensor given by

$$C_1 I = \begin{bmatrix} I_{xx1} & 0 & 0 \\ 0 & I_{yy1} & 0 \\ 0 & 0 & I_{zz1} \end{bmatrix}.$$

Assume that link 2 has all its mass, m_2, located at a point at the end-effector. Assume that gravity is directed downward (opposite $\hat{Z}_1$).

6.9 [37] Derive the dynamic equations for the three-link manipulator with one prismatic joint shown in Fig. 3.9. Link 1 has an inertia tensor given by

$$C_1 I = \begin{bmatrix} I_{xx1} & 0 & 0 \\ 0 & I_{yy1} & 0 \\ 0 & 0 & I_{zz1} \end{bmatrix}.$$

Link 2 has point mass m_2 located at the origin of its link frame. Link 3 has an inertia tensor given by

$$C_3 I = \begin{bmatrix} I_{xx3} & 0 & 0 \\ 0 & I_{yy3} & 0 \\ 0 & 0 & I_{zz3} \end{bmatrix}.$$

Assume that gravity is directed opposite $\hat{Z}_1$ and that viscous friction of magnitude v_i is active at each joint.

6.10 [35] Derive the dynamic equations in Cartesian space for the manipulator of Exercise 6.8. Write the equations in frame {2}.

6.11 [20] A certain one-link manipulator has

$$C_1 I = \begin{bmatrix} I_{xx1} & 0 & 0 \\ 0 & I_{yy1} & 0 \\ 0 & 0 & I_{zz1} \end{bmatrix}.$$

Assume that this is just the inertia of the link itself. If the motor armature has a moment of inertia I_m and the gear ratio is 100, what is the total inertia as seen from the motor shaft [1]?

6.12 [20] The single-degree-of-freedom "manipulator" in Fig. 6.9 has total mass $m = 1$, with the center of mass at

$$^1P_C = \begin{bmatrix} 2 \\ 0 \\ 0 \end{bmatrix},$$

and has inertia tensor

$$^C I_1 = \begin{bmatrix} 1 & 0 & 0 \\ 0 & 2 & 0 \\ 0 & 0 & 2 \end{bmatrix}.$$

From rest at $t = 0$, the joint angle θ_1 moves in accordance with the time function

$$\theta_1(t) = bt + ct^2$$

in radians. Give the angular acceleration of the link and the linear acceleration of the center of mass in terms of frame {1} as a function of t.

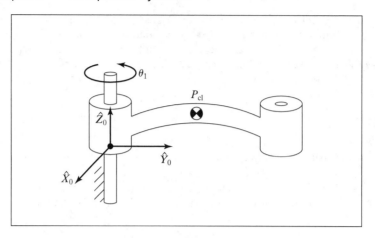

FIGURE 6.9: One-link "manipulator" of Exercise 6.12.

6.13 [40] Construct the Cartesian dynamic equations for the two-link nonplanar manipulator shown in Fig. 6.8. Assume that all the mass of the links can be considered as a point mass located at the distal (outermost) end of the link. The mass values are m_1 and m_2, and the link lengths are l_1 and l_2. This manipulator is like the first two links of the arm in Exercise 3.3. Also assume that viscous friction is acting at each joint with coefficients v_1 and v_2. Write the Cartesian dynamics in frame {3}, which is located at the tip of the manipulator and has the same orientation as link frame {2}.

6.14 [18] The following equations were derived for a 2-DOF RP manipulator:

$$\tau_1 = m_1(d_1^2 + d_2)\ddot{\theta}_1 + m_2 d_2^2 \ddot{\theta}_1 + 2m_2 d_2 \dot{d}_2 \dot{\theta}_1$$

$$+g\cos(\theta_1)[m_1(d_1 + d_2\dot{\theta}_1) + m_2(d_2 + \dot{d}_2)]$$

$$\tau_2 = m_1 \dot{d}_2 \ddot{\theta}_1 + m_2 \ddot{d}_2 - m_1 d_1 \dot{d}_2 - m_2 d_2 \dot{\theta}^2 + m_2(d_2 + 1)g\sin(\theta_1).$$

Some of the terms are obviously incorrect. Indicate the incorrect terms.

6.15 [28] Derive the dynamic equations for the RP manipulator of Example 6.5, using the Newton–Euler procedure instead of the Lagrangian technique.

6.16 [25] Derive the equations of motion for the PR manipulator shown in Fig. 6.10. Neglect friction, but include gravity. (Here, $\hat{X}_0$ is upward.) The inertia tensors of the links are diagonal, with moments $I_{xx1}, I_{yy1}, I_{zz1}$ and $I_{xx2}, I_{yy2}, I_{zz2}$. The centers of mass for the links are given by

$$^1P_{C_1} = \begin{bmatrix} 0 \\ 0 \\ -l_1 \end{bmatrix},$$

$$^2P_{C_2} = \begin{bmatrix} 0 \\ 0 \\ 0 \end{bmatrix}.$$

6.17 [40] The velocity-related terms appearing in the manipulator dynamic equation can be written as a matrix-vector product—that is,

$$V(\Theta, \dot{\Theta}) = V_m(\Theta, \dot{\Theta})\dot{\Theta},$$

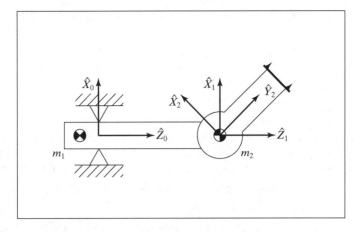

FIGURE 6.10: PR manipulator of Exercise 6.16.

where the m subscript stands for "matrix form." Show that an interesting relationship exists between the time derivative of the manipulator mass matrix and $V_m(\cdot)$, namely,

$$\dot{M}(\Theta) = 2V_m(\Theta, \dot{\Theta}) - S,$$

where S is some skew-symmetric matrix.

6.18 [15] Give two properties that any reasonable friction model (i.e., the term $F(\Theta, \dot{\Theta})$ in (6.114)) would possess.

6.19 [28] Do Exercise 6.5, using Lagrange's equations.

6.20 [28] Derive the dynamic equations of the 2-DOF manipulator of Section 6.7, using a Lagrangian formulation.

PROGRAMMING EXERCISE (PART 6)

1. Derive the dynamic equations of motion for the three-link manipulator (from Example 3.3). That is, expand Section 6.7 for the three-link case. The following numerical values describe the manipulator:

$$l_1 = l_2 = 0.5\text{m},$$

$$m_1 = 4.6\text{Kg},$$

$$m_2 = 2.3\text{Kg},$$

$$m_3 = 1.0\text{Kg},$$

$$g = 9.8\text{m/s}^2.$$

For the first two links, we assume that the mass is all concentrated at the distal end of the link. For link 3, we assume that the center of mass is located at the origin of frame {3}—that is, at the proximal end of the link. The inertia tensor for link 3 is

$$^{C_s}I = \begin{bmatrix} 0.05 & 0 & 0 \\ 0 & 0.1 & 0 \\ 0 & 0 & 0.1 \end{bmatrix} \text{Kg-m}^2.$$

The vectors that locate each center of mass relative to the respective link frame are

$$^1P_{C_1} = l_1\hat{X}_1,$$

$$^2P_{C_2} = l_2\hat{X}_2,$$

$$^3P_{C_3} = 0.$$

2. Write a simulator for the three-link manipulator. A simple Euler-integration routine is sufficient for performing the numerical integration (as in Section 6.12). To keep your code modular, it might be helpful to define the routine

```
Procedure UPDATE(VAR tau: vec3; VAR period: real; VAR
theta, thetadot: vec3);
```

where "tau" is the torque command to the manipulator (always zero for this assignment), "period" is the length of time you wish to advance time (in seconds), and "theta" and "thetadot" are the *state* of the manipulator. Theta and thetadot are updated by "period" seconds each time you call UPDATE. Note that "period" would typically be longer than the integration step size, Δt, used in the numerical integration. For example, although the step size for numerical integration might be 0.001 second, you might wish to print out the manipulator position and velocity only each 0.1 seconds.

To test your simulation, set the joint-torque commands to zero (for all time) and perform these tests:

   (a) Set the initial position of the manipulator to

$$[\theta_1\ \theta_2\ \theta_3] = [-90\ 0\ 0].$$

   Simulate for a few seconds. Is the motion of the manipulator what you would expect?

   (b) Set the initial position of the manipulator to

$$[\theta_1\ \theta_2\ \theta_3] = [30\ 30\ 10].$$

   Simulate for a few seconds. Is the motion of the manipulator what you would expect?

   (c) Introduce some viscous friction at each joint of the simulated manipulator—that is, add a term to the dynamics of each joint in the form $\tau_f = v\dot{\theta}$, where $v = 5.0$ newton-meter-seconds for each joint. Repeat test (b) above. Is the motion what you would expect?

MATLAB EXERCISE 6A

This exercise focuses on the inverse-dynamics analysis (in a resolved-rate control framework—see MATLAB Exercise 5) for the planar 2-DOF 2R robot. This robot is the first two R-joints and first two moving links of the planar 3-DOF 3R robot. (See Figures 3.6 and 3.7; the DH parameters are given in the first two rows of Figure 3.8.)

For the planar 2R robot, calculate the required joint torques (i.e., solve the inverse-dynamics problem) to provide the commanded motion at every time step in a resolved-rate control scheme. You can use either numerical Newton–Euler recursion or the analytical equations from the results of Exercise 6.2, or both.

Given: $L_1 = 1.0$ m, $L_2 = 0.5$ m; Both links are solid steel with mass density $\rho = 7806$ kg/m^3; both have the width and thickness dimensions $w = t = 5$ cm. The revolute joints are assumed to be perfect, connecting the links at their very edges (not physically possible).

The initial angles are $\Theta = \begin{Bmatrix} \theta_1 \\ \theta_2 \end{Bmatrix} = \begin{Bmatrix} 10° \\ 90° \end{Bmatrix}$.

The (constant) commanded Cartesian velocity is $^0\dot{X} = {}^0\begin{Bmatrix} \dot{x} \\ \dot{y} \end{Bmatrix} = {}^0\begin{Bmatrix} 0 \\ 0.5 \end{Bmatrix}$ (m/s).

Simulate motion for 1 sec, with a control time step of 0.01 sec.
Present five plots (each set on a separate graph, please):

1. the two joint angles (degrees) $\Theta = \{\theta_1 \ \theta_2\}^T$ vs. time;
2. the two joint rates (rad/s) $\dot{\Theta} = \{\dot{\theta}_1 \ \dot{\theta}_2\}^T$ vs. time;
3. the two joint accelerations (rad/s^2) $\ddot{\Theta} = \{\ddot{\theta}_1 \ \ddot{\theta}_2\}^T$ vs. time;
4. the three Cartesian components of $^0_H T$, $X = \{x \ y \ \phi\}^T$ (rad is fine for ϕ so it will fit) vs. time;
5. the two inverse dynamics joint torques (Nm) $T = \{\tau_1 \ \tau_2\}^T$ vs. time.

Carefully label (by hand is fine!) each component on each plot. Also, label the axis names and units.

Perform this simulation twice. The first time, ignore gravity (the motion plane is normal to the effect of gravity); the second time, consider gravity g in the negative Y direction.

MATLAB EXERCISE 6B

This exercise focuses on the inverse-dynamics solution for the planar 3-DOF, 3R robot (of Figures 3.6 and 3.7; the DH parameters are given in Figure 3.8) for a motion snapshot in time only. The following fixed-length parameters are given: $L_1 = 4$, $L_2 = 3$, and $L_3 = 2$ (m). For dynamics, we must also be given mass and moment-of-inertia information: $m_1 = 20$, $m_2 = 15$, $m_3 = 10$ (kg), $^C I_{ZZ1} = 0.5$, $^C I_{ZZ2} = 0.2$, and $^C I_{ZZ3} = 0.1$ (kgm^2). Assume that the CG of each link is in its geometric center. Also, assume that gravity acts in the $-Y$ direction in the plane of motion. For this exercise, ignore actuator dynamics and the joint gearing.

a) Write a MATLAB program to implement the recursive Newton–Euler inverse-dynamics solution (i.e., given the commanded motion, calculate the required driving joint torques) for the following motion snapshot in time:

$$\Theta = \begin{Bmatrix} \theta_1 \\ \theta_2 \\ \theta_3 \end{Bmatrix} = \begin{Bmatrix} 10° \\ 20° \\ 30° \end{Bmatrix} \quad \dot{\Theta} = \begin{Bmatrix} \dot{\theta}_1 \\ \dot{\theta}_2 \\ \dot{\theta}_3 \end{Bmatrix} = \begin{Bmatrix} 1 \\ 2 \\ 3 \end{Bmatrix} (rad/s) \ddot{\Theta} = \begin{Bmatrix} \ddot{\theta}_1 \\ \ddot{\theta}_2 \\ \ddot{\theta}_3 \end{Bmatrix} = \begin{Bmatrix} 0.5 \\ 1 \\ 1.5 \end{Bmatrix} (rad/s^2)$$

b) Check your results in (a) by means of the Corke MATLAB Robotics Toolbox. Try functions $rne()$ and $gravload()$.

MATLAB EXERCISE 6C

This exercise focuses on the forward-dynamics solution for the planar 3-DOF, 3R robot (parameters from MATLAB Exercise 6B) for motion over time. In this case, ignore gravity (i.e., assume that gravity acts in a direction normal to the plane of motion). Use the Corke MATLAB Robotics Toolbox to solve the forward-dynamics problem (i.e., given the commanded driving joint torques, calculate the resulting robot motion) for the following constant joint torques and the given initial joint angles and initial joint rates:

$$T = \begin{Bmatrix} \tau_1 \\ \tau_2 \\ \tau_3 \end{Bmatrix} = \begin{Bmatrix} 20 \\ 5 \\ 1 \end{Bmatrix} \text{ (Nm, constant)} \quad \Theta_0 = \begin{Bmatrix} \theta_{10} \\ \theta_{20} \\ \theta_{30} \end{Bmatrix} = \begin{Bmatrix} -60° \\ 90° \\ 30° \end{Bmatrix}$$

$$\dot{\Theta}_0 = \begin{Bmatrix} \dot{\theta}_{10} \\ \dot{\theta}_{20} \\ \dot{\theta}_{30} \end{Bmatrix} = \begin{Bmatrix} 0 \\ 0 \\ 0 \end{Bmatrix} \text{ (rad/s)}$$

Perform this simulation for 4 seconds. Try function *fdyn()*.

Present two plots for the resulting robot motion (each set on a separate graph, please):

1. the three joint angles (degrees) $\Theta = \{\theta_1 \ \theta_2 \ \theta_3\}^T$ vs. time;
2. the three joint rates (rad/s) $\dot{\Theta} = \{\dot{\theta}_1 \ \dot{\theta}_2 \ \dot{\theta}_3\}^T$ vs. time.

Carefully label (by hand is fine!) each component on each plot. Also, label the axis names and units.

CHAPTER 7

Trajectory generation

7.1 INTRODUCTION

In this chapter, we concern ourselves with methods of computing a trajectory that describes the desired motion of a manipulator in multidimensional space. Here, **trajectory** refers to a time history of position, velocity, and acceleration for each degree of freedom.

This problem includes the human-interface problem of how we wish to *specify* a trajectory or path through space. In order to make the description of manipulator motion easy for a human user of a robot system, the user should not be required to write down complicated functions of space and time to specify the task. Rather, we must allow the capability of specifying trajectories with simple descriptions of the desired motion, and let the system figure out the details. For example, the user might want to be able to specify nothing more than the desired goal position and orientation of the end-effector and leave it to the system to decide on the exact shape of the path to get there, the duration, the velocity profile, and other details.

We also are concerned with how trajectories are *represented* in the computer after they have been planned. Finally, there is the problem of actually computing the trajectory from the internal representation—or *generating* the trajectory. Generation occurs at *run time*; in the most general case, position, velocity, and acceleration are computed. These trajectories are computed on digital computers, so the trajectory points are computed at a certain rate, called the **path-update rate**. In typical manipulator systems, this rate lies between 60 and 2000 Hz.

7.2 GENERAL CONSIDERATIONS IN PATH DESCRIPTION AND GENERATION

For the most part, we will consider motions of a manipulator as motions of the tool frame, $\{T\}$, relative to the station frame, $\{S\}$. This is the same manner in which an eventual user of the system would think, and designing a path description and generation system in these terms will result in a few important advantages.

When we specify paths as motions of the tool frame relative to the station frame, we decouple the motion description from any particular robot, end-effector, or workpieces. This results in a certain modularity and would allow the same path description to be used with a different manipulator—or with the same manipulator, but a different tool size. Further, we can specify and plan motions relative to a moving workstation (perhaps a conveyor belt) by planning motions relative to the station frame as always and, at run time, causing the definition of $\{S\}$ to be changing with time.

As shown in Fig. 7.1, the basic problem is to move the manipulator from an initial position to some desired final position—that is, we wish to move the tool frame from its current value, $\{T_{\text{initial}}\}$, to a desired final value, $\{T_{\text{final}}\}$. Note that, in general, this motion involves both a change in orientation and a change in the position of the tool relative to the station.

Sometimes it is necessary to specify the motion in much more detail than by simply stating the desired final configuration. One way to include more detail in a path description is to give a sequence of desired **via points** (intermediate points between the initial and final positions). Thus, in completing the motion, the tool frame must pass through a set of intermediate positions and orientations as described by the via points. Each of these via points is actually a frame that specifies both the position and orientation of the tool relative to the station. The name **path points** includes all the via points plus the initial and final points. Remember that, although we generally use the term "points," these are actually frames, which give both position and orientation. Along with these *spatial* constraints on the motion, the user could also wish to specify *temporal* attributes of the motion. For example, the time elapsed between via points might be specified in the description of the path.

Usually, it is desirable for the motion of the manipulator to be *smooth*. For our purposes, we will define a smooth function as a function that is continuous and has a continuous first derivative. Sometimes a continuous second derivative is also

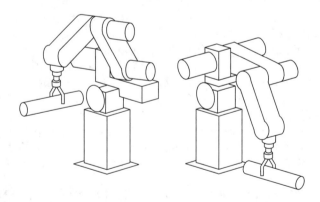

FIGURE 7.1: In executing a trajectory, a manipulator moves from its initial position to a desired goal position in a smooth manner.

desirable. Rough, jerky motions tend to cause increased wear on the mechanism and cause vibrations by exciting resonances in the manipulator. In order to guarantee smooth paths, we must put some sort of constraints on the spatial and temporal qualities of the path *between* the via points.

At this point, there are many choices that may be made and, consequently, a great variety in the ways that paths might be specified and planned. Any smooth functions of time that pass through the via points could be used to specify the exact path shape. In this chapter, we will discuss a couple of simple choices for these functions. Other approaches can be found in [1, 2] and [13–16].

7.3   JOINT-SPACE SCHEMES

In this section, we consider methods of path generation in which the path shapes (in space and in time) are described in terms of functions of joint angles.

Each path point is usually specified in terms of a desired position and orientation of the tool frame, {T}, relative to the station frame, {S}. Each of these via points is "converted" into a set of desired joint angles by application of the inverse kinematics. Then a smooth function is found for each of the n joints that pass through the via points and end at the goal point. The time required for each segment is the same for each joint so that all joints will reach the via point at the same time, thus resulting in the desired Cartesian position of {T} at each via point. Other than specifying the same duration for each joint, the determination of the desired joint angle function for a particular joint does not depend on the functions for the other joints.

Hence, joint-space schemes achieve the desired position and orientation at the via points. In between via points, the shape of the path, although rather simple in joint space, is complex if described in Cartesian space. Joint-space schemes are usually the easiest to compute, and, because we make no continuous correspondence between joint space and Cartesian space, there is essentially no problem with singularities of the mechanism.

Cubic polynomials

Consider the problem of moving the tool from its initial position to a goal position in a certain amount of time. Inverse kinematics allow the set of joint angles that correspond to the goal position and orientation to be calculated. The initial position of the manipulator is also known in the form of a set of joint angles. What is required is a function for each joint whose value at t_0 is the initial position of the joint and whose value at t_f is the desired goal position of that joint. As shown in Fig. 7.2, there are many smooth functions, $\theta(t)$, that might be used to interpolate the joint value.

In making a single smooth motion, at least four constraints on $\theta(t)$ are evident. Two constraints on the function's value come from the selection of initial and final values:

$$\theta(0) = \theta_0,$$
$$\theta(t_f) = \theta_f. \tag{7.1}$$

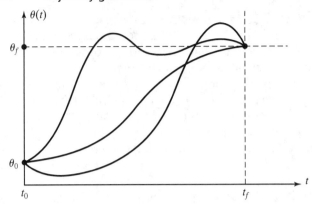

FIGURE 7.2: Several possible path shapes for a single joint.

An additional two constraints are that the function be continuous in velocity, which in this case means that the initial and final velocity are zero:

$$\dot{\theta}(0) = 0,$$
$$\dot{\theta}(t_f) = 0. \qquad (7.2)$$

These four constraints can be satisfied by a polynomial of at least third degree. (A cubic polynomial has four coefficients, so it can be made to satisfy the four constraints given by (7.1) and (7.2).) These constraints uniquely specify a particular cubic. A cubic has the form

$$\theta(t) = a_0 + a_1 t + a_2 t^2 + a_3 t^3, \qquad (7.3)$$

so the joint velocity and acceleration along this path are clearly

$$\dot{\theta}(t) = a_1 + 2a_2 t + 3a_3 t^2,$$
$$\ddot{\theta}(t) = 2a_2 + 6a_3 t. \qquad (7.4)$$

Combining (7.3) and (7.4) with the four desired constraints yields four equations in four unknowns:

$$\theta_0 = a_0,$$
$$\theta_f = a_0 + a_1 t_f + a_2 t_f^2 + a_3 t_f^3,$$
$$0 = a_1, \qquad (7.5)$$
$$0 = a_1 + 2a_2 t_f + 3a_3 t_f^2.$$

Solving these equations for the a_i, we obtain

$$a_0 = \theta_0,$$
$$a_1 = 0,$$

$$a_2 = \frac{3}{t_f^2}(\theta_f - \theta_0),$$ (7.6)

$$a_3 = -\frac{2}{t_f^3}(\theta_f - \theta_0).$$

Using (7.6), we can calculate the cubic polynomial that connects any initial joint-angle position with any desired final position. This solution is for the case when the joint starts and finishes at zero velocity.

EXAMPLE 7.1

A single-link robot with a rotary joint is motionless at $\theta = 15$ degrees. It is desired to move the joint in a smooth manner to $\theta = 75$ degrees in 3 seconds. Find the coefficients of a cubic that accomplishes this motion and brings the manipulator to rest at the goal. Plot the position, velocity, and acceleration of the joint as a function of time.

Plugging into (7.6), we find that

$$a_0 = 15.0,$$
$$a_1 = 0.0,$$
$$a_2 = 20.0,$$ (7.7)
$$a_3 = -4.44.$$

Using (7.3) and (7.4), we obtain

$$\theta(t) = 15.0 + 20.0t^2 - 4.44t^3,$$
$$\dot{\theta}(t) = 40.0t - 13.33t^2,$$ (7.8)
$$\ddot{\theta}(t) = 40.0 - 26.66t.$$

Figure 7.3 shows the position, velocity, and acceleration functions for this motion sampled at 40 Hz. Note that the velocity profile for any cubic function is a parabola and that the acceleration profile is linear.

Cubic polynomials for a path with via points

So far, we have considered motions described by a desired duration and a final goal point. In general, we wish to allow paths to be specified that include intermediate via points. If the manipulator is to come to rest at each via point, then we can use the cubic solution of Section 7.3.

Usually, we wish to be able to pass through a via point without stopping, and so we need to generalize the way in which we fit cubics to the path constraints.

As in the case of a single goal point, each via point is usually specified in terms of a desired position and orientation of the tool frame relative to the station frame. Each of these via points is "converted" into a set of desired joint angles by application of the inverse kinematics. We then consider the problem of computing cubics that connect the via-point values for each joint together in a smooth way.

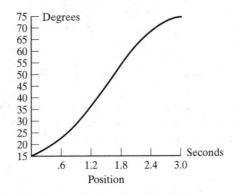

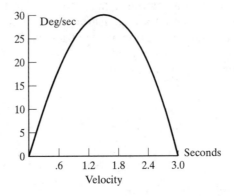

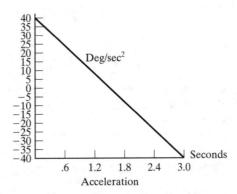

FIGURE 7.3: Position, velocity, and acceleration profiles for a single cubic segment that starts and ends at rest.

If desired velocities of the joints at the via points are known, then we can construct cubic polynomials as before; now, however, the velocity constraints at each end are not zero, but rather, some known velocity. The constraints of (7.3) become

$$\dot{\theta}(0) = \dot{\theta}_0,$$

$$\dot{\theta}(t_f) = \dot{\theta}_f. \tag{7.9}$$

The four equations describing this general cubic are

$$\theta_0 = a_0,$$

$$\theta_f = a_0 + a_1 t_f + a_2 t_f^2 + a_3 t_f^3,$$

$$\dot{\theta}_0 = a_1,$$ (7.10)

$$\dot{\theta}_f = a_1 + 2a_2 t_f + 3a_3 t_f^2.$$

Solving these equations for the a_i, we obtain

$$a_0 = \theta_0,$$

$$a_1 = \dot{\theta}_0,$$

$$a_2 = \frac{3}{t_f^2}(\theta_f - \theta_0) - \frac{2}{t_f}\dot{\theta}_0 - \frac{1}{t_f}\dot{\theta}_f,$$ (7.11)

$$a_3 = -\frac{2}{t_f^3}(\theta_f - \theta_0) + \frac{1}{t_f^2}(\dot{\theta}_f + \dot{\theta}_0).$$

Using (7.11), we can calculate the cubic polynomial that connects any initial and final positions with any initial and final velocities.

If we have the desired joint velocities at each via point, then we simply apply (7.11) to each segment to find the required cubics. There are several ways in which the desired velocity at the via points might be specified:

1. The user specifies the desired velocity at each via point in terms of a Cartesian linear and angular velocity of the tool frame at that instant.
2. The system automatically chooses the velocities at the via points by applying a suitable heuristic in either Cartesian space or joint space.
3. The system automatically chooses the velocities at the via points in such a way as to cause the acceleration at the via points to be continuous.

In the first option, Cartesian desired velocities at the via points are "mapped" to desired joint rates by using the inverse Jacobian of the manipulator evaluated at the via point. If the manipulator is at a singular point at a particular via point, then the user is not free to assign an arbitrary velocity at this point. It is a useful capability of a path-generation scheme to be able to meet a desired velocity that the user specifies, but it would be a burden to require that the user always make these specifications. Therefore, a convenient system should include either option 2 or 3 (or both).

In option 2, the system automatically chooses reasonable intermediate velocities, using some kind of heuristic. Consider the path specified by the via points shown for some joint, θ, in Fig. 7.4.

In Fig. 7.4, we have made a reasonable choice of joint velocities at the via points, as indicated with small line segments representing tangents to the curve at each via point. This choice is the result of applying a conceptually and computationally simple heuristic. Imagine the via points connected with straight line segments. If the

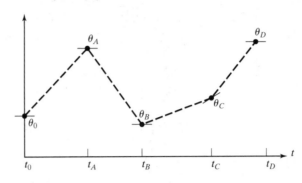

FIGURE 7.4: Via points with desired velocities at the points indicated by tangents.

slope of these lines changes sign at the via point, choose zero velocity; if the slope of these lines does not change sign, choose the average of the two slopes as the via velocity. In this way, from specification of the desired via points alone, the system can choose the velocity at each point.

In option 3, the system chooses velocities in such a way that acceleration is continuous at the via point. To do this, a new approach is needed. In this kind of spline, set of data[1] we replace the two velocity constraints at the connection of two cubics with the two constraints that velocity be continuous and acceleration be continuous.

EXAMPLE 7.2

Solve for the coefficients of two cubics that are connected in a two-segment spline with continuous acceleration at the intermediate via point. The initial angle is θ_0, the via point is θ_v, and the goal point is θ_g.

The first cubic is

$$\theta(t) = a_{10} + a_{11}t + a_{12}t^2 + a_{13}t^3, \tag{7.12}$$

and the second is

$$\theta(t) = a_{20} + a_{21}t + a_{22}t^2 + a_{23}t^3. \tag{7.13}$$

Each cubic will be evaluated over an interval starting at $t = 0$ and ending at $t = t_{fi}$, where $i = 1$ or $i = 2$.

The constraints we wish to enforce are

$$\theta_0 = a_{10},$$
$$\theta_v = a_{10} + a_{11}t_{f1} + a_{12}t_{f1}^2 + a_{13}t_{f1}^3,$$
$$\theta_v = a_{20},$$
$$\theta_g = a_{20} + a_{21}t_{f2} + a_{22}t_{f2}^2 + a_{23}t_{f2}^3, \tag{7.14}$$
$$0 = a_{11},$$

---

[1] In our usage, the term "spline" simply means a function of time.

$$0 = a_{21} + 2a_{22}t_{f2} + 3a_{23}t_{f2}^2,$$

$$a_{11} + 2a_{12}t_{f1} + 3a_{12}t_{f1}^2 = a_{21},$$

$$2a_{12} + 6a_{13}t_{f1} = 2a_{22}.$$

These constraints specify a linear-equation problem having eight equations and eight unknowns. Solving for the case $t_f = t_{f1} = t_{f2}$, we obtain

$$a_{10} = \theta_0,$$

$$a_{11} = 0,$$

$$a_{12} = \frac{12\theta_v - 3\theta_g - 9\theta_0}{4t_f^2},$$

$$a_{13} = \frac{-8\theta_v + 3\theta_g + 5\theta_0}{4t_f^3},$$

$$a_{20} = \theta_v, \qquad\qquad\qquad (7.15)$$

$$a_{21} = \frac{3\theta_g - 3\theta_0}{4t_f},$$

$$a_{22} = \frac{-12\theta_v + 6\theta_g + 6\theta_0}{4t_f^2},$$

$$a_{23} = \frac{8\theta_v - 5\theta_g - 3\theta_0}{4t_f^3}.$$

For the general case, involving n cubic segments, the equations that arise from insisting on continuous acceleration at the via points can be cast in matrix form, which is solved to compute the velocities at the via points. The matrix turns out to be tridiagonal and easily solved [4].

Higher-order polynomials

Higher-order polynomials are sometimes used for path segments. For example, if we wish to be able to specify the position, velocity, *and* acceleration at the beginning and end of a path segment, a quintic polynomial is required, namely,

$$\theta(t) = a_0 + a_1 t + a_2 t^2 + a_3 t^3 + a_4 t^4 + a_5 t^5, \qquad\qquad (7.16)$$

where the constraints are given as

$$\theta_0 = a_0,$$

$$\theta_f = a_0 + a_1 t_f + a_2 t_f^2 + a_3 t_f^3 + a_4 t_f^4 + a_5 t_f^5,$$

$$\dot{\theta}_0 = a_1,$$

$$\dot{\theta}_f = a_1 + 2a_2t_f + 3a_3t_f^2 + 4a_4t_f^3 + 5a_5t_f^4, \tag{7.17}$$

$$\ddot{\theta}_0 = 2a_2,$$

$$\ddot{\theta}_f = 2a_2 + 6a_3t_f + 12a_4t_f^2 + 20a_5t_f^3.$$

These constraints specify a linear set of six equations with six unknowns, whose solution is

$$a_0 = \theta_0,$$

$$a_1 = \dot{\theta}_0,$$

$$a_2 = \frac{\ddot{\theta}_0}{2},$$

$$a_3 = \frac{20\theta_f - 20\theta_0 - (8\dot{\theta}_f + 12\dot{\theta}_0)t_f - (3\ddot{\theta}_0 - \ddot{\theta}_f)t_f^2}{2t_f^3}, \tag{7.18}$$

$$a_4 = \frac{30\theta_0 - 30\theta_f + (14\dot{\theta}_f + 16\dot{\theta}_0)t_f + (3\ddot{\theta}_0 - 2\ddot{\theta}_f)t_f^2}{2t_f^4},$$

$$a_5 = \frac{12\theta_f - 12\theta_0 - (6\dot{\theta}_f + 6\dot{\theta}_0)t_f - (\ddot{\theta}_0 - \ddot{\theta}_f)t_f^2}{2t_f^5}.$$

Various algorithms are available for computing smooth functions (polynomial or otherwise) that pass through a given set of data points [3, 4]. Complete coverage is beyond the scope of this book.

Linear function with parabolic blends

Another choice of path shape is linear. That is, we simply interpolate linearly to move from the present joint position to the final position, as in Fig. 7.5. Remember that, although the motion of each joint in this scheme is linear, the end-effector in general does not move in a straight line in space.

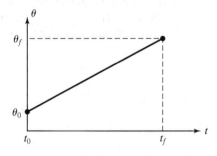

FIGURE 7.5: Linear interpolation requiring infinite acceleration.

However, straightforward linear interpolation would cause the velocity to be discontinuous at the beginning and end of the motion. To create a smooth path with continuous position and velocity, we start with the linear function but add a parabolic *blend* region at each path point.

During the blend portion of the trajectory, constant acceleration is used to change velocity smoothly. Figure 7.6 shows a simple path constructed in this way. The linear function and the two parabolic functions are "splined" together so that the entire path is continuous in position and velocity.

In order to construct this single segment, we will assume that the parabolic blends both have the same duration; therefore, the same constant acceleration (modulo a sign) is used during both blends. As indicated in Fig. 7.7, there are many solutions to the problem—but note that the answer is always symmetric about the halfway point in time, t_h, and about the halfway point in position, θ_h. The velocity at the end of the blend region must equal the velocity of the linear section, and so we have

$$\ddot{\theta}t_b = \frac{\theta_h - \theta_b}{t_h - t_b}, \tag{7.19}$$

where θ_b is the value of θ at the end of the blend region, and $\ddot{\theta}$ is the acceleration acting during the blend region. The value of θ_b is given by

$$\theta_b = \theta_0 + \tfrac{1}{2}\ddot{\theta}t_b^2. \tag{7.20}$$

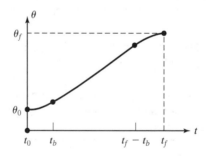

FIGURE 7.6: Linear segment with parabolic blends.

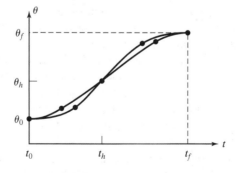

FIGURE 7.7: Linear segment with parabolic blends.

Combining (7.19) and (7.20) and $t = 2t_h$, we get

$$\ddot{\theta}t_b^2 - \ddot{\theta}tt_b + (\theta_f - \theta_0) = 0, \tag{7.21}$$

where t is the desired duration of the motion. Given any θ_f, θ_0, and t, we can follow any of the paths given by the choices of $\ddot{\theta}$ and t_b that satisfy (7.21). Usually, an acceleration, $\ddot{\theta}$, is chosen, and (7.21) is solved for the corresponding t_b. The acceleration chosen must be sufficiently high, or a solution will not exist. Solving (7.21) for t_b in terms of the acceleration and other known parameters, we obtain

$$t_b = \frac{t}{2} - \frac{\sqrt{\ddot{\theta}^2 t^2 - 4\ddot{\theta}(\theta_f - \theta_0)}}{2\ddot{\theta}}. \tag{7.22}$$

The constraint on the acceleration used in the blend is

$$\ddot{\theta} \geq \frac{4(\theta_f - \theta_0)}{t^2} \tag{7.23}$$

When equality occurs in (7.23) the linear portion has shrunk to zero length and the path is composed of two blends that connect with equivalent slope. As the acceleration used becomes larger and larger, the length of the blend region becomes shorter and shorter. In the limit, with infinite acceleration, we are back to the simple linear-interpolation case.

EXAMPLE 7.3

For the same single-segment path discussed in Example 7.1, show two examples of a linear path with parabolic blends.

Figure 7.8(a) shows one possibility where $\ddot{\theta}$ was chosen quite high. In this case we quickly accelerate, then coast at constant velocity, and then decelerate. Figure 7.8(b) shows a trajectory where acceleration is kept quite low, so that the linear section almost disappears.

Linear function with parabolic blends for a path with via points

We now consider linear paths with parabolic blends for the case in which there are an arbitrary number of via points specified. Figure 7.9 shows a set of joint-space via points for some joint θ. Linear functions connect the via points, and parabolic blend regions are added around each via point.

We will use the following notation: Consider three neighboring path points, which we will call points j, k, and l. The duration of the blend region at path point k is t_k. The duration of the linear portion between points j and k is t_{jk}. The overall duration of the segment connecting points j and k is t_{djk}. The velocity during the linear portion is $\dot{\theta}_{jk}$, and the acceleration during the blend at point j is $\ddot{\theta}_j$. See Fig. 7.9 for an example.

As with the single-segment case, there are many possible solutions, depending on the value of acceleration used at each blend. Given all the path points θ_k, the desired durations t_{djk}, and the magnitude of acceleration to use at each path point

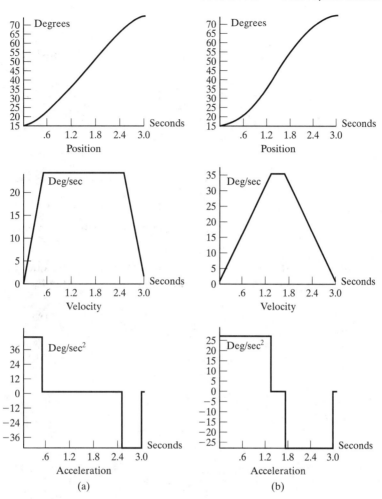

FIGURE 7.8: Position, velocity, and acceleration profiles for linear interpolation with parabolic blends. The set of curves on the left is based on a higher acceleration during the blends than is that on the right.

$|\ddot{\theta}_k|$, we can compute the blend times t_k. For interior path points, this follows simply from the equations

$$\dot{\theta}_{jk} = \frac{\theta_k - \theta_j}{t_{djk}},$$

$$\ddot{\theta}_k = SGN(\dot{\theta}_{kl} - \dot{\theta}_{jk})|\ddot{\theta}_k|,$$

$$t_k = \frac{\dot{\theta}_{kl} - \dot{\theta}_{jk}}{\ddot{\theta}_k}, \tag{7.24}$$

$$t_{jk} = t_{djk} - \frac{1}{2}t_j - \frac{1}{2}t_k.$$

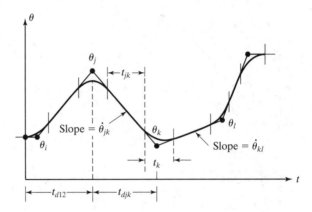

FIGURE 7.9: Multisegment linear path with blends.

The first and last segments must be handled slightly differently, because an entire blend region at one end of the segment must be counted in the total segment's time duration.

For the first segment, we solve for t_1 by equating two expressions for the velocity during the linear phase of the segment:

$$\frac{\theta_2 - \theta_1}{t_{12} - \frac{1}{2}t_1} = \ddot{\theta}_1 t_1. \tag{7.25}$$

This can be solved for t_1, the blend time at the initial point; then $\dot{\theta}_{12}$ and t_{12} are easily computed:

$$\ddot{\theta}_1 = SGN(\theta_2 - \theta_1)|\ddot{\theta}_1|,$$

$$t_1 = t_{d12} - \sqrt{t_{d12}^2 - \frac{2(\theta_2 - \theta_1)}{\ddot{\theta}_1}},$$

$$\dot{\theta}_{12} = \frac{\theta_2 - \theta_1}{t_{d12} - \frac{1}{2}t_1}, \tag{7.26}$$

$$t_{12} = t_{d12} - t_1 - \frac{1}{2}t_2.$$

Likewise, for the last segment (the one connecting points $n - 1$ and n), we have

$$\frac{\theta_{n-1} - \theta_n}{t_{d(n-1)n} - \frac{1}{2}t_n} = \ddot{\theta}_n t_n, \tag{7.27}$$

which leads to the solution

$$\ddot{\theta}_n = SGN(\theta_{n-1} - \theta_n)|\ddot{\theta}_n|,$$

$$t_n = t_{d(n-1)n} - \sqrt{t_{d(n-1)n}^2 + \frac{2(\theta_n - \theta_{n-1})}{\ddot{\theta}_n}},$$

$$\dot{\theta}_{(n-1)n} = \frac{\theta_n - \theta_{n-1}}{t_{d(n-1)n} - \frac{1}{2}t_n},$$ (7.28)

$$t_{(n-1)n} = t_{d(n-1)n} - t_n - \frac{1}{2}t_{n-1}.$$

Using (7.24) through (7.28), we can solve for the blend times and velocities for a multisegment path. Usually, the user specifies only the via points and the desired duration of the segments. In this case, the system uses default values for acceleration for each joint. Sometimes, to make things even simpler for the user, the system will calculate durations based on default velocities. At all blends, sufficiently large acceleration must be used so that there is sufficient time to get into the linear portion of the segment before the next blend region starts.

EXAMPLE 7.4

The trajectory of a particular joint is specified as follows: Path points in degrees: 10, 35, 25, 10. The duration of these three segments should be 2, 1, 3 seconds, respectively. The magnitude of the default acceleration to use at all blend points is 50 degrees/second². Calculate all segment velocities, blend times, and linear times.

For the first segment, we apply (7.26) to find

$$\ddot{\theta}_1 = 50.0.$$ (7.29)

Applying (7.26) to calculate the blend time at the initial point, we get

$$t_1 = 2 - \sqrt{4 - \frac{2(35 - 10)}{50.0}} = 0.27.$$ (7.30)

The velocity, $\dot{\theta}_{12}$, is calculated from (7.26) as

$$\dot{\theta}_{12} = \frac{35 - 10}{2 - 0.5(0.27)} = 13.50.$$ (7.31)

The velocity, $\dot{\theta}_{23}$, is calculated from (7.24) as

$$\dot{\theta}_{23} = \frac{25 - 35}{1} = -10.0.$$ (7.32)

Next, we apply (7.24) to find

$$\ddot{\theta}_2 = -50.0.$$ (7.33)

Then t_2 is calculated from (7.24), and we get

$$t_2 = \frac{-10.0 - 13.50}{-50.0} = 0.47.$$ (7.34)

The linear-portion length of segment 1 is then calculated from (7.26):

$$t_{12} = 2 - 0.27 - \frac{1}{2}(0.47) = 1.50.$$ (7.35)

Next, from (7.29), we have

$$\ddot{\theta}_4 = 50.0.$$ (7.36)

So, for the last segment, (7.28) is used to compute t_4, and we have

$$t_4 = 3 - \sqrt{9 + \frac{2(10 - 25)}{50.0}} = 0.102. \tag{7.37}$$

The velocity, $\dot{\theta}_{34}$, is calculated from (7.28) as

$$\dot{\theta}_{34} = \frac{10 - 25}{3 - 0.050} = -5.10. \tag{7.38}$$

Next, (7.24) is used to obtain

$$\ddot{\theta}_3 = 50.0. \tag{7.39}$$

Then t_3 is calculated from (7.24):

$$t_3 = \frac{-5.10 - (-10.0)}{50} = 0.098. \tag{7.40}$$

Finally, from (7.24), we compute

$$t_{23} = 1 - \tfrac{1}{2}(0.47) - \tfrac{1}{2}(0.098) = 0.716, \tag{7.41}$$

$$t_{34} = 3 - \tfrac{1}{2}(0.098) - 0.012 = 2.849. \tag{7.42}$$

The results of these computations constitute a "plan" for the trajectory. At execution time, these numbers would be used by the **path generator** to compute values of $\theta, \dot{\theta}$, and $\ddot{\theta}$ at the path-update rate.

In these linear-parabolic-blend splines, note that the via points are not actually reached unless the manipulator comes to a stop. Often, when acceleration capability is sufficiently high, the paths will come quite close to the desired via point. If we wish to actually pass through a point, by coming to a stop, the via point is simply repeated in the path specification.

If the user wishes to specify that the manipulator pass *exactly* through a via point without stopping, this specification can be accommodated by using the same formulation as before, but with the following addition: The system automatically replaces the via point through which we wish the manipulator to pass with two *pseudo via points*, one on each side of the original (as in Fig. 7.10). Then path generation takes place as before. The original via point will now lie in the linear region of the path connecting the two pseudo via points. In addition to requesting that the manipulator pass exactly through a via point, the user can also request that it pass through with a certain velocity. If the user does not specify this velocity, the system chooses it by means of a suitable heuristic. The term **through point** might be used (rather than via point) to specify a path point *through* which we force the manipulator to pass exactly.

7.4   CARTESIAN-SPACE SCHEMES

As was mentioned in Section 7.3, paths computed in joint space can ensure that via and goal points are attained, even when these path points were specified by means of

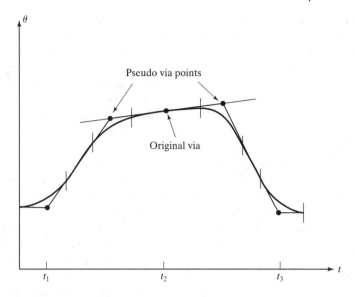

FIGURE 7.10: Use of pseudo via points to create a "through" point.

Cartesian frames. However, the spatial shape of the path taken by the end-effector is not a straight line through space; rather, it is some complicated shape that depends on the particular kinematics of the manipulator being used. In this section, we consider methods of path generation in which the path shapes are described in terms of functions that compute Cartesian position and orientation as functions of time. In this way, we can also specify the spatial shape of the path between path points. The most common path shape is a straight line, but circular, sinusoidal, or other path shapes could be used.

Each path point is usually specified in terms of a desired position and orientation of the tool frame relative to the station frame. In Cartesian-based path-generation schemes, the functions splined together to form a trajectory are functions of time that represent Cartesian variables. These paths can be *planned* directly from the user's definition of path points, which are $\{T\}$ specifications relative to $\{S\}$, without first performing inverse kinematics. However, Cartesian schemes are more computationally expensive to execute, because, at run time, inverse kinematics must be solved at the path-update rate—that is, after the path is generated in Cartesian space, as a last step the inverse kinematic calculation is performed to calculate desired joint angles.

Several schemes for generating Cartesian paths have been proposed in literature from the research and industrial robotics community [1, 2]. In the following section, we introduce one scheme as an example. In this scheme, we are able to use the same linear/parabolic spliner that we developed for the joint-space case.

Cartesian straight-line motion

Often, we would like to be able to specify easily a spatial path that causes the tip of the tool to move through space in a straight line. Obviously, if we specify many closely separated via points lying on a straight line, then the tool tip will appear to follow a straight line, regardless of the choice of smooth function that interconnects the via points. However, it is much more convenient if the tool follows straight-line paths between even widely separated via points. This mode of path specification and execution is called **Cartesian straight-line motion**. Defining motions in terms of straight lines is a subset of the more general capability of **Cartesian motion**, in which arbitrary functions of Cartesian variables as functions of time could be used to specify a path. In a system that allowed general Cartesian motion, such path shapes as ellipses or sinusoids could be executed.

In planning and generating Cartesian straight-line paths, a spline of linear functions with parabolic blends is appropriate. During the linear portion of each segment, all three components of position change in a linear fashion, and the end-effector will move along a linear path in space. However, if we are specifying the orientation as a rotation matrix at each via point, we cannot linearly interpolate its elements, because doing so would not necessarily result in a valid rotation matrix at all times. A rotation matrix must be composed of orthonormal columns, and this condition would not be guaranteed if it were constructed by linear interpolation of matrix elements between two valid matrices. Instead, we will use another representation of orientation.

As stated in Chapter 2, the so-called **angle–axis** representation can be used to specify an orientation with three numbers. If we combine this representation of orientation with the 3×1 Cartesian-position representation, we have a 6×1 representation of Cartesian position and orientation. Consider a via point specified relative to the station frame as $^S_A T$. That is, the frame $\{A\}$ specifies a via point with position of the end-effector given by $^S P_{AORG}$, and orientation of the end-effector given by $^S_A R$. This rotation matrix can be converted to the angle–axis representation $ROT(^S \hat{K}_A, \theta_{SA})$—or simply $^S K_A$. We will use the symbol χ to represent this 6×1 vector of Cartesian position and orientation. Thus, we have

$$^S \chi_A = \begin{bmatrix} ^S P_{AORG} \\ ^S K_A \end{bmatrix}, \tag{7.43}$$

where $^S K_A$ is formed by scaling the unit vector $^S \hat{K}_A$ by the amount of rotation, θ_{SA}. If every path point is specified in this representation, we then need to describe spline functions that smoothly vary these six quantities from path point to path point as functions of time. If linear splines with parabolic blends are used, the path shape between via points will be linear. When via points are passed, the linear and angular velocity of the end-effector are changed smoothly.

Note that, unlike some other Cartesian-straight-line-motion schemes that have been proposed, this method does not guarantee that rotations occur about a single "equivalent axis" in moving from point to point. Rather, our scheme is a simple one that provides smooth orientation changes and allows the use of the same mathematics we have already developed for planning joint-interpolated trajectories.

One slight complication arises from the fact that the angle–axis representation of orientation is not unique—that is,

$$(^S \hat{K}_A, \theta_{SA}) = (^S \hat{K}_A, \theta_{SA} + n360°), \tag{7.44}$$

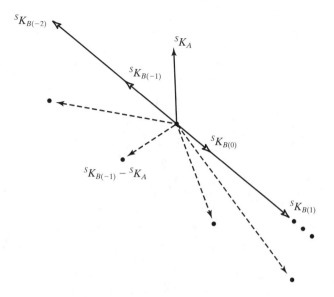

FIGURE 7.11: Choosing angle–axis representation to minimize rotation.

where n is any positive or negative integer. In going from a via point $\{A\}$ to a via point $\{B\}$, the total amount of rotation should be minimized. If our representation of the orientation of $\{A\}$ is given as ^SK_A, we must choose the particular ^SK_B such that $|^SK_B - ^SK_A|$ is minimized. For example, Fig. 7.11 shows four different possible ^SK_B's and their relation to the given ^SK_A. The difference vectors (broken lines) are compared to learn which ^SK_B which will result in minimum rotation—in this case, $^SK_{B(-1)}$.

Once we select the six values of χ for each via point, we can use the same mathematics we have already developed for generating splines that are composed of linear and parabolic sections. However, we must add one more constraint: The blend times for each degree of freedom must be the same. This will ensure that the resultant motion of all the degrees of freedom will be a straight line in space. Because all blend times must be the same, the acceleration used during the blend for each degree of freedom will differ. Hence, we specify a duration of blend, and, using (7.24), we compute the needed acceleration (instead of the other way around). The blend time can be chosen so that a certain upper bound on acceleration is not exceeded.

Many other schemes for representing and interpolating the orientation portion of a Cartesian path can be used. Among these are the use of some of the other 3×1 representations of orientation introduced in Section 2.8. For example, some industrial robots move along Cartesian straight-line paths in which interpolation of orientation is done by means of a representation similar to Z–Y–Z Euler angles.

7.5 GEOMETRIC PROBLEMS WITH CARTESIAN PATHS

Because a continuous correspondence is made between a path shape described in Cartesian space and joint positions, Cartesian paths are prone to various problems relating to workspace and singularities.

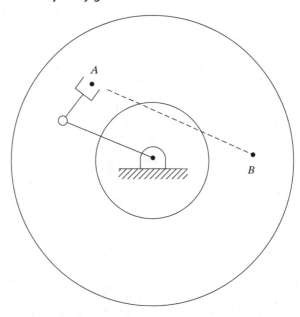

FIGURE 7.12: Cartesian-path problem of type 1.

Problems of type 1: intermediate points unreachable

Although the initial location of the manipulator and the final goal point are both within the manipulator workspace, it is quite possible that not all points lying on a straight line connecting these two points are in the workspace. As an example, consider the planar two-link robot shown in Fig. 7.12 and its associated workspace. In this case, link 2 is shorter than link 1, so the workspace contains a hole in the middle whose radius is the difference between link lengths. Drawn on the workspace is a start point A and a goal point B. Moving from A to B would be no problem in joint space, but if a Cartesian straight-line motion were attempted, intermediate points along the path would not be reachable. This is an example of a situation in which a joint-space path could easily be executed, but a Cartesian straight-line path would fail.[2]

Problems of type 2: high joint rates near singularity

We saw in Chapter 5 that there are locations in the manipulator's workspace where it is impossible to choose finite joint rates that yield the desired velocity of the end-effector in Cartesian space. It should not be surprising, therefore, that there are certain paths (described in Cartesian terms) which are impossible for the manipulator to perform. If, for example, a manipulator is following a Cartesian straight-line path and approaches a singular configuration of the mechanism, one or more joint velocities might increase toward infinity. Because velocities of the

---

[2]Some robot systems would notify the user of a problem before moving the manipulator; in others, motion would start along the path until some joint reaches its limit, at which time manipulator motion would be halted.

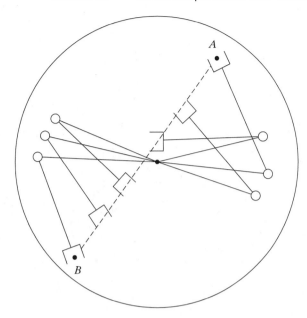

FIGURE 7.13: Cartesian-path problem of type 2.

mechanism are upper bounded, this situation usually results in the manipulator's deviating from the desired path.

As an example, Fig. 7.13 shows a planar two-link (with equal link lengths) moving along a path from point A to point B. The desired trajectory is to move the end tip of the manipulator at constant linear velocity along the straight-line path. In the figure, several intermediate positions of the manipulator have been drawn to help visualize its motion. All points along the path are reachable, but as the robot goes past the middle portion of the path, the velocity of joint one is very high. The closer the path comes to the joint-one axis, the faster this rate will be. One approach is to scale down the overall velocity of the path to a speed where all joints stay within their velocity capabilities. In this way, the desired temporal attributes of the path might be lost, but at least the spatial aspect of the trajectory definition is adhered to.

Problems of type 3: start and goal reachable in different solutions

A third kind of problem that could arise is shown in Fig. 7.14. Here, a planar two-link with equal link lengths has joint limits that restrict the number of solutions with which it can reach a given point in space. In particular, a problem will arise if the goal point cannot be reached in the same physical solution as the robot is in at the start point. In Fig. 7.14, the manipulator can reach all points of the path in some solution, but not in any one solution. In this situation, the manipulator trajectory planning system can detect this problem without ever attempting to move the robot along the path and can signal an error to the user.

To handle these problems with paths specified in Cartesian space, most industrial manipulator-control systems support both joint-space and Cartesian-space

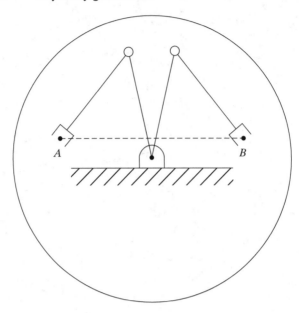

FIGURE 7.14: Cartesian-path problem of type 3.

path generation. The user quickly learns that, because of the difficulties with Cartesian paths, joint-space paths should be used as the default, and Cartesian-space paths should be used only when actually needed by the application.

7.6 PATH GENERATION AT RUN TIME

At **run time**, the path-generator routine constructs the trajectory, usually in terms of θ, $\dot{\theta}$, and $\ddot{\theta}$, and feeds this information to the manipulator's control system. This path generator computes the trajectory at the path-update rate.

Generation of joint-space paths

The result of having planned a path by using any of the splining methods mentioned in Section 7.3 is a set of data for each segment of the trajectory. These data are used by the path generator at run time to calculate θ, $\dot{\theta}$, and $\ddot{\theta}$.

In the case of cubic splines, the path generator simply computes (7.3) as t is advanced. When the end of one segment is reached, a new set of cubic coefficients is recalled, t is set back to zero, and the generation continues.

In the case of linear splines with parabolic blends, the value of time, t, is checked on each update to determine whether we are currently in the linear or the blend portion of the segment. In the linear portion, the trajectory for each joint is calculated as

$$\theta = \theta_j + \dot{\theta}_{jk}t,$$
$$\dot{\theta} = \dot{\theta}_{jk}, \qquad\qquad (7.45)$$
$$\ddot{\theta} = 0,$$

where t is the time since the jth via point and $\dot{\theta}_{jk}$ was calculated at path-planning time from (7.24). In the blend region, the trajectory for each joint is calculated as

$$t_{inb} = t - (\tfrac{1}{2}t_j + t_{jk}),$$
$$\theta = \theta_j + \dot{\theta}_{jk}(t - t_{inb}) + \tfrac{1}{2}\ddot{\theta}_k t_{inb}^2, \tag{7.46}$$
$$\dot{\theta} = \dot{\theta}_{jk} + \ddot{\theta}_k t_{inb},$$
$$\ddot{\theta} = \ddot{\theta}_k,$$

where $\dot{\theta}_{jk}, \ddot{\theta}_k, t_j$, and t_{jk} were calculated at path-planning time by equations (7.24) through (7.28). This continues, with t being reset to $\tfrac{1}{2}t_k$ when a new linear segment is entered, until we have worked our way through all the data sets representing the path segments.

Generation of Cartesian-space paths

For the Cartesian-path scheme presented in Section 7.4, we use the path generator for the linear spline with parabolic blends path. However, the values computed represent the Cartesian position and orientation rather than joint-variable values, so we rewrite (7.45) and (7.46) with the symbol x representing a component of the Cartesian position and orientation vector. In the linear portion of the segment, each degree of freedom in χ is calculated as

$$x = x_j + \dot{x}_{jk}t,$$
$$\dot{x} = \dot{x}_{jk}, \tag{7.47}$$
$$\ddot{x} = 0,$$

where t is the time since the jth via point and $\dot{x}_{jk}$ was calculated at path-planning time by using an equation analogous to (7.24). In the blend region, the trajectory for each degree of freedom is calculated as

$$t_{inb} = t - (\tfrac{1}{2}t_j + t_{jk}),$$
$$x = x_j + \dot{x}_{jk}(t - t_{inb}) + \tfrac{1}{2}\ddot{x}_k t_{inb}^2, \tag{7.48}$$
$$\dot{x} = \dot{x}_{jk} + \ddot{x}_k t_{inb},$$
$$\ddot{x} = \ddot{x}_k,$$

where the quantities $\dot{x}_{jk}, \ddot{x}_k, t_j$, and t_{jk} were computed at plan time, just as in the joint-space case.

Finally, this Cartesian trajectory (χ, $\dot{\chi}$, and $\ddot{\chi}$) must be converted into equivalent joint-space quantities. A complete analytical solution to this problem would use the inverse kinematics to calculate joint positions, the inverse Jacobian for velocities, and the inverse Jacobian plus its derivative for accelerations [5]. A simpler way often used in practice is as follows: At the path-update rate, we convert χ into its equivalent frame representation, $^S_G T$. We then use the SOLVE routine (see Section 4.8) to calculate the required vector of joint angles, Θ. Numerical differentiation is then used to compute $\dot{\Theta}$ and $\ddot{\Theta}$.[3] Thus, the algorithm is

$$\chi \to {}^S_G T,$$

$$\Theta(t) = SOLVE({}^S_G T),$$

$$\dot{\Theta}(t) = \frac{\Theta(t) - \Theta(t - \delta t)}{\delta t}, \qquad (7.49)$$

$$\ddot{\Theta}(t) = \frac{\dot{\Theta}(t) - \dot{\Theta}(t - \delta t)}{\delta t}.$$

Then Θ, $\dot{\Theta}$, and $\ddot{\Theta}$ are supplied to the manipulator's control system.

7.7   DESCRIPTION OF PATHS WITH A ROBOT PROGRAMMING LANGUAGE

In Chapter 12, we will discuss **robot programming languages** further. Here, we will illustrate how various types of paths that we have discussed in this chapter might be specified in a robot language. In these examples, we use the syntax of **AL**, a robot programming language developed at Stanford University [6].

The symbols A, B, C, and D stand for variables of type "frame" in the AL-language examples that follow. These frames specify path points that we will assume to have been taught or textually described to the system. Assume that the manipulator begins in position A. To move the manipulator in joint-space mode along linear-parabolic-blend paths, we could say

```
move ARM to C with duration = 3*seconds;
```
To move to the same position and orientation in a straight line we could say

```
move ARM to C linearly with duration = 3*seconds;
```
where the keyword "linearly" denotes that Cartesian straight-line motion is to be used. If duration is not important, the user can omit this specification, and the system will use a default velocity—that is,

```
move ARM to C;
```
A via point can be added, and we can write

```
move ARM to C via B;
```
or a whole set of via points might be specified by

```
move ARM to C via B,A,D;
```
Note that in

```
move ARM to C via B with duration = 6*seconds;
```
the duration is given for the entire motion. The system decides how to split this duration between the two segments. It is possible in AL to specify the duration of a single segment—for example, by

```
move ARM to C via B where duration = 3*seconds;
```
The first segment which leads to point B will have a duration of 3 seconds.

7.8   PLANNING PATHS WHEN USING THE DYNAMIC MODEL

Usually, when paths are planned, we use a default or a maximum acceleration at each blend point. Actually, the amount of acceleration that the manipulator is capable

---

[3]This differentiation can be done noncausally for preplanned paths, resulting in better-quality $\dot{\Theta}$ and $\ddot{\Theta}$. Also, many control systems do not require a $\ddot{\Theta}$ input, and so it would not be computed.

of at any instant is a function of the dynamics of the arm and the actuator limits. Most actuators are not characterized by a fixed maximum torque or acceleration, but rather by a torque–speed curve.

When we plan a path assuming there is a maximum acceleration at each joint or along each degree of freedom, we are making a tremendous simplification. In order to be careful not to exceed the actual capabilities of the device, this maximum acceleration must be chosen conservatively. Therefore, we are not making full use of the speed capabilities of the manipulator in paths planned by the methods introduced in this chapter.

We might ask the following question: Given a desired spatial path of the end-effector, find the timing information (which turns a description of a spatial path into a trajectory) such that the manipulator reaches the goal point in minimum time. Such problems have been solved by numerical means [7, 8]. The solution takes both the rigid-body dynamics and actuator speed–torque constraint curves into account.

7.9 COLLISION-FREE PATH PLANNING

It would be extremely convenient if we could simply tell the robot system what the desired goal point of the manipulator motion is and let the system determine where and how many via points are required so that the goal is reached without the manipulator's hitting any obstacles. In order to do this, the system must have models of the manipulator, the work area, and all potential obstacles in the area. A second manipulator could even be working in the same area; in that case, each arm would have to be considered a moving obstacle for the other.

Systems that plan collision-free paths are not available commercially. Research in this area has led to two competing principal techniques and to several variations and combinations thereof. One approach solves the problem by forming a connected-graph representation of the free space and then searching the graph for a collision-free path [9–11, 17, 18]. Unfortunately, these techniques have exponential complexity in the number of joints in the device. The second approach is based on creating artificial potential fields around obstacles, which cause the manipulator(s) to avoid the obstacles while they are drawn toward an artificial attractive pole at the goal point [12]. Unfortunately, these methods generally have a local view of the environment and are subject to becoming "stuck" at local minima of the artificial field.

BIBLIOGRAPHY

[1] R.P. Paul and H. Zong, "Robot Motion Trajectory Specification and Generation," 2nd International Symposium on Robotics Research, Kyoto, Japan, August 1984.

[2] R. Taylor, "Planning and Execution of Straight Line Manipulator Trajectories," in *Robot Motion*, Brady et al., Editors, MIT Press, Cambridge, MA, 1983.

[3] C. DeBoor, *A Practical Guide to Splines*, Springer-Verlag, New York, 1978.

[4] D. Rogers and J.A. Adams, *Mathematical Elements for Computer Graphics*, McGraw-Hill, New York, 1976.

[5] B. Gorla and M. Renaud, *Robots Manipulateurs*, Cepadues-Editions, Toulouse, 1984.

[6] R. Goldman, *Design of an Interactive Manipulator Programming Environment*, UMI Research Press, Ann Arbor, MI, 1985.

[7] J. Bobrow, S. Dubowsky, and J. Gibson, "On the Optimal Control of Robotic Manipulators with Actuator Constraints," *Proceedings of the American Control Conference*, June 1983.

[8] K. Shin and N. McKay, "Minimum-Time Control of Robotic Manipulators with Geometric Path Constraints," *IEEE Transactions on Automatic Control*, June 1985.

[9] T. Lozano-Perez, "Spatial Planning: A Configuration Space Approach," AI Memo 605, MIT Artificial Intelligence Laboratory, Cambridge, MA, 1980.

[10] T. Lozano-Perez, "A Simple Motion Planning Algorithm for General Robot Manipulators," *IEEE Journal of Robotics and Automation*, Vol. RA-3, No. 3, June 1987.

[11] R. Brooks, "Solving the Find-Path Problem by Good Representation of Free Space," *IEEE Transactions on Systems, Man, and Cybernetics*, SMC-13:190–197, 1983.

[12] O. Khatib, "Real-Time Obstacle Avoidance for Manipulators and Mobile Robots," *The International Journal of Robotics Research*, Vol. 5, No. 1, Spring 1986.

[13] R.P. Paul, "Robot Manipulators: Mathematics, Programming, and Control," MIT Press, Cambridge, MA, 1981.

[14] R. Castain and R.P. Paul, "An Online Dynamic Trajectory Generator," *The International Journal of Robotics Research*, Vol. 3, 1984.

[15] C.S. Lin and P.R. Chang, "Joint Trajectory of Mechanical Manipulators for Cartesian Path Approximation," *IEEE Transactions on Systems, Man, and Cybernetics*, Vol. SMC-13, 1983.

[16] C.S. Lin, P.R. Chang, and J.Y.S. Luh, "Formulation and Optimization of Cubic Polynomial Joint Trajectories for Industrial Robots," *IEEE Transactions on Automatic Control*, Vol. AC-28, 1983.

[17] L. Kavraki, P. Svestka, J.C. Latombe, and M. Overmars, "Probabilistic Roadmaps for Path Planning in High-Dimensional Configuration Spaces," *IEEE Transactions on Robotics and Automation*, 12(4): 566–580, 1996.

[18] J. Barraquand, L. Kavraki, J.C. Latombe, T.Y. Li, R. Motwani, and P. Raghavan, "A Random Sampling Scheme for Path Planning," *International Journal of Robotics Research*, 16(6): 759–774, 1997.

EXERCISES

7.1 [8] How many individual cubics are computed when a six-jointed robot moves along a cubic spline path through two via points and stops at a goal point? How many coefficients are stored to describe these cubics?

7.2 [13] A single-link robot with a rotary joint is motionless at $\theta = -5°$. It is desired to move the joint in a smooth manner to $\theta = 80°$ in 4 seconds. Find the coefficients of a cubic which accomplishes this motion and brings the arm to rest at the goal. Plot the position, velocity, and acceleration of the joint as a function of time.

7.3 [14] A single-link robot with a rotary joint is motionless at $\theta = -5°$. It is desired to move the joint in a smooth manner to $\theta = 80°$ in 4 seconds and stop smoothly. Compute the corresponding parameters of a linear trajectory with parabolic blends. Plot the position, velocity, and acceleration of the joint as a function of time.

7.4 [30] Write a path-planning software routine that implements (7.24) through (7.28) in a general way for paths described by an arbitrary number of path points. For example, this routine could be used to solve Example 7.4.

7.5 [18] Sketch graphs of position, velocity, and acceleration for the two-segment continuous-acceleration spline given in Example 7.2. Sketch them for a joint for which $\theta_0 = 5.0°, \theta_v = 15.0°, \theta_g = 40.0°$, and each segment lasts 1.0 second.

7.6 [18] Sketch graphs of position, velocity, and acceleration for a two-segment spline where each segment is a cubic, using the coefficients as given in (7.11). Sketch them for a joint where $\theta_0 = 5.0°$ for the initial point, $\theta_v = 15.0°$ is a via point, and $\theta_g = 40.0°$ is the goal point. Assume that each segment has a duration of 1.0 second and that the velocity at the via point is to be 17.5 degrees/second.

7.7 [20] Calculate $\dot{\theta}_{12}, \dot{\theta}_{23}, t_1, t_2$, and t_3 for a two-segment linear spline with parabolic blends. (Use (7.24) through (7.28).) For this joint, $\theta_1 = 5.0°, \theta_2 = 15.0°, \theta_3 = 40.0°$. Assume that $t_{d12} = t_{d23} = 1.0$ second and that the default acceleration to use during blends is 80 degrees/second^2. Sketch plots of position, velocity, and acceleration of θ.

7.8 [18] Sketch graphs of position, velocity, and acceleration for the two-segment continuous-acceleration spline given in Example 7.2. Sketch them for a joint for which $\theta_0 = 5.0°, \theta_v = 15.0°, \theta_g = -10.0°$, and each segment lasts 2.0 seconds.

7.9 [18] Sketch graphs of position, velocity, and acceleration for a two-segment spline where each segment is a cubic, using the coefficients as given in (7.11). Sketch them for a joint where $\theta_0 = 5.0°$ for the initial point, $\theta_v = 15.0°$ is a via point, and $\theta_g = -10.0°$ is the goal point. Assume that each segment has a duration of 2.0 seconds and that the velocity at the via point is to be 0.0 degrees/second.

7.10 [20] Calculate $\dot{\theta}_{12}, \dot{\theta}_{23}, t_1, t_2$, and t_3 for a two-segment linear spline with parabolic blends. (Use (7.24) through (7.28).) For this joint, $\theta_1 = 5.0°, \theta_2 = 15.0°, \theta_3 = -10.0°$. Assume that $t_{d12} = t_{d23} = 2.0$ seconds and that the default acceleration to use during blends is 60 degrees/second^2. Sketch plots of position, velocity, and acceleration of θ.

7.11 [6] Give the 6×1 Cartesian position and orientation representation $^S\chi_G$ that is equivalent to ^S_GT where $^S_GR = ROT(\hat{Z}, 30°)$ and $^SP_{GORG} = [10.0 \ 20.0 \ 30.0]^T$.

7.12 [6] Give the ^S_GT that is equivalent to the 6×1 Cartesian position and orientation representation $^S\chi_G = [5.0 \ -20.0 \ 10.0 \ 45.0 \ 0.0 \ 0.0]^T$.

7.13 [30] Write a program that uses the dynamic equations from Section 6.7 (the two-link planar manipulator) to compute the time history of torques needed to move the arm along the trajectory of Exercise 7.8. What are the maximum torques required and where do they occur along the trajectory?

7.14 [32] Write a program that uses the dynamic equations from Section 6.7 (the two-link planar manipulator) to compute the time history of torques needed to move the arm along the trajectory of Exercise 7.8. Make separate plots of the joint torques required due to inertia, velocity terms, and gravity.

7.15 [22] Do Example 7.2 when $t_{f1} \neq t_{f2}$.

7.16 [25] We wish to move a single joint from θ_0 to θ_f starting from rest, ending at rest, in time t_f. The values of θ_0 and θ_f are given, but we wish to compute t_f so that $\|\dot{\theta}(t)\| < \dot{\theta}_{max}$ and $\|\ddot{\theta}(t)\| < \ddot{\theta}_{max}$ for all t, where $\dot{\theta}_{max}$ and $\ddot{\theta}_{max}$ are given positive constants. Use a single cubic segment, and give an expression for t_f and for the cubic's coefficients.

7.17 [10] A single cubic trajectory is given by

$$\theta(t) = 10 + 90t^2 - 60t^3$$

and is used over the time interval from $t = 0$ to $t = 1$. What are the starting and final positions, velocities, and accelerations?

7.18 [12] A single cubic trajectory is given by

$$\theta(t) = 10 + 90t^2 - 60t^3$$

and is used over the time interval from $t = 0$ to $t = 2$. What are the starting and final positions, velocities, and accelerations?

7.19 [13] A single cubic trajectory is given by

$$\theta(t) = 10 + 5t + 70t^2 - 45t^3$$

and is used over the time interval from $t = 0$ to $t = 1$. What are the starting and final positions, velocities, and accelerations?

7.20 [15] A single cubic trajectory is given by

$$\theta(t) = 10 + 5t + 70t^2 - 45t^3$$

and is used over the time interval from $t = 0$ to $t = 2$. What are the starting and final positions, velocities, and accelerations?

PROGRAMMING EXERCISE (PART 7)

1. Write a joint-space, cubic-splined path-planning system. One routine that your system should include is

   ```
   Procedure CUBCOEF (VAR th0, thf, thdot0, thdotf: real; VAR cc:
   vec4);
   ```

   where

   $$\text{th0} = \text{initial position of } \theta \text{ at beginning of segment,}$$
   $$\text{thf} = \text{final position of } \theta \text{ at segment end,}$$
   $$\text{thdot0} = \text{initial velocity of segment,}$$
   $$\text{thdotf} = \text{final velocity of segment.}$$

   These four quantities are inputs, and "cc", an array of the four cubic coefficients, is the output.

   Your program should accept up to (at least) five via-point specifications—in the form of tool frame, $\{T\}$, relative to station frame, $\{S\}$—in the usual user form: (x, y, ϕ). To keep life simple, all segments will have the same duration. Your system should solve for the coefficients of the cubics, using some reasonable heuristic for assigning joint velocities at the via points. *Hint*: See option 2 in Section 7.3.

2. Write a path-generator system that calculates a trajectory in joint space based on sets of cubic coefficients for each segment. It must be able to generate the multisegment path you planned in Problem 1. A duration for the segments will be specified by the user. It should produce position, velocity, and acceleration information at the path-update rate, which will also be specified by the user.

3. The manipulator is the same three-link used previously. The definitions of the $\{T\}$ and $\{S\}$ frames are the same as before:

   $$^W_T T = [x \ y \ \theta] = [0.1 \ 0.2 \ 30.0],$$
   $$^B_S T = [x \ y \ \theta] = [0.0 \ 0.0 \ 0.0].$$

Using a duration of 3.0 seconds per segment, plan and execute the path that starts with the manipulator at position

$$[x_1 \ y_1 \ \phi_1] = [0.758 \ 0.173 \ 0.0],$$

moves through the via points

$$[x_2 \ y_2 \ \phi_2] = [0.6 \ -0.3 \ 45.0]$$

and

$$[x_3 \ y_3 \ \phi_3] = [-0.4 \ 0.3 \ 120.0],$$

and ends at the goal point (in this case, same as initial point)

$$[x_4 \ y_4 \ \phi_4] = [0.758 \ 0.173 \ 0.0].$$

Use a path-update rate of 40 Hz, but print the position only every 0.2 seconds. Print the positions out in terms of Cartesian user form. You don't have to print out velocities or accelerations, though you might be interested in doing so.

MATLAB EXERCISE 7

The goal of this exercise is to implement polynomial joint-space trajectory-generation equations for a single joint. (Multiple joints would require n applications of the result.) Write a MATLAB program to implement the joint-space trajectory generation for the three cases that follow. Report your results for the specific assignments given; for each case, give the polynomial functions for the joint angle, angular velocity, angular acceleration, and angular jerk (the time derivative of acceleration). For each case, plot the results. (Arrange the plots vertically with angle, velocity, acceleration, and then jerk, all with the same time scale—check out the *subplot* MATLAB function to accomplish this.) Don't just plot out results—give some discussion; do your results make sense? Here are the three cases:

a) *Third-order polynomial.* Force the angular velocity to be zero at the start and at the finish. Given $\theta_s = 120°$ (start), $\theta_f = 60°$ (finish), and $t_f = 1$ sec.

b) *Fifth-order polynomial.* Force the angular velocity and acceleration to be zero at the start and at the finish. Given $\theta_s = 120°, \theta_f = 60°$, and $t_f = 1$ sec. Compare your results (functions and plots) with this same example, but using a single third-order polynomial, as in problem (a).

c) *Two third-order polynomials with via point.* Force the angular velocity to be zero at the start and at the finish. Don't force the angular velocity to be zero at the via point—you must match velocity and acceleration at this point for the two polynomials meeting at that point in time. Demonstrate that this condition is satisfied. Given $\theta_s = 60°$ (start), $\theta_v = 120°$ (via), $\theta_f = 30°$ (finish), and $t_1 = t_2 = 1$ sec (relative time steps—i.e., $t_f = 2$ sec).

d) Check the results of (a) and (b) by means of the Corke MATLAB Robotics Toolbox. Try function *jtraj()*.

CHAPTER 8

Manipulator-mechanism design

8.1 INTRODUCTION

In previous chapters, we have seen that the particular structure of a manipulator influences kinematic and dynamic analysis. For example, some kinematic configurations will be easy to solve; others will have no closed-form kinematic solution. Likewise, the complexity of the dynamic equations can vary greatly with the kinematic configuration and the mass distribution of the links. In coming chapters, we will see that manipulator control depends not only on the rigid-body dynamics, but also upon the friction and flexibility of the drive systems.

The tasks that a manipulator can perform will also vary greatly with the particular design. Although we have generally dealt with the robot manipulator as an abstract entity, its performance is ultimately limited by such pragmatic factors as load capacity, speed, size of workspace, and repeatability. For certain applications, the overall manipulator size, weight, power consumption, and cost will be significant factors.

This chapter discusses some of the issues involved in the design of the manipulator. In general, methods of design and even the evaluation of a finished design are partially subjective topics. It is difficult to narrow the spectrum of design choices with many hard and fast rules.

The elements of a robot system fall roughly into four categories:

1. The manipulator, including its internal or **proprioceptive** sensors;
2. the end-effector, or **end-of-arm tooling**;
3. external sensors and effectors, such as vision systems and part feeders; and
4. the controller.

The breadth of engineering disciplines encompassed forces us to restrict our attention only to the design of the manipulator itself.

In developing a manipulator design, we will start by examining the factors likely to have the greatest overall effect on the design and then consider more detailed questions. Ultimately, however, designing a manipulator is an iterative process. More often than not, problems that arise in the solving of a design detail will force rethinking of previous higher level design decisions.

8.2 BASING THE DESIGN ON TASK REQUIREMENTS

Although robots are nominally "universally programmable" machines capable of performing a wide variety of tasks, economies and practicalities dictate that different manipulators be designed for particular types of tasks. For example, large robots capable of handling payloads of hundreds of pounds do not generally have the capability to insert electronic components into circuit boards. As we shall see, not only the size, but the number of joints, the arrangement of the joints, and the types of actuation, sensing, and control will all vary greatly with the sort of task to be performed.

Number of degrees of freedom

The number of degrees of freedom in a manipulator should match the number required by the task. Not all tasks require a full six degrees of freedom.

The most common such circumstance occurs when the end-effector has an axis of symmetry. Figure 8.1 shows a manipulator positioning a grinding tool in two different ways. In this case, the orientation of the tool with respect to the axis of the tool, $\hat{Z}_T$, is immaterial, because the grinding wheel is spinning at several hundred RPM. To say that we can position this 6-DOF robot in an infinity of ways for this task (rotation about $\hat{Z}_T$ is a free variable), we say that the robot is **redundant** for this task. Arc welding, spot welding, deburring, glueing, and polishing provide other examples of tasks that often employ end-effectors with at least one axis of symmetry.

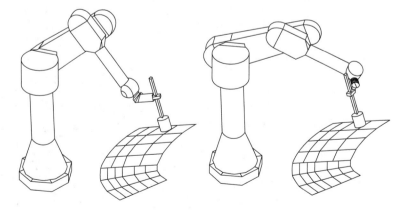

FIGURE 8.1: A 6-DOF manipulator with a symmetric tool contains a redundant degree of freedom.

In analyzing the symmetric-tool situation, it is sometimes helpful to imagine a *fictitious joint* whose axis lies along the axis of symmetry. In positioning any end-effector to a specific pose, we need a total of six degrees of freedom. Because one of these six is our fictitious joint, the actual manipulator need not have more than five degrees of freedom. If a 5-DOF robot were used in the application of Fig. 8.1, then we would be back to the usual case in which only a finite number of different solutions are available for positioning the tool. Quite a large percentage of existing industrial robots are 5-DOF, in recognition of the relative prevalence of symmetric-tool applications.

Some tasks are performed in domains that, fundamentally, have fewer than six degrees of freedom. Placement of components on circuit boards provides a common example of this. Circuit boards generally are planar and contain parts of various heights. Positioning parts on a planar surface requires three degrees of freedom (x, y, and θ); in order to lift and insert the parts, a fourth motion normal to the plane is added (z).

Robots with fewer than six degrees of freedom can also perform tasks in which some sort of active positioning device presents the parts. In welding pipes, for example, a tilt/roll platform, shown in Fig. 8.2, often presents the parts to be welded. In counting the number of degrees of freedom between the pipes and the end-effector, the tilt/roll platform accounts for two. This, together with the fact that arc welding is a symmetric-tool task, means that, in theory, a 3-DOF manipulator could be used. In practice, realities such as the need to avoid collisions with the workpiece generally dictate the use of a robot with more degrees of freedom.

Parts with an axis of symmetry also reduce the required degrees of freedom for the manipulator. For example, cylindrical parts can in many cases be picked up and inserted independent of the orientation of the gripper with respect to the axis of the cylinder. Note, however, that after the part is grasped, the orientation of the part about its symmetric axis must fail to matter for *all* subsequent operations, because its orientation is not guaranteed.

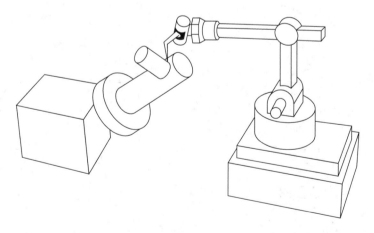

FIGURE 8.2: A tilt/roll platform provides two degrees of freedom to the overall manipulator system.

Workspace

In performing tasks, a manipulator has to reach a number of workpieces or fixtures. In some cases, these can be positioned as needed to suit the workspace of the manipulator. In other cases, a robot can be installed in a fixed environment with rigid workspace requirements. **Workspace** is also sometimes called **work volume** or **work envelope**.

The overall scale of the task sets the required workspace of the manipulator. In some cases, the details of the shape of the workspace and the location of workspace singularities will be important considerations.

The intrusion of the manipulator itself in the workspace can sometimes be a factor. Depending on the kinematic design, operating a manipulator in a given application could require more or less space around the fixtures in order to avoid collisions. Restricted environments can affect the choice of kinematic configuration.

Load capacity

The **load capacity** of a manipulator depends upon the sizing of its structural members, power-transmission system, and actuators. The load placed on actuators and drive system is a function of the configuration of the robot, the percentage of time supporting a load, and dynamic loading due to inertial- and velocity-related forces.

Speed

An obvious goal in design has been for faster and faster manipulators. High speed offers advantages in many applications when a proposed robotic solution must compete on economic terms with hard automation or human workers. For some applications, however, the process itself limits the speed rather than the manipulator. This is the case with many welding and spray-painting applications.

An important distinction is that between the maximum end-effector speed and the overall **cycle time** for a particular task. For pick-and-place applications, the manipulator must accelerate and decelerate to and from the *pick* and *place* locations within some positional accuracy bounds. Often, the acceleration and deceleration phases take up most of the cycle time. Hence, acceleration capability, not just peak speed, is very important.

Repeatability and accuracy

High repeatability and accuracy, although desirable in any manipulator design, are expensive to achieve. For example, it would be absurd to design a paint-spraying robot to be accurate to within 0.001 inches when the spray spot diameter is 8 inches ±2 inches. To a large extent, accuracy of a particular model of industrial robot depends upon the details of its manufacture rather than on its design. High accuracy is achieved by having good knowledge of the link (and other) parameters. Making it possible are accurate measurements after manufacture or careful attention to tolerances during manufacturing.

8.3 KINEMATIC CONFIGURATION

Once the required number of degrees of freedom has been decided upon, a particular configuration of joints must be chosen to realize those freedoms. For serial kinematic linkages, the number of joints equals the required number of degrees of freedom. Most manipulators are designed so that the last $n-3$ joints orient the end-effector and have axes that intersect at the **wrist point**, and the first three joints position this wrist point. Manipulators with this design could be said to be composed of a **positioning structure** followed by an **orienting structure** or **wrist**. As we saw in Chapter 4, these manipulators always have closed-form kinematic solutions. Although other configurations exist that possess closed-form kinematic solutions, almost every industrial manipulator belongs to this **wrist-partitioned** class of mechanisms. Furthermore, the positioning structure is almost without exception designed to be kinematically simple, having link twists equal to $0°$ or $±90°$ and having many of the link lengths and offsets equal to zero.

It has become customary to classify manipulators of the wrist-partitioned, kinematically simple class according to the design of their first three joints (the positioning structure). The following paragraphs briefly describe the most common of these classifications.

Cartesian

A **Cartesian manipulator** has perhaps the most straightforward configuration. As shown in Fig. 8.3, joints 1 through 3 are prismatic, mutually orthogonal, and correspond to the $\hat{X}$, $\hat{Y}$, and $\hat{Z}$ Cartesian directions. The inverse kinematic solution for this configuration is trivial.

This configuration produces robots with very stiff structures. As a consequence, very large robots can be built. These large robots, often called **gantry robots**, resemble overhead gantry cranes. Gantry robots sometimes manipulate entire automobiles or inspect entire aircraft.

The other advantages of Cartesian manipulators stem from the fact that the first three joints are *decoupled*. This makes them simpler to design and prevents kinematic singularities due to the first three joints.

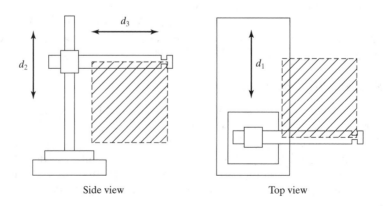

Side view   Top view

FIGURE 8.3: A Cartesian manipulator.

Their primary disadvantage is that all of the feeders and fixtures associated with an application must lie "inside" the robot. Consequently, application workcells for Cartesian robots become very machine dependent. The size of the robot's support structure limits the size and placement of fixtures and sensors. These limitations make retrofitting Cartesian robots into existing workcells extremely difficult.

Articulated

Figure 8.4 shows an **articulated manipulator**, sometimes also called a **jointed, elbow**, or **anthropomorphic** manipulator. A manipulator of this kind typically consists of two "shoulder" joints (one for rotation about a vertical axis and one for elevation out of the horizontal plane), an "elbow" joint (whose axis is usually parallel to the shoulder elevation joint), and two or three wrist joints at the end of the manipulator. Both the PUMA 560 and the Motoman L-3, which we studied in earlier chapters, fall into this class.

Articulated robots minimize the intrusion of the manipulator structure into the workspace, making them capable of reaching into confined spaces. They require much less overall structure than Cartesian robots, making them less expensive for applications needing smaller workspaces.

SCARA

The **SCARA**[1] configuration, shown in Fig. 8.5, has three parallel revolute joints (allowing it to move and orient in a plane), with a fourth prismatic joint for moving the end-effector normal to the plane. The chief advantage is that the first three joints don't have to support any of the weight of the manipulator or the load. In addition, link 0 can easily house the actuators for the first two joints. The actuators can be made very large, so the robot can move very fast. For example, the Adept

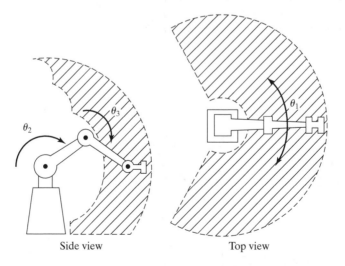

Side view   Top view

FIGURE 8.4: An articulated manipulator.

---

[1]SCARA stands for "selectively compliant assembly robot arm."

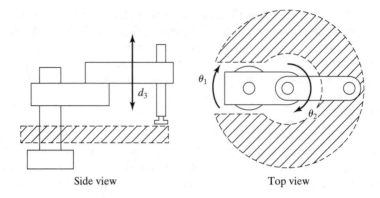

FIGURE 8.5: A SCARA manipulator.

One SCARA manipulator can move at up to 30 feet per second, about 10 times faster than most articulated industrial robots [1]. This configuration is best suited to planar tasks.

Spherical

The spherical configuration in Fig. 8.6 has many similarities to the articulated manipulator, but with the elbow joint replaced by a prismatic joint. This design is better suited to some applications than is the elbow design. The link that moves prismatically might telescope—or even "stick out the back" when retracted.

Cylindrical

Cylindrical manipulators (Fig. 8.7) consist of a prismatic joint for translating the arm vertically, a revolute joint with a vertical axis, another prismatic joint orthogonal to the revolute joint axis, and, finally, a wrist of some sort.

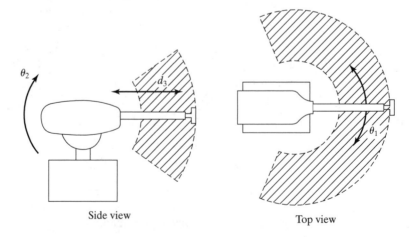

FIGURE 8.6: A spherical manipulator.

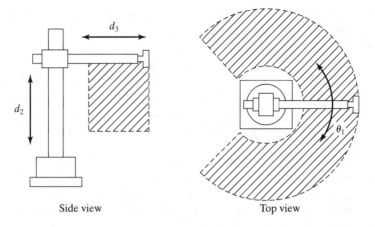

Side view   Top view

FIGURE 8.7: A cylindrical manipulator.

Wrists

The most common wrist configurations consist of either two or three revolute joints with orthogonal, intersecting axes. The first of the wrist joints usually forms joint 4 of the manipulator.

A configuration of three orthogonal axes will guarantee that any orientation can be achieved (assuming no joint-angle limits) [2]. As was stated in Chapter 4, any manipulator with three consecutive intersecting axes will possess a closed-form kinematic solution. Therefore, a three-orthogonal-axis wrist can be located at the end of the manipulator in any desired orientation with no penalty. Figure 8.8 is a schematic of one possible design of such a wrist, which uses several sets of bevel gears to drive the mechanism from remotely located actuators.

In practice, it is difficult to build a three-orthogonal-axis wrist not subject to rather severe joint-angle limitations. An interesting design used in several robots

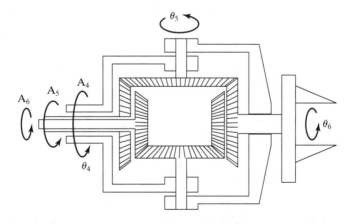

FIGURE 8.8: An orthogonal-axis wrist driven by remotely located actuators via three concentric shafts.

manufactured by Cincinatti Milacron (Fig. 1.4) employs a wrist that has three intersecting but nonorthogonal axes. In this design (called the "three roll wrist"), all three joints of the wrist can rotate continuously without limits. The nonorthogonality of the axes creates, however, a set of orientations that are impossible to reach with this wrist. This set of unattainable orientations is described by a cone within which the third axis of the wrist cannot lie. (See Exercise 8.11.) However, the wrist can be mounted to link 3 of the manipulator in such a way that the link structure occupies this cone and so would be block access anyway. Figure 8.9 shows two drawings of such a wrist [24].

Some industrial robots have wrists that do *not* have intersecting axes. This implies that a closed-form kinematic solution might not exist. If, however, the wrist is mounted on an articulated manipulator in such a way that the joint-4 axis is parallel to the joint-2 and -3 axes, as in Fig. 8.10, there *will* be a closed-form kinematic solution. Likewise, a nonintersecting-axis wrist mounted on a Cartesian robot yields a closed-form-solvable manipulator.

Typically, 5-DOF welding robots use two-axis wrists oriented as shown in Fig. 8.11. Note that, if the robot has a symmetric tool, this "fictitious joint" must follow the rules of wrist design. That is, in order to reach all orientations, the tool must be mounted with its axis of symmetry orthogonal to the joint-5 axis. In the worst case, when the axis of symmetry is parallel to the joint-5 axis, the fictitious sixth axis is in a permanently singular configuration.

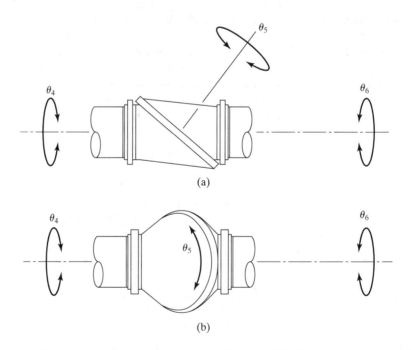

(a)

(b)

FIGURE 8.9: Two views of a nonorthogonal-axis wrist [24]. From *International Encyclopedia of Robotics*, by R. Dorf and S. Nof (editors). From "Wrists" by M. Rosheim, John C. Wiley and Sons, Inc., New York, NY ©1988. Reprinted by permission.

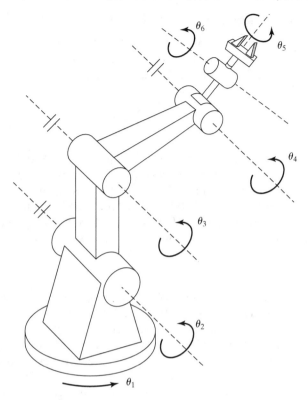

FIGURE 8.10: A manipulator with a wrist whose axes do not intersect. However, this robot does possess a closed-form kinematic solution.

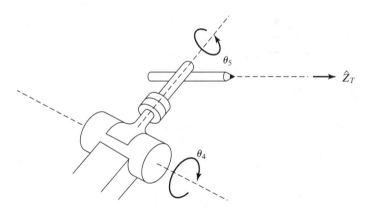

FIGURE 8.11: Typical wrist design of a 5-DOF welding robot.

8.4   QUANTITATIVE MEASURES OF WORKSPACE ATTRIBUTES

Manipulator designers have proposed several interesting quantitative measures of various workspace attributes.

Efficiency of design in terms of generating workspace

Some designers noticed that it seemed to take more material to build a Cartesian manipulator than to build an articulated manipulator of similar workspace volume. To get a quantitative handle on this, we first define the **length sum** of a manipulator as

$$L = \sum_{i=1}^{N}(a_{i-1} + d_i), \tag{8.1}$$

where a_{i-1} and d_i are the link length and joint offset as defined in Chapter 3. Thus, the length sum of a manipulator gives a rough measure of the "length" of the complete linkage. Note that, for prismatic joints, d_i must here be interpreted as a constant equal to the length of travel between the joint-travel limits.

In [3], the **structural length index**, Q_L, is defined as the ratio of the manipulator's length sum to the cube root of the workspace volume—that is,

$$Q_L = L/\sqrt[3]{w}, \tag{8.2}$$

where L is given in (8.1) and W is the volume of the manipulator's workspace. Hence, Q_L attempts to index the relative amount of structure (linkage length) required by different configurations to generate a given work volume. Thus, a good design would be one in which a manipulator with a small length sum nonetheless possessed a large workspace volume. Good designs have a low Q_L.

Considering just the positioning structure of a Cartesian manipulator (and therefore the workspace of the wrist point), the value of Q_L is minimized when all three joints have the same length of travel. This minimal value is $Q_L = 3.0$. On the other hand, an ideal articulated manipulator, such as the one in Fig. 8.4, has $Q_L = \frac{1}{\sqrt[3]{4\pi/3}} \cong 0.62$. This helps quantify our earlier statement that articulated manipulators are superior to other configurations in that they have minimal intrusion into their own workspace. Of course, in any real manipulator structure, the figure just given would be made somewhat larger by the effect of joint limits in reducing the workspace volume.

EXAMPLE 8.1

A SCARA manipulator like that of Fig. 8.5 has links 1 and 2 of equal length $l/2$, and the range of motion of the prismatic joint 3 is given by d_3. Assume for simplicity that the joint limits are absent, and find Q_L. What value of d_3 minimizes Q_L and what is this minimal value?

The length sum of this manipulator is $L = l/2 + l/2 + d_3 = l + d_3$, and the workspace volume is that of a right cylinder of radius l and height d_3; therefore,

$$Q_L = \frac{l + d_3}{\sqrt[3]{\pi l^2 d_3}}. \tag{8.3}$$

Minimizing Q_L as a function of the ratio d_3/l gives $d_3 = l/2$ as optimal [3]. The corresponding minimal value of Q_L is 1.29.

Designing well-conditioned workspaces

At singular points, a manipulator effectively loses one or more degrees of free-dom, so certain tasks may not be able to be performed at that point. In fact, in the neighborhood of singular points (including workspace-boundary singularities), actions of the manipulator could fail to be **well-conditioned**. In some sense, the farther the manipulator is away from singularities, the better able it is to move uniformly and apply forces uniformly in all directions. Several measures have been suggested for quantifying this effect. The use of such measures at design time might yield a manipulator design with a maximally large well-conditioned subspace of the workspace.

Singular configurations are given by

$$\det(J(\Theta)) = 0, \tag{8.4}$$

so it is natural to use the determinant of the Jacobian in a measure of manipulator dexterity. In [4], the **manipulability measure**, w, is defined as

$$w = \sqrt{\det(J(\Theta)J^T(\Theta))}, \tag{8.5}$$

which, for a nonredundant manipulator, reduces to

$$w = |\det(J(\Theta))|. \tag{8.6}$$

A good manipulator design has large areas of its workspace characterized by high values of w.

Whereas velocity analysis motivated (8.6), other researchers have proposed manipulability measures based on acceleration analysis or force-application capa-bility. Asada [5] suggested an examination of the eigenvalues of the Cartesian mass matrix

$$M_x(\Theta) = J^{-T}(\Theta)M(\Theta)J^{-1}(\Theta) \tag{8.7}$$

as a measure of how well the manipulator can accelerate in various Cartesian directions. He suggests a graphic representation of this measure as an **inertia ellipsoid**, given by

$$X^T M_x(\Theta)X = 1, \tag{8.8}$$

the equation of an n-dimensional ellipse, where n is the dimension of X. The axes of the ellipsoid given in (8.8) lie in the directions of the eigenvectors of $M_x(\Theta)$, and the reciprocals of the square roots of the corresponding eigenvalues provide the lengths of the axes of the ellipsoid. Well-conditioned points in the manipulator workspace are characterized by inertia ellipsoids that are spherical (or nearly so).

Figure 8.12 shows graphically the properties of a planar two-link manipulator. In the center of the workspace, the manipulator is well conditioned, as is indicated by nearly circular ellipsoids. At workspace boundaries, the ellipses flatten, indicating the manipulator's difficulty in accelerating in certain directions.

Other measures of workspace conditioning have been proposed in [6–8, 25].

8.5   REDUNDANT AND CLOSED-CHAIN STRUCTURES

In general, the scope of this book is limited to manipulators that are serial-chain linkages of six or fewer joints. In this section, however, we briefly discuss manipulators outside of this class.

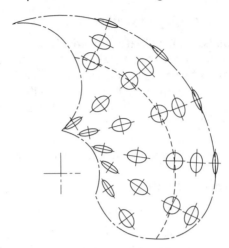

FIGURE 8.12: Workspace of a 2-DOF planar arm, showing inertia ellipsoids, from [5] (© 1984 IEEE). The dashed line indicates a locus of isotropic points in the workspace. Reprinted by permission.

Micromanipulators and other redundancies

General spatial positioning capability requires only six degrees of freedom, but there are advantages to having even more controllable joints.

One use for these extra freedoms is already finding some practical application [9,10] and is of growing interest in the research community: a **micromanipulator**. A micromanipulator is generally formed by several fast, precise degrees of freedom located near the distal end of a "conventional" manipulator. The conventional manipulator takes care of large motions, while the micromanipulator, whose joints generally have a small range of motion, accomplishes fine motion and force control.

Additional joints can also help a mechanism avoid singular configurations, as is suggested in [11, 12]. For example, any three-degree-of-freedom wrist will suffer from singular configurations (when all three axes lie in a plane), but a four-degree-of-freedom wrist can effectively avoid such configurations [13–15].

Figure 8.13 shows two configurations suggested [11, 12] for seven-degree-of-freedom manipulators.

A major potential use of redundant robots is in avoiding collisions while operating in cluttered work environments. As we have seen, a six-degree-of-freedom manipulator can reach a given position and orientation in only a finite number of ways. The addition of a seventh joint allows an infinity of ways, permitting the desire to avoid obstacles to influence the choice.

Closed-loop structures

Although we have considered only serial-chain manipulators in our analysis, some manipulators contain **closed-loop structures**. For example, the Motoman L-3 robot described in Chapters 3 and 4 possesses closed-loop structures in the drive mechanism of joints 2 and 3. Closed-loop structures offer a benefit: increased stiffness of

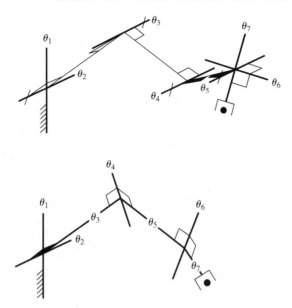

FIGURE 8.13: Two suggested seven-degree-of-freedom manipulator designs [3].

the mechanism [16]. On the other hand, closed-loop structures generally reduce the allowable range of motion of the joints and thus decrease the workspace size.

Figure 8.14 depicts a **Stewart mechanism**, a closed-loop alternative to the serial 6-DOF manipulator. The position and orientation of the "end-effector" is controlled by the lengths of the six linear actuators which connect it to the base. At the base end, each actuator is connected by a two-degree-of-freedom universal joint. At the end-effector, each actuator is attached with a three-degree-of-freedom ball-and-socket joint. It exhibits characteristics common to most closed-loop mechanisms: it can be made very stiff, but the links have a much more limited range of motion than do serial linkages. The Stewart mechanism, in particular, demonstrates an interesting reversal in the nature of the forward and inverse kinematic solutions: the inverse solution is quite simple, whereas the forward solution is typically quite complex, sometimes lacking a closed-form formulation. (See Exercises 8.7 and 8.12.)

In general, the number of degrees of freedom of a closed-loop mechanism is not obvious. The total number of degrees of freedom can be computed by means of **Grübler's** formula [17],

$$F = 6(l - n - 1) + \sum_{i=1}^{n} f_i,$$ (8.9)

where F is the total number of degrees of freedom in the mechanism, l is the number of links (including the base), n is the total number of joints, and f_i is the number of degrees of freedom associated with the ith joint. A planar version of Grübler's formula (when all objects are considered to have three degrees of freedom if unconstrained) is obtained by replacing the 6 in (8.9) with a 3.

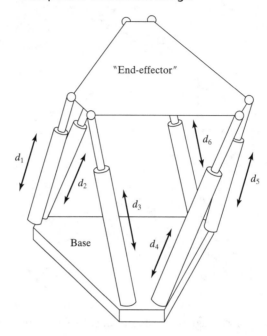

FIGURE 8.14: The Stewart mechanism is a six-degree-of-freedom fully parallel manipulator.

EXAMPLE 8.2

Use Grübler's formula to verify that the Stewart mechanism (Fig. 8.14) indeed has six degrees of freedom.

The number of joints is 18 (6 universal, 6 ball and socket, and 6 prismatic in the actuators). The number of links is 14 (2 parts for each actuator, the end-effector, and the base). The sum of all the joint freedoms is 36. Using Grübler's formula, we can verify that the total number of degrees of freedom is six:

$$F = 6(14 - 18 - 1) + 36 = 6. \tag{8.10}$$

8.6 ACTUATION SCHEMES

Once the general kinematic structure of a manipulator has been chosen, the next most important matter of concern is the actuation of the joints. Typically, the actuator, reduction, and transmission are closely coupled and must be designed together.

Actuator location

The most straightforward choice of actuator location is at or near the joint it drives. If the actuator can produce enough torque or force, its output can attach directly to the joint. This arrangement, known as a **direct-drive** configuration [18], offers the advantages of simplicity in design and superior controllability—that is, with no transmission or reduction elements between the actuator and the joint, the joint motions can be controlled with the same fidelity as the actuator itself.

Unfortunately, many actuators are best suited to relatively high speeds and low torques and therefore require a **speed-reduction system**. Furthermore, actuators tend to be rather heavy. If they can be located remotely from the joint and toward the base of the manipulator, the overall inertia of the manipulator can be reduced considerably. This, in turn, reduces the size needed for the actuators. To realize these benefits, a **transmission system** is needed to transfer the motion from the actuator to the joint.

In a joint-drive system with a remotely mounted actuator, the reduction system could be placed either at the actuator or at the joint. Some arrangements combine the functions of transmission and reduction. Aside from added complexity, the major disadvantage of reduction and transmission systems is that they introduce additional friction and flexibility into the mechanism. When the reduction is at the joint, the transmission will be working at higher speeds and lower torques. Lower torque means that flexibility will be less of a problem. However, if the weight of the reducer is significant, some of the advantage of remotely mounted actuators is lost.

In Chapter 3, details were given for the actuation scheme of the Yasukawa Motoman L-3, which is typical of a design in which actuators are mounted remotely and resulting joint motions are coupled. Equations (3.16) show explicitly how actuator motions cause joint motions. Note, for example, that motion of actuator 2 causes motion of joints 2, 3, and 4.

The optimal distribution of reduction stages throughout the transmission will depend ultimately on the flexibility of the transmission, the weight of the reduction system, the friction associated with the reduction system, and the ease of incorporating these components into the overall manipulator design.

Reduction and transmission systems

Gears are the most common element used for reduction. They can provide for large reductions in relatively compact configurations. Gear pairs come in various configurations for parallel shafts (spur gears), orthogonal intersecting shafts (bevel gears), skew shafts (worm gears or cross helical gears), and other configurations. Different types of gears have different load ratings, wear characteristics, and frictional properties.

The major disadvantages of using gearing are added **backlash** and friction. Backlash, which arises from the imperfect meshing of gears, can be defined as the maximum angular motion of the output gear when the input gear remains fixed. If the gear teeth are meshed tightly to eliminate backlash, there can be excessive amounts of friction. Very precise gears and very precise mounting minimize these problems, but also increase cost.

The **gear ratio**, η, describes the speed-reducing and torque-increasing effects of a gear pair. For speed-reduction systems, we will define $\eta > 1$; then the relationships between input and output speeds and torques are given by

$$\dot{\theta}_o = (1/\eta)\dot{\theta}_i$$
$$\tau_o = \eta\tau_i, \tag{8.11}$$

where $\dot{\theta}_o$ and $\dot{\theta}_i$ are output and input speeds, respectively, and τ_o and τ_i are output and input torques, respectively.

The second broad class of reduction elements includes flexible bands, cables, and belts. Because all of these elements must be flexible enough to bend around pulleys, they also tend to be flexible in the longitudinal direction. The flexibility of these elements is proportional to their length. Because these systems are flexible, there must be some mechanism for preloading the loop to ensure that the belt or cable stays engaged on the pulley. Large preloads can add undue strain to the flexible element and introduce excessive friction.

Cables or flexible bands can be used either in a closed loop or as single-ended elements that are always kept in tension by some sort of preload. In a joint that is spring loaded in one direction, a single-ended cable could be used to pull against it. Alternately, two active single-ended systems can oppose each other. This arrangement eliminates the problem of excessive preloads but adds more actuators.

Roller chains are similar to flexible bands but can bend around relatively small pulleys while retaining a high stiffness. As a result of wear and high loads on the pins connecting the links, toothed-belt systems are more compact than roller chains for certain applications.

Band, cable, belt, and chain drives have the ability to combine transmission with reduction. As is shown in Fig. 8.15, when the input pulley has radius r_1 and the output pulley has radius r_2, the "gear" ratio of the transmission system is

$$\eta = \frac{r_2}{r_1}. \tag{8.12}$$

Lead screws or ball-bearing screws provide another popular method of getting a large reduction in a compact package (Fig. 8.16). Lead screws are very stiff and can support very large loads, and have the property that they transform rotary motion into linear motion. Ball-bearing screws are similar to lead screws, but instead of having the nut threads riding directly on the screw threads, a recirculating circuit of ball bearings rolls between the sets of threads. Ball-bearings screws have very low friction and are usually backdrivable.

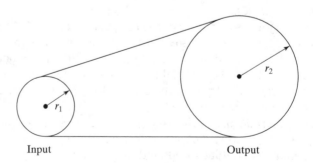

Input   Output

FIGURE 8.15: Band, cable, belt, and chain drives have the ability to combine transmission with reduction.

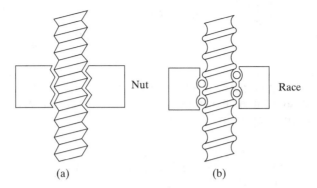

FIGURE 8.16: Lead screws (a) and ball-bearing screws (b) combine a large reduction and transformation from rotary to linear motion.

8.7 STIFFNESS AND DEFLECTIONS

An important goal for the design of most manipulators is overall stiffness of the structure and the drive system. Stiff systems provide two main benefits. First, because typical manipulators do not have sensors to measure the tool frame location directly, it is calculated by using the forward kinematics based on sensed joint positions. For an accurate calculation, the links cannot sag under gravity or other loads. In other words, we wish our Denavit–Hartenberg description of the linkages to remain fixed under various loading conditions. Second, flexibilities in the structure or drive train will lead to **resonances**, which have an undesirable effect on manipulator performance. In this section, we consider issues of stiffness and the resulting deflections under loads. We postpone further discussion of resonances until Chapter 9.

Flexible elements in parallel and in series

As can be easily shown (see Exercise 8.21), the combination of two flexible members of stiffness k_1 and k_2 "connected in parallel" produces the net stiffness

$$k_{\text{parallel}} = k_1 + k_2;\qquad(8.13)$$

"connected in series," the combination produces the net stiffness

$$\frac{1}{k_{\text{series}}} = \frac{1}{k_1} + \frac{1}{k_2}.\qquad(8.14)$$

In considering transmission systems, we often have the case of one stage of reduction or transmission in series with a following stage of reduction or transmission; hence, (8.14) becomes useful.

Shafts

A common method for transmitting rotary motion is through shafts. The torsional stiffness of a round shaft can be calculated [19] as

$$k = \frac{G\pi d^4}{32l},\qquad(8.15)$$

where d is the shaft diameter, l is the shaft length, and G is the shear modulus of elasticity (about 7.5×10^{10} Nt/m^2 for steel, and about a third as much for aluminum).

Gears

Gears, although typically quite stiff, introduce compliance into the drive system. An approximate formula to estimate the stiffness of the output gear (assuming the input gear is fixed) is given in [20] as

$$k = C_g b r^2, \tag{8.16}$$

where b is the face width of the gears, r is the radius of the output gear, and $C_g = 1.34 \times 10^{10}$ Nt/m^2 for steel.

Gearing also has the effect of changing the effective stiffness of the drive system by a factor of η^2. If the stiffness of the transmission system prior to the reduction (i.e., on the input side) is k_i, so that

$$\tau_i = k_i \delta \theta_i, \tag{8.17}$$

and the stiffness of the output side of the reduction is k_o, so that

$$\tau_o = k_o \delta \theta_o, \tag{8.18}$$

then we can compute the relationship between k_i and k_o (under the assumption of a perfectly rigid gear pair) as

$$k_o = \frac{\tau_o}{\delta \theta_o} = \frac{\eta k_i \delta \theta_i}{(1/\eta) \delta \theta_i} = \eta^2 k_i. \tag{8.19}$$

Hence, a gear reduction has the effect of increasing the stiffness by the square of the gear ratio.

EXAMPLE 8.3

A shaft with torsional stiffness equal to 500.0 Nt-m/radian is connected to the input side of a gear set with $\eta = 10$, whose output gear (when the input gear is fixed) exhibits a stiffness of 5000.0 Nt m/radian. What is the output stiffness of the combined drive system?

Using (8.14) and (8.19), we have

$$\frac{1}{k_{series}} = \frac{1}{5000.0} + \frac{1}{10^2(500.0)}, \tag{8.20}$$

or

$$k_{series} = \frac{50000}{11} \cong 4545.4 \text{ Nt m/radian.} \tag{8.21}$$

When a relatively large speed reduction is the last element of a multielement transmission system, the stiffnesses of the preceding elements can generally be ignored.

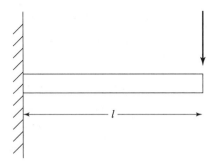

FIGURE 8.17: Simple cantilever beam used to model the stiffness of a link to an end load.

Belts

In such a belt drive as that of Fig. 8.15, stiffness is given by

$$k = \frac{AE}{l},\qquad(8.22)$$

where A is the cross-sectional area of the belt, E is the modulus of elasticity of the belt, and l is the length of the free belt between pulleys plus one-third of the length of the belt in contact with the pulleys [19].

Links

As a rough approximation of the stiffness of a link, we might model a single link as a cantilever beam and calculate the stiffness at the end point, as in Fig. 8.17. For a round hollow beam, this stiffness is given by [19]

$$k = \frac{3\pi E(d_o^4 - d_i^4)}{64 l^3},\qquad(8.23)$$

where d_i and d_o are the inner and outer diameters of the tubular beam, l is the length, and E is the modulus of elasticity (about 2×10^{11} Nt/m^2 for steel, and about a third as much for aluminum). For a square-cross-section hollow beam, this stiffness is given by

$$k = \frac{E(w_o^4 - w_i^4)}{4 l^3},\qquad(8.24)$$

where w_i and w_o are the outer and inner widths of the beam (i.e., the wall thickness is $w_o - w_i$).

EXAMPLE 8.4

A square-cross-section link of dimensions $5 \times 5 \times 50$ cm with a wall thickness of 1 cm is driven by a set of rigid gears with $\eta = 10$, and the input of the gears is driven by a shaft having diameter 0.5 cm and length 30 cm. What deflection is caused by a force of 100 Nt at the end of the link?

Using (8.24), we calculate the stiffness of the link as

$$k_{link} = \frac{(2 \times 10^{11})(0.05^4 - 0.04^4)}{4(0.5)} \cong 3.69 \times 10^5. \tag{8.25}$$

Hence, for a load of 100 Nt, there is a deflection in the link itself of

$$\delta x = \frac{100}{k_{link}} \cong 2.7 \times 10^{-4} \text{ m}, \tag{8.26}$$

or 0.027 cm.

Additionally, 100 Nt at the end of a 50-cm link is placing a torque of 50 Nt-m on the output gear. The gears are rigid, but the flexibility of the input shaft is

$$k_{shaft} = \frac{(7.5 \times 10^{10})(3.14)(5 \times 10^{-3})^4}{(32)(0.3)} \cong 15.3 \text{ Nt m/radian}, \tag{8.27}$$

which, viewed from the output gear, is

$$k'_{shaft} = (15.3)(10^2) = 1530.0 \text{ Nt-m/radian}. \tag{8.28}$$

Loading with 50 Nt-m causes an angular deflection of

$$\delta\theta = \frac{50.0}{1530.0} \cong 0.0326 \text{ radian}, \tag{8.29}$$

so the total linear deflection at the tip of the link is

$$\delta x \cong 0.027 + (0.0326)(50) = 0.027 + 1.630 = 1.657 \text{ cm}. \tag{8.30}$$

In our solution, we have assumed that the shaft and link are made of steel. The stiffness of both members is linear in E, the modulus of elasticity, so, for aluminum elements, we can multiply our result by about 3.

In this section, we have examined some simple formulas for estimating the stiffness of gears, shafts, belts, and links. They are meant to give some guidance in sizing structural members and transmission elements. However, in practical applications, many sources of flexibility are very difficult to model. Often, the drive train introduces significantly more flexibility than the link of a manipulator. Furthermore, many sources of flexibility in the drive system have not been considered here (bearing flexibility, flexibility of the actuator mounting, etc.). Generally, any attempt to predict stiffness analytically results in an overly high prediction, because many sources are not considered.

Finite-element techniques can be used to predict the stiffness (and other properties) of more realistic structural elements more accurately. This is an entire field in itself [21] and is beyond the scope of this book.

Actuators

Among various actuators, **hydraulic cylinders** or **vane actuators** were originally the most popular for use in manipulators. In a relatively compact package, they can produce enough force to drive joints without a reduction system. The speed of operation depends upon the pump and accumulator system, usually located remotely from the manipulator. The position control of hydraulic systems is well understood and relatively straightforward. All of the early industrial robots and many modern large industrial robots use hydraulic actuators.

Unfortunately, hydraulics require a great deal of equipment, such as pumps, accumulators, hoses, and servo valves. Hydraulic systems also tend to be inherently messy, making them unsuitable for some applications. With the advent of more advanced robot-control strategies, in which actuator forces must be applied accurately, hydraulics proved disadvantageous, because of the friction contributed by their seals.

Pneumatic cylinders possess all the favorable attributes of hydraulics, and they are cleaner than hydraulics—air seeps out instead of hydraulic fluid. However, pneumatic actuators have proven difficult to control accurately, because of the compressibility of air and the high friction of the seals.

Electric motors are the most popular actuator for manipulators. Although they don't have the power-to-weight ratio of hydraulics or pneumatics, their controllability and ease of interface makes them attractive for small-to-medium-sized manipulators.

Direct current (DC) brush motors (Fig. 8.18) are the most straightforward to interface and control. The current is conducted to the windings of the rotor via brushes, which make contact with the revolving commutator. Brush wear and friction can be problems. New magnetic materials have made high peak torques possible. The limiting factor on the torque output of these motors is the overheating

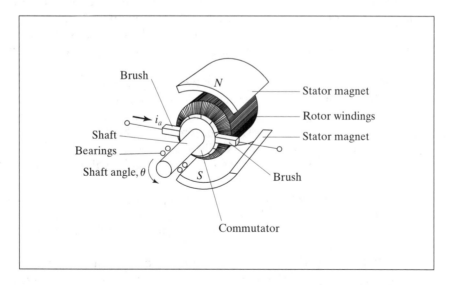

FIGURE 8.18: DC brush motors are among the actuators occurring most frequently in manipulator design. Franklin, Powell, Emami-Naeini, *Feedback Control of Dynamic Systems*, © 1988, Addison-Wesley, Reading, MA. Reprinted with permission.

of the windings. For short duty cycles, high torques can be achieved, but only much lower torques can be sustained over long periods of time.

Brushless motors solve brush wear and friction problems. Here, the windings remain stationary and the magnetic field piece rotates. A sensor on the rotor detects the shaft angle and is then used by external electronics to perform the commutation. Another advantage of brushless motors is that the winding is on the outside, attached to the motor case, affording it much better cooling. Sustained torque ratings tend to be somewhat higher than for similar-sized brush motors.

Alternating current (AC) motors and stepper motors have been used infrequently in industrial robotics. Difficulty of control of the former and low torque ability of the latter have limited their use.

8.8 POSITION SENSING

Virtually all manipulators are servo-controlled mechanisms—that is, the force or torque command to an actuator is based on the error between the sensed position of the joint and the desired position. This requires that each joint have some sort of position-sensing device.

The most common approach is to locate a position sensor directly on the shaft of the actuator. If the drive train is stiff and has no backlash, the true joint angles can be calculated from the actuator shaft positions. Such **co-located** sensor and actuator pairs are easiest to control.

The most popular position-feedback device is the **rotary optical encoder**. As the encoder shaft turns, a disk containing a pattern of fine lines interrupts a light beam. A photodetector turns these light pulses into a binary waveform. Typically, there are two such channels, with wave pulse trains 90 degrees out of phase. The shaft angle is determined by counting the number of pulses, and the direction of rotation is determined by the relative phase of the two square waves. Additionally, encoders generally emit an **index pulse** at one location, which can be used to set a home position in order to compute an absolute angular position.

Resolvers are devices that output two analog signals—one the sine of the shaft angle, the other the cosine. The shaft angle is computed from the relative magnitude of the two signals. The resolution is a function of the quality of the resolver and the amount of noise picked up in the electronics and cabling. Resolvers are often more reliable than optical encoders, but their resolution is lower. Typically, resolvers cannot be placed directly at the joint without additional gearing to improve the resolution.

Potentiometers provide the most straightforward form of position sensing. Connected in a bridge configuration, they produce a voltage proportional to the shaft position. Difficulties with resolution, linearity, and noise susceptibility limit their use.

Tachometers are sometimes used to provide an analog signal proportional to the shaft velocity. In the absence of such velocity sensors, the velocity feedback is derived by taking differences of sensed position over time. This **numerical differentiation** can introduce both noise and a time lag. Despite these potential problems, most manipulators are without direct velocity sensing.

8.9   FORCE SENSING

A variety of devices have been designed to measure forces of contact between a manipulator's end-effector and the environment that it touches. Most such sensors make use of sensing elements called **strain gauges**, of either the semiconductor or the metal-foil variety. These strain gauges are bonded to a metal structure and produce an output proportional to the strain in the metal. In this type of force-sensor design, the issues the designer must address include the following:

1. How many sensors are needed to resolve the desired information?
2. How are the sensors mounted relative to each other on the structure?
3. What structure allows good sensitivity while maintaining stiffness?
4. How can protection against mechanical overload be built into the device?

There are three places where such sensors are usually placed on a manipulator:

1. At the joint actuators. These sensors measure the torque or force output of the actuator/reduction itself. These are useful for some control schemes, but usually do not provide good sensing of contact between the end-effector and the environment.
2. Between the end-effector and last joint of the manipulator. These sensors are usually referred to as **wrist sensors**. They are mechanical structures instrumented with strain gauges, which can measure the forces and torques acting on the end-effector. Typically, these sensors are capable of measuring from three to six components of the force/torque vector acting on the end-effector.
3. At the "fingertips" of the end-effector. Usually, these **force-sensing fingers** have built-in strain gauges to measure from one to four components of force acting at each fingertip.

As an example, Fig. 8.19 is a drawing of the internal structure of a popular style of wrist-force sensor designed by Scheinman [22]. Bonded to the cross-bar structure of the device are eight pairs of semiconductor strain gauges. Each pair is wired in a voltage-divider arrangement. Each time the wrist is queried, eight analog voltages are digitized and read into the computer. Calibration schemes have been designed with which to arrive at a constant 6×8 **calibration matrix** that maps these eight strain measurements into the force–torque vector, $\mathcal{F}$, acting on the end-effector. The sensed force–torque vector can be transformed to a reference frame of interest, as we saw in Example 5.8.

Force-sensor design issues

Use of strain gauges to measure force relies on measuring the deflection of a stressed **flexure**. Therefore, one of the primary design trade-offs is between the stiffness and the sensitivity of the sensor. A stiffer sensor is inherently less sensitive.

The stiffness of the sensor also affects the construction of **overload protection**. Strain gauges can be damaged by impact loading and therefore must be protected against such overloads. Transducer damage can be prevented by having **limit stops**,

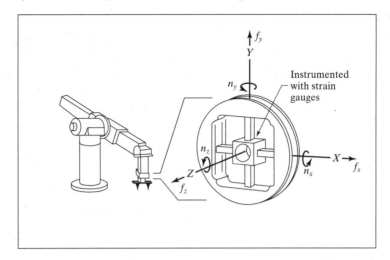

FIGURE 8.19: The internal structure of a typical force-sensing wrist.

which prevent the flexures from deflecting past a certain point. Unfortunately, a very stiff sensor might deflect only a few ten-thousandths of an inch. Manufacturing limit stops with such small clearances is very difficult. Consequently, for many types of transducers, a certain amount of flexibility *must* be built-in in order to make possible effective limit stops.

Eliminating **hysteresis** is one of the most cumbersome restrictions in the sensor design. Most metals used as flexures, if not overstrained, have very little hysteresis. However, bolted, press-fit, or welded joints near the flexure introduce hysteresis. Ideally, the flexure and the material near it are made from a single piece of metal.

It is also important to use differential measurements to increase the linearity and disturbance rejection of torque sensors. Different physical configurations of transducers can eliminate influences due to temperature effects and off-axis forces.

Foil gauges are relatively durable, but they produce a very small resistance change at full strain. Eliminating noise in the strain-gauge cabling and amplification electronics is of crucial importance for a good dynamic range.

Semiconductor strain gauges are much more susceptible to damage through overload. In their favor, they produce a resistance change about seventy times that of foil gauges for a given strain. This makes the task of signal processing much simpler for a given dynamic range.

BIBLIOGRAPHY

[1] W. Rowe, Editor, *Robotics Technical Directory 1986*, Instrument Society of America, Research Triangle Park, NC, 1986.

[2] R. Vijaykumar and K. Waldron, "Geometric Optimization of Manipulator Structures for Working Volume and Dexterity," *International Journal of Robotics Research*, Vol. 5, No. 2, 1986.

[3] K. Waldron, "Design of Arms," in *The International Encyclopedia of Robotics*, R. Dorf and S. Nof, Editors, John Wiley and Sons, New York, 1988.

[4] T. Yoshikawa, "Manipulability of Robotic Mechanisms," *The International Journal of Robotics Research*, Vol. 4, No. 2, MIT Press, Cambridge, MA, 1985.

[5] H. Asada, "Dynamic Analysis and Design of Robot Manipulators Using Inertia Ellipsoids," *Proceedings of the IEEE International Conference on Robotics*, Atlanta, March 1984.

[6] J.K. Salisbury and J. Craig, "Articulated Hands: Force Control and Kinematic Issues," *The International Journal of Robotics Research*, Vol. 1, No. 1, 1982.

[7] O. Khatib and J. Burdick, "Optimization of Dynamics in Manipulator Design: The Operational Space Formulation," *International Journal of Robotics and Automation*, Vol. 2, No. 2, IASTED, 1987.

[8] T. Yoshikawa, "Dynamic Manipulability of Robot Manipulators," *Proceedings of the IEEE International Conference on Robotics and Automation*, St. Louis, March 1985.

[9] J. Trevelyan, P. Kovesi, and M. Ong, "Motion Control for a Sheep Shearing Robot," *The 1st International Symposium of Robotics Research*, MIT Press, Cambridge, MA, 1984.

[10] P. Marchal, J. Cornu, and J. Detriche, "Self Adaptive Arc Welding Operation by Means of an Automatic Joint Following System," *Proceedings of the 4th Symposium on Theory and Practice of Robots and Manipulators*, Zaburow, Poland, September 1981.

[11] J.M. Hollerbach, "Optimum Kinematic Design for a Seven Degree of Freedom Manipulator," *Proceedings of the 2nd International Symposium of Robotics Research*, Kyoto, Japan, August 1984.

[12] K. Waldron and J. Reidy, "A Study of Kinematically Redundant Manipulator Structure," *Proceedings of the IEEE Robotics and Automation Conference*, San Francisco, April 1986.

[13] V. Milenkovic, "New Nonsingular Robot Wrist Design," *Proceedings of the Robots 11 / 17th ISIR Conference*, SME, 1987.

[14] E. Rivin, *Mechanical Design of Robots*, McGraw-Hill, New York, 1988.

[15] T. Yoshikawa, "Manipulability of Robotic Mechanisms," in *Proceedings of the 2nd International Symposium on Robotics Research*, Kyoto, Japan, 1984.

[16] M. Leu, V. Dukowski, and K. Wang, "An Analytical and Experimental Study of the Stiffness of Robot Manipulators with Parallel Mechanisms," in *Robotics and Manufacturing Automation*, M. Donath and M. Leu, Editors, ASME, New York, 1985.

[17] K. Hunt, *Kinematic Geometry of Mechanisms*, Cambridge University Press, Cambridge, MA, 1978.

[18] H. Asada and K. Youcef-Toumi, *Design of Direct Drive Manipulators*, MIT Press, Cambridge, MA, 1987.

[19] J. Shigley, *Mechanical Engineering Design*, 3rd edition, McGraw-Hill, New York, 1977.

[20] D. Welbourne, "Fundamental Knowledge of Gear Noise—A Survey," *Proceedings of the Conference on Noise and Vibrations of Engines and Transmissions*, Institute of Mechanical Engineers, Cranfield, UK, 1979.

[21] O. Zienkiewicz, *The Finite Element Method*, 3rd edition, McGraw-Hill, New York, 1977.

[22] V. Scheinman, "Design of a Computer Controlled Manipulator," M.S. Thesis, Mechanical Engineering Department, Stanford University, 1969.

[23] K. Lau, N. Dagalakis, and D. Meyers, "Testing," in *The International Encyclopedia of Robotics*, R. Dorf and S. Nof, Editors, John Wiley and Sons, New York, 1988.

[24] M. Roshiem, "Wrists," in *The International Encyclopedia of Robotics*, R. Dorf and S. Nof, Editors, John Wiley and Sons, New York, 1988.

[25] A. Bowling and O. Khatib, "Robot Acceleration Capability: The Actuation Efficiency Measure," *Proceedings of the IEEE International Conference on Robotics and Automation*, San Francisco, April 2000.

EXERCISES

8.1 [15] A robot is to be used for positioning a laser cutting device. The laser produces a pinpoint, nondivergent beam. For general cutting tasks, how many degrees of freedom does the positioning robot need? Justify your answer.

8.2 [15] Sketch a possible joint configuration for the laser-positioning robot of Exercise 8.1, assuming that it will be used primarily for cutting at odd angles through 1-inch-thick, 8 × 8-foot plates.

8.3 [17] For a spherical robot like that of Fig. 8.6, if joints 1 and 2 have no limits and joint 3 has lower limit l and upper limit u, find the structural length index, Q_L, for the wrist point of this robot.

8.4 [25] A steel shaft of length 30 cm and diameter 0.2 cm drives the input gear of a reduction having $\eta = 8$. The output gear drives a steel shaft having length 30 cm and diameter 0.3 cm. If the gears introduce no compliance of their own, what is the overall stiffness of the transmission system?

8.5 [20] In Fig. 8.20, a link is driven through a shaft after a gear reduction. Model the link as rigid with mass of 10 Kg located at a point 30 cm from the shaft axis. Assume that the gears are rigid and that the reduction, η, is large. The shaft is steel and must be 30 cm long. If the design specifications call for the center of link mass to undergo accelerations of 2.0 g, what should the shaft diameter be to limit dynamic deflections to 0.1 radian at the joint angle?

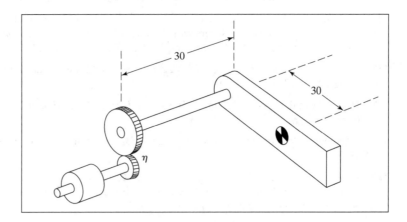

FIGURE 8.20: A link actuated through a shaft after a gear reduction.

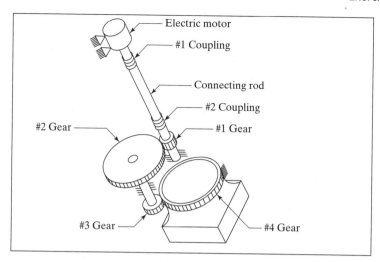

FIGURE 8.21: Simplified version of the drive train of joint 4 of the PUMA 560 manipulator (from [23]). From *International Encyclopedia of Robotics*, by R. Dorf and S. Nof, editors. From "Testing," by K. Law, N. Dagalakis, and D. Myers.

8.6 [15] If the output gear exhibits a stiffness of 1000 Nt-m/radian with input gear locked and the shaft has stiffness of 300 Nt-m/radian, what is the combined stiffness of the drive system in Fig. 8.20?

8.7 [43] Pieper's criteria for serial-link manipulators state that the manipulator will be solvable if three consecutive axes intersect at a single point or are parallel. This is based on the idea that inverse kinematics can be decoupled by looking at the position of the wrist point independently from the orientation of the wrist frame. Propose a similar result for the Stewart mechanism in Fig. 8.14, to allow the forward kinematic solution to be similarly decoupled.

8.8 [20] In the Stewart mechanism of Fig. 8.14, if the 2-DOF universal joints at the base were replaced with 3-DOF ball-and-socket joints, what would the total number of degrees of freedom of the mechanism be? Use Grübler's formula.

8.9 [22] Figure 8.21 shows a simplified schematic of the drive system of joint 4 of the PUMA 560 [23]. The torsional stiffness of the couplings is 100 Nt-m/radian each, that of the shaft is 400 Nt-m/radian, and each of the reduction pairs has been measured to have output stiffness of 2000 Nt-m/radian with its input gears fixed. Both the first and second reductions have $\eta = 6$.[2] Assuming the structure and bearing are perfectly rigid, what is the stiffness of the joint (i.e., when the motor's shaft is locked)?

8.10 [25] What is the error if one approximates the answer to Exercise 8.9 by considering just the stiffness of the final speed-reduction gearing?

8.11 [20] Figure 4.14 shows an orthogonal-axis wrist and a nonorthogonal wrist. The orthogonal-axis wrist has link twists of magnitude 90°; the nonorthogonal wrist has link twists of ϕ and $180° - \phi$ in magnitude. Describe the set of orientations that are *unattainable* with the nonorthogonal mechanism. Assume that all axes can turn 360° and that links can pass through one another if need be (i.e., workspace is not limited by self-collision).

---

[2]None of the numerical values in this exercise is meant to be realistic!

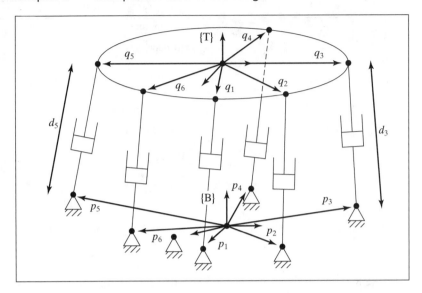

FIGURE 8.22: Stewart mechanism of Exercise 8.12.

8.12 [18] Write down a general inverse-kinematic solution for the Stewart mechanism
shown in Fig. 8.22. Given the location of $\{T\}$ relative to the base frame $\{B\}$, solve
for the joint-position variables d_1 through d_6. The ^Bp_i are 3×1 vectors which
locate the base connections of the linear actuators relative to frame $\{B\}$. The
^Tq_i are 3×1 vectors which locate the upper connections of the linear actuators
relative to the frame $\{T\}$.

8.13 [20] The planar two-link of example 5.3 has the determinant of its Jacobian given
by

$$\det(J(\Theta)) = l_1 l_2 s_2. \tag{8.31}$$

If the sum of the two link lengths, $l_1 + l_2$, is constrained to be equal to a constant,
what should the relative lengths be in order to maximize the manipulator's
manipulability as defined by (8.6)?

8.14 [28] For a SCARA robot, given that the sum of the link lengths of link 1 and link
2 must be constant, what is the optimal choice of relative length in terms of the
manipulability index given in (8.6)? Solving Exercise 8.13 first could be helpful.

8.15 [35] Show that the manipulability measure defined in (8.6) is also equal to the
product of the eigenvalues of $J(\Theta)$.

8.16 [15] What is the torsional stiffness of a 40-cm aluminum rod with radius 0.1 cm?

8.17 [5] What is the effective "gear" reduction, η, of a belt system having an input
pulley of radius 2.0 cm and an output pulley of radius 12.0 cm?

8.18 [10] How many degrees of freedom are required in a manipulator used to
place cylindrical-shaped parts on a flat plane? The cylindrical parts are perfectly
symmetrical about their main axes.

8.19 [25] Figure 8.23 shows a three-fingered hand grasping an object. Each finger has
three single-degree-of-freedom joints. The contact points between fingertips and
the object are modeled as "point contact"—that is, the position is fixed, but the
relative orientation is free in all three degrees of freedom. Hence, these point
contacts can be replaced by 3-DOF ball-and-socket joints for the purposes of

analysis. Apply Grübler's formula to compute how many degrees of freedom the overall system possesses.

8.20 [23] Figure 8.24 shows an object connected to the ground with three rods. Each rod is connected to the object with a 2-DOF universal joint and to the ground with a 3-DOF ball-and-socket joint. How many degrees of freedom does the system possess?

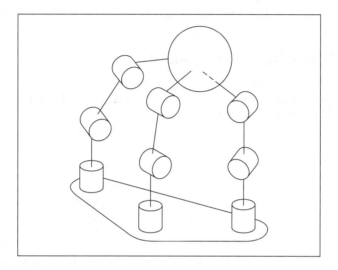

FIGURE 8.23: A three-fingered hand in which each finger has three degrees of freedom grasps an object with "point contact."

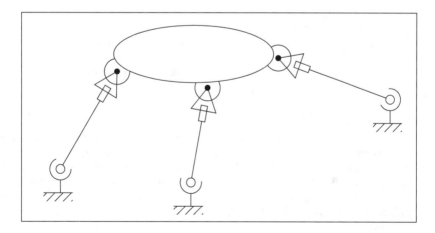

FIGURE 8.24: Closed loop mechanism of Exercise 8.20.

8.21 [18] Verify that, if two transmission systems are connected serially, then the equivalent stiffness of the overall system is given by (8.14). It is perhaps simplest to think of the serial connection of two linear springs having stiffness coefficients

k_1 and k_2 and of the resulting equations:

$$f = k_1 \delta x_1,$$
$$f = k_2 \delta x_2, \tag{8.32}$$
$$f = k_{sum}(\delta x_1 + \delta x_2).$$

8.22 [20] Derive a formula for the stiffness of a belt-drive system in terms of the pulley radii (r_1 and r_2) and the center-to-center distance between the pulleys, d_c. Start from (8.22).

PROGRAMMING EXERCISE (PART 8)

1. Write a program to compute the determinant of a 3×3 matrix.
2. Write a program to move the simulated three-link robot in 20 steps in a straight line and constant orientation from

$$_3^0 T = \begin{bmatrix} 0.25 \\ 0.0 \\ 0.0 \end{bmatrix}$$

to

$$_3^0 T = \begin{bmatrix} 0.95 \\ 0.0 \\ 0.0 \end{bmatrix}$$

in increments of 0.05 meter. At each location, compute the manipulability measure for the robot at that configuration (i.e., the determinant of the Jacobian). List, or, better yet, make a plot of the values as a function of the position along the $\hat{X}_0$ axis. Generate the preceding data for two cases:

(a) $l_1 = l_2 = 0.5$ meter, and
(b) $l_1 = 0.625$ meter, $l_2 = 0.375$ meter.

Which manipulator design do you think is better? Explain your answer.

MATLAB EXERCISE 8

Section 8.5 introduced the concept of kinematically redundant robots. This exercise deals with the resolved-rate control simulation for a kinematically redundant robot. We will focus on the planar 4-DOF 4R robot with one degree of kinematic redundancy (four joints to provide three Cartesian motions: two translations and one rotation). This robot is obtained by adding a fourth R-joint and a fourth moving link L_4 to the planar 3-DOF, 3R robot (of Figures 3.6 and 3.7; the DH parameters can be extended by adding one row to Figure 3.8).

For the planar 4R robot, derive analytical expressions for the 3×4 Jacobian matrix; then, perform resolved-rate control simulation in MATLAB (as in MATLAB Exercise 5). The form of the velocity equation is again $^k \dot{X} = {}^k J \dot{\Theta}$; however, this equation cannot be inverted by means of the normal matrix inverse, because the Jacobian matrix is nonsquare (three equations, four unknowns, infinite solutions to $\dot{\Theta}$). Therefore, let us use the Moore–Penrose pseudoinverse J^* of the Jacobian matrix: $J^* = J^T (JJ^T)^{-1}$. For the resulting commanded relative joint rates for the resolved-rate algorithm, $\dot{\Theta} = {}^k J^{*k} \dot{X}$, choose the minimum-norm solution from the infinite possibilities (i.e., this specific $\dot{\Theta}$ is as small as possible to satisfy the commanded Cartesian velocities $^k\dot{X}$).

This solution represents the particular solution only—that is, there exists a homogeneous solution to optimize performance (such as avoiding manipulator singularities or avoiding joint limits) in addition to satisfying the commanded Cartesian motion. Performance optimization is beyond the scope of this exercise.

Given: $L_1 = 1.0\ m$, $L_2 = 1.0$ m, $L_3 = 0.2\ m$, $L_4 = 0.2\ m$.
The initial angles are:

$$\Theta = \begin{Bmatrix} \theta_1 \\ \theta_2 \\ \theta_3 \\ \theta_4 \end{Bmatrix} = \begin{Bmatrix} -30° \\ 70° \\ 30° \\ 40° \end{Bmatrix}.$$

The (constant) commanded Cartesian velocity is

$$^0\dot{X} = {}^0\begin{Bmatrix} \dot{x} \\ \dot{y} \\ \omega_z \end{Bmatrix} = {}^0\begin{Bmatrix} -0.2 \\ -0.2 \\ 0.2 \end{Bmatrix} \text{ (m/s, rad/s)}.$$

Simulate resolved-rate motion, for the particular solution only, for 3 sec, with a control time step of 0.1 sec. Also, in the same loop, animate the robot to the screen during each time step, so that you can watch the simulated motion to verify that it is correct.

a) Present four plots (each set on a separate graph, please):

1. the four joint angles (degrees) $\Theta = \{\theta_1\ \theta_2\ \theta_3\ \theta_4\}^T$ vs. time;
2. the four joint rates (rad/s) $\dot{\Theta} = \{\dot{\theta}_1\ \dot{\theta}_2\ \dot{\theta}_3\ \dot{\theta}_4\}^T$ vs. time;
3. the joint-rate Euclidean norm $\|\dot{\Theta}\|$ (vector magnitude) vs. time;
4. the three Cartesian components of $^0_H T$, $X = \{x\ y\ \phi\}^T$ (rad is fine for ϕ so that it will fit) vs. time.

Carefully label (by hand is fine!) each component on each plot. Also, label the axis names and units.

b) Check your Jacobian matrix results for the initial and final joint-angle sets by means of the Corke MATLAB Robotics Toolbox. Try function *jacob0()*. **Caution:** The toolbox Jacobian functions are for motion of {4} with respect to {0}, not for {H} with respect to {0} as in the problem assignment. The preceding function gives the Jacobian result in {0} coordinates; *jacobn()* would give results in {4} coordinates.

CHAPTER 9

Linear control of manipulators

9.1   INTRODUCTION

Armed with the previous material, we now have the means to calculate joint-position time histories that correspond to desired end-effector motions through space. In this chapter, we begin to discuss how to cause the manipulator actually to perform these desired motions.

The control methods that we will discuss fall into the class called **linear-control** systems. Strictly speaking, the use of linear-control techniques is valid only when the system being studied can be modeled mathematically by *linear* differential equations. For the case of manipulator control, such linear methods must essentially be viewed as approximate methods, for, as we have seen in Chapter 6, the dynamics of a manipulator are more properly represented by a *nonlinear* differential equation. Nonetheless, we will see that it is often reasonable to make such approximations, and it also is the case that these linear methods are the ones most often used in current industrial practice.

Finally, consideration of the linear approach will serve as a basis for the more complex treatment of nonlinear control systems in Chapter 10. Although we approach linear control as an approximate method for manipulator control, the justification for using linear controllers is not only empirical. In Chapter 10, we will prove that a certain linear controller leads to a reasonable control system even without resorting to a linear approximation of manipulator dynamics. Readers familiar with linear-control systems might wish to skip the first four sections of the current chapter.

9.2   FEEDBACK AND CLOSED-LOOP CONTROL

We will model a manipulator as a mechanism that is instrumented with sensors at each joint to measure the joint angle and that has an actuator at each joint to apply a torque on the neighboring (next higher) link.[1]. Although other physical arrangements of sensors are sometimes used, the vast majority of robots have a position sensor at each joint. Sometimes velocity sensors (tachometers) are also present at the joints. Various actuation and transmission schemes are prevalent in industrial robots, but many of these can be modeled by supposing that there is a single actuator at each joint.

We wish to cause the manipulator joints to follow prescribed position trajectories, but the actuators are commanded in terms of torque, so we must use some kind of **control system** to compute appropriate actuator commands that will realize this desired motion. Almost always, these torques are determined by using **feedback** from the joint sensors to compute the torque required.

Figure 9.1 shows the relationship between the trajectory generator and the physical robot. The robot accepts a vector of joint torques, τ, from the control system. The manipulator's sensors allow the controller to read the vectors of joint positions, Θ, and joint velocities, $\dot{\Theta}$. All signal lines in Fig. 9.1 carry $N \times 1$ vectors (where N is the number of joints in the manipulator).

Let's consider what algorithm might be implemented in the block labeled "control system" in Fig. 9.1. One possibility is to use the dynamic equation of the robot (as studied in Chapter 6) to calculate the torques required for a particular trajectory. We are given Θ_d, $\dot{\Theta}_d$, and $\ddot{\Theta}_d$ by the trajectory generator, so we could use (6.59) to compute

$$\tau = M(\Theta_d)\ddot{\Theta}_d + V(\Theta_d, \dot{\Theta}_d) + G(\Theta_d). \tag{9.1}$$

This computes the torques that our model dictates would be required to realize the desired trajectory. If our dynamic model were complete and accurate and no "noise" or other disturbances were present, continuous use of (9.1) along the desired trajectory would realize the desired trajectory. Unfortunately, imperfection in the dynamic model and the inevitable presence of disturbances make such a scheme impractical for use in real applications. Such a control technique is termed an **open-loop** scheme, because there is no use made of the feedback from the joint sensors

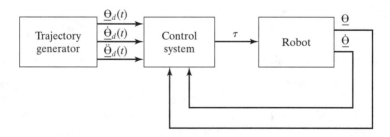

FIGURE 9.1: High-level block diagram of a robot-control system.

---

[1]Remember, all remarks made concerning rotational joints hold analogously for linear joints, and vice versa

(i.e., (9.1) is a function only of the desired trajectory, Θ_d, and its derivatives, and *not* a function of Θ, the actual trajectory).

Generally, the only way to build a high-performance control system is to make use of feedback from joint sensors, as indicated in Fig. 9.1. Typically, this feedback is used to compute any **servo error** by finding the difference between the desired and the actual position and that between the desired and the actual velocity:

$$E = \Theta_d - \Theta,$$

$$\dot{E} = \dot{\Theta}_d - \dot{\Theta}. \tag{9.2}$$

The control system can then compute how much torque to require of the actuators as some function of the servo error. Obviously, the basic idea is to compute actuator torques that would tend to reduce servo errors. A control system that makes use of feedback is called a **closed-loop** system. The "loop" closed by such a control system around the manipulator is apparent in Fig. 9.1.

The central problem in designing a control system is to ensure that the resulting closed-loop system meets certain performance specifications. The most basic such criterion is that the system remain **stable**. For our purposes, we will define a system to be stable if the errors remain "small" when executing various desired trajectories even in the presence of some "moderate" disturbances. It should be noted that an improperly designed control system can sometimes result in **unstable** performance, in which servo errors are enlarged instead of reduced. Hence, the first task of a control engineer is to prove that his or her design yields a stable system; the second is to prove that the closed-loop performance of the system is satisfactory. In practice, such "proofs" range from mathematical proofs based on certain assumptions and models to more empirical results, such as those obtained through simulation or experimentation.

Figure 9.1, in which all signals lines represent $N \times 1$ vectors, summarizes the fact that the manipulator-control problem is a **multi-input, multi-output (MIMO)** control problem. In this chapter, we take a simple approach to constructing a control system by treating each joint as a separate system to be controlled. Hence, for an N-jointed manipulator, we will design N independent **single-input, single-output (SISO)** control systems. This is the design approach presently adopted by most industrial-robot suppliers. This **independent joint control** approach is an approximate method in that the equations of motion (developed in Chapter 6) are not independent, but rather are highly coupled. Later, this chapter will present justification for the linear approach, at least for the case of highly geared manipulators.

9.3 SECOND-ORDER LINEAR SYSTEMS

Before considering the manipulator control problem, let's step back and start by considering a simple mechanical system. Figure 9.2 shows a block of mass m attached to a spring of stiffness k and subject to friction of coefficient b. Figure 9.2 also indicates the zero position and positive sense of x, the block's position. Assuming a frictional force proportional to the block's velocity, a free-body diagram of the forces acting on the block leads directly to the equation of motion,

$$m\ddot{x} + b\dot{x} + kx = 0. \tag{9.3}$$

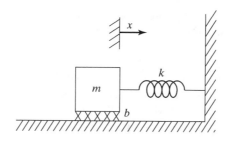

FIGURE 9.2: Spring–mass system with friction.

Hence, the open-loop dynamics of this one-degree-of-freedom system are described by a second-order linear constant-coefficient differential equation [1]. The solution to the differential equation (9.3) is a time function, $x(t)$, that specifies the motion of the block. This solution will depend on the block's **initial conditions**—that is, its initial position and velocity.

We will use this simple mechanical system as an example with which to review some basic control system concepts. Unfortunately, it is impossible to do justice to the field of control theory with only a brief introduction here. We will discuss the control problem, assuming no more than that the student is familiar with simple differential equations. Hence, we will not use many of the popular tools of the control-engineering trade. For example, **Laplace transforms** and other common techniques neither are a prerequisite nor are introduced here. A good reference for the field is [4].

Intuition suggests that the system of Fig. 9.2 might exhibit several different characteristic motions. For example, in the case of a very weak spring (i.e., k small) and very heavy friction (i.e., b large) one imagines that, if the block were perturbed, it would return to its resting position in a very slow, sluggish manner. However, with a very stiff spring and very low friction, the block might oscillate several times before coming to rest. These different possibilities arise because the character of the solution to (9.3) depends upon the values of the parameters m, b, and k.

From the study of differential equations [1], we know that the form of the solution to an equation of the form of (9.3) depends on the roots of its **characteristic equation**,

$$ms^2 + bs + k = 0. \tag{9.4}$$

This equation has the roots

$$s_1 = -\frac{b}{2m} + \frac{\sqrt{b^2 - 4mk}}{2m},$$

$$s_2 = -\frac{b}{2m} - \frac{\sqrt{b^2 - 4mk}}{2m}. \tag{9.5}$$

The location of s_1 and s_2 (sometimes called the **poles** of the system) in the real–imaginary plane dictate the nature of the motions of the system. If s_1 and s_2 are real, then the behavior of the system is sluggish and nonoscillatory. If s_1 and s_2 are complex (i.e., have an imaginary component) then the behavior of the system is oscillatory. If we include the special limiting case between these two behaviors, we have three classes of response to study:

1. **Real and Unequal Roots.** This is the case when $b^2 > 4\,mk$; that is, friction dominates, and sluggish behavior results. This response is called **overdamped**.
2. **Complex Roots.** This is the case when $b^2 < 4\,mk$; that is, stiffness dominates, and oscillatory behavior results. This response is called **underdamped**.
3. **Real and Equal Roots.** This is the special case when $b^2 = 4\,mk$; that is, friction and stiffness are "balanced," yielding the fastest possible nonoscillatory response. This response is called **critically damped**.

The third case (critical damping) is generally a desirable situation: the system nulls out nonzero initial conditions and returns to its nominal position as rapidly as possible, yet without oscillatory behavior.

Real and unequal roots

It can easily be shown (by direct substitution into (9.3)) that the solution, $x(t)$, giving the motion of the block in the case of real, unequal roots has the form

$$x(t) = c_1 e^{s_1 t} + c_2 e^{s_2 t}, \qquad (9.6)$$

where s_1 and s_2 are given by (9.5). The coefficients c_1 and c_2 are constants that can be computed for any given set of initial conditions (i.e., initial position and velocity of the block).

Figure 9.3 shows an example of pole locations and the corresponding time response to a nonzero initial condition. When the poles of a second-order system are real and unequal, the system exhibits sluggish or overdamped motion.

In cases where one of the poles has a much greater magnitude than the other, the pole of larger magnitude can be neglected, because the term corresponding to it will decay to zero rapidly in comparison to the other, **dominant pole**. This same notion of dominance extends to higher order systems—for example, often a

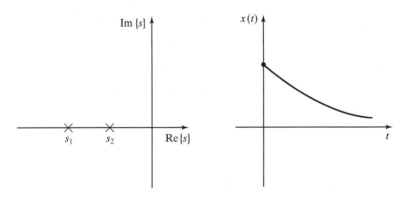

FIGURE 9.3: Root location and response to initial conditions for an overdamped system.

third-order system can be studied as a second-order system by considering only two dominant poles.

EXAMPLE 9.1

Determine the motion of the system in Fig. 9.2 if parameter values are $m = 1, b = 5$, and $k = 6$ and the block (initially at rest) is released from the position $x = -1$.

The characteristic equation is

$$s^2 + 5s + 6 = 0, \tag{9.7}$$

which has the roots $s_1 = -2$ and $s_2 = -3$. Hence, the response has the form

$$x(t) = c_1 e^{-2t} + c_2 e^{-3t}. \tag{9.8}$$

We now use the given initial conditions, $x(0) = -1$ and $\dot{x}(0) = 0$, to compute c_1 and c_2. To satisfy these conditions at $t = 0$, we must have

$$c_1 + c_2 = -1$$

and

$$-2c_1 - 3c_2 = 0, \tag{9.9}$$

which are satisfied by $c_1 = -3$ and $c_2 = 2$. So, the motion of the system for $t \geq 0$ is given by

$$x(t) = -3e^{-2t} + 2e^{-3t}. \tag{9.10}$$

Complex roots

For the case where the characteristic equation has complex roots of the form

$$s_1 = \lambda + \mu i,$$

$$s_2 = \lambda - \mu i, \tag{9.11}$$

it is still the case that the solution has the form

$$x(t) = c_1 e^{s_1 t} + c_2 e^{s_2 t}. \tag{9.12}$$

However, equation (9.12) is difficult to use directly, because it involves imaginary numbers explicitly. It can be shown (see Exercise 9.1) that **Euler's formula**,

$$e^{ix} = \cos x + i \sin x, \tag{9.13}$$

allows the solution (9.12) to be manipulated into the form

$$x(t) = c_1 e^{\lambda t} \cos(\mu t) + c_2 e^{\lambda t} \sin(\mu t). \tag{9.14}$$

As before, the coefficients c_1 and c_2 are constants that can be computed for any given set of initial conditions (i.e., initial position and velocity of the block). If we write the constants c_1 and c_2 in the form

$$c_1 = r \cos \delta,$$

$$c_2 = r \sin \delta, \tag{9.15}$$

then (9.14) can be written in the form

$$x(t) = re^{\lambda t} \cos(\mu t - \delta), \tag{9.16}$$

where

$$r = \sqrt{c_1^2 + c_2^2},$$

$$\delta = \text{Atan2}(c_2, c_1). \tag{9.17}$$

In this form, it is easier to see that the resulting motion is an oscillation whose amplitude is exponentially decreasing toward zero.

Another common way of describing oscillatory second-order systems is in terms of **damping ratio** and **natural frequency**. These terms are defined by the parameterization of the characteristic equation given by

$$s^2 + 2\zeta\omega_n s + \omega_n^2 = 0, \tag{9.18}$$

where ζ is the damping ratio (a dimensionless number between 0 and 1) and ω_n is the natural frequency.[2] Relationships between the pole locations and these parameters are

$$\lambda = -\zeta\omega_n$$

and

$$\mu = \omega_n\sqrt{1 - \zeta^2}. \tag{9.19}$$

In this terminology, μ, the imaginary part of the poles, is sometimes called the **damped natural frequency**. For a damped spring–mass system such as the one in Fig. 9.2, the damping ratio and natural frequency are, respectively,

$$\zeta = \frac{b}{2\sqrt{km}},$$

$$\omega_n = \sqrt{k/m}. \tag{9.20}$$

When no damping is present ($b = 0$ in our example), the damping ratio becomes zero; for critical damping ($b^2 = 4km$), the damping ratio is 1.

Figure 9.4 shows an example of pole locations and the corresponding time response to a nonzero initial condition. When the poles of a second-order system are complex, the system exhibits oscillatory or underdamped motion.

EXAMPLE 9.2

Find the motion of the system in Fig. 9.2 if parameter values are $m = 1$, $b = 1$, and $k = 1$ and the block (initially at rest) is released from the position $x = -1$.

The characteristic equation is

$$s^2 + s + 1 = 0, \tag{9.21}$$

---

[2]The terms *damping ratio* and *natural frequency* can also be applied to overdamped systems, in which case $\zeta > 1.0$.

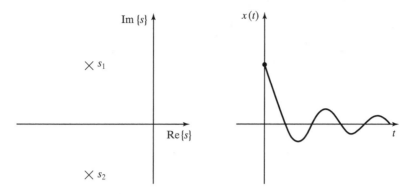

FIGURE 9.4: Root location and response to initial conditions for an underdamped system.

which has the roots $s_i = -\frac{1}{2} \pm \frac{\sqrt{3}}{2}i$. Hence, the response has the form

$$x(t) = e^{-\frac{t}{2}} \left(c_1 \cos \frac{\sqrt{3}}{2}t + c_2 \sin \frac{\sqrt{3}}{2}t \right). \tag{9.22}$$

We now use the given initial conditions, $x(0) = -1$ and $\dot{x}(0) = 0$, to compute c_1 and c_2. To satisfy these conditions at $t = 0$, we must have

$$c_1 = -1$$

and

$$-\frac{1}{2}c_1 - \frac{\sqrt{3}}{2}c_2 = 0, \tag{9.23}$$

which are satisfied by $c_1 = -1$ and $c_2 = \frac{\sqrt{3}}{3}$. So, the motion of the system for $t \geq 0$ is given by

$$x(t) = e^{-\frac{t}{2}} \left(-\cos \frac{\sqrt{3}}{2}t - \frac{\sqrt{3}}{3} \sin \frac{\sqrt{3}}{2}t \right). \tag{9.24}$$

This result can also be put in the form of (9.16), as

$$x(t) = \frac{2\sqrt{3}}{3} e^{-\frac{t}{2}} \cos \left(\frac{\sqrt{3}}{2}t + 120° \right). \tag{9.25}$$

Real and equal roots

By substitution into (9.3), it can be shown that, in the case of real and equal roots (i.e., **repeated roots**), the solution has the form

$$x(t) = c_1 e^{s_1 t} + c_2 t e^{s_2 t}, \tag{9.26}$$

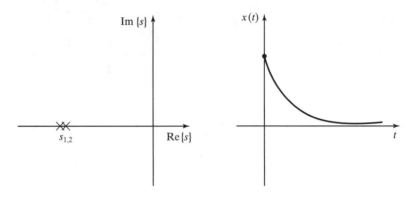

FIGURE 9.5: Root location and response to initial conditions for a critically damped system.

where, in this case, $s_1 = s_2 = -\frac{b}{2m}$, so (9.26) can be written

$$x(t) = (c_1 + c_2 t)e^{-\frac{b}{2m}t}. \tag{9.27}$$

In case it is not clear, a quick application of **l'Hôpital's rule** [2] shows that, for any c_1, c_2, and a,

$$\lim_{t \to \infty} (c_1 + c_2 t)e^{-at} = 0. \tag{9.28}$$

Figure 9.5 shows an example of pole locations and the corresponding time response to a nonzero initial condition. When the poles of a second-order system are real and equal, the system exhibits critically damped motion, the fastest possible nonoscillatory response.

EXAMPLE 9.3

Work out the motion of the system in Fig. 9.2 if parameter values are $m = 1$, $b = 4$, and $k = 4$ and the block (initially at rest) is released from the position $x = -1$.

The characteristic equation is

$$s^2 + 4s + 4 = 0, \tag{9.29}$$

which has the roots $s_1 = s_2 = -2$. Hence, the response has the form

$$x(t) = (c_1 + c_2 t)e^{-2t}. \tag{9.30}$$

We now use the given initial conditions, $x(0) = -1$ and $\dot{x}(0) = 0$, to calculate c_1 and c_2. To satisfy these conditions at $t = 0$, we must have

$$c_1 = -1$$

and

$$-2c_1 + c_2 = 0, \tag{9.31}$$

which are satisfied by $c_1 = -1$ and $c_2 = -2$. So, the motion of the system for $t \geq 0$ is given by

$$x(t) = (-1 - 2t)e^{-2t}. \tag{9.32}$$

In Examples 9.1 through 9.3, all the systems were stable. For any passive physical system like that of Fig. 9.2, this will be the case. Such mechanical systems always have the properties

$$m > 0,$$

$$b > 0, \tag{9.33}$$

$$k > 0.$$

In the next section, we will see that the action of a control system is, in effect, to change the value of one or more of these coefficients. It will then be necessary to consider whether the resulting system is stable.

9.4 CONTROL OF SECOND-ORDER SYSTEMS

Suppose that the natural response of our second-order mechanical system is not what we wish it to be. Perhaps it is underdamped and oscillatory, and we would like it to be critically damped; or perhaps the spring is missing altogether $(k = 0)$, so the system never returns to $x = 0$ if disturbed. Through the use of sensors, an actuator, and a control system, we can modify the system's behavior as desired.

Figure 9.6 shows a damped spring–mass system with the addition of an actuator with which it is possible to apply a force f to the block. A free-body diagram leads to the equation of motion,

$$m\ddot{x} + b\dot{x} + kx = f. \tag{9.34}$$

Let's also assume that we have sensors capable of detecting the block's position and velocity. We now propose a **control law** which computes the force that should be applied by the actuator as a function of the sensed feedback:

$$f = -k_p x - k_v \dot{x}. \tag{9.35}$$

Figure 9.7 is a block diagram of the closed-loop system, where the portion to the left of the dashed line is the control system (usually implemented in a computer) and that to the right of the dashed line is the physical system. Implicit in the figure are interfaces between the control computer and the output actuator commands and the input sensor information.

The control system we have proposed is a **position-regulation** system—it simply attempts to maintain the position of the block in one fixed place regardless

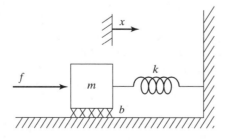

FIGURE 9.6: A damped spring–mass system with an actuator.

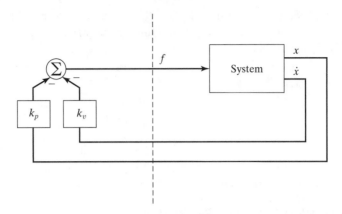

FIGURE 9.7: A closed-loop control system. The control computer (to the left of the dashed line) reads sensor input and writes actuator output commands.

of disturbance forces applied to the block. In a later section, we will construct a **trajectory-following** control system, which can cause the block to follow a desired position trajectory.

By equating the open-loop dynamics of (9.34) with the control law of (9.35), we can derive the closed-loop dynamics as

$$m\ddot{x} + b\dot{x} + kx = -k_p x - k_v \dot{x},\tag{9.36}$$

or

$$m\ddot{x} + (b + k_v)\dot{x} + (k + k_p)x = 0,\tag{9.37}$$

or

$$m\ddot{x} + b'\dot{x} + k'x = 0,\tag{9.38}$$

where $b' = b + k_v$ and $k' = k + k_p$. From (9.37) and (9.38), it is clear that, by setting the **control gains**, k_v and k_p, we can cause the closed-loop system to appear to have any second system behavior that we wish. Often, gains would be chosen to obtain critical damping (i.e., $b' = 2\sqrt{mk'}$) and some desired **closed-loop stiffness** given directly by k'.

Note that k_v and k_p could be positive or negative, depending on the parameters of the original system. However, if b' or k' became negative, the result would be an unstable control system. This instability will be obvious if one writes down the solution of the second-order differential equation (in the form of (9.6), (9.14), or (9.26)). It also makes intuitive sense that, if b' or k' is negative, servo errors tend to get magnified rather than reduced.

EXAMPLE 9.4

If the parameters of the system in Fig. 9.6 are $m = 1$, $b = 1$, and $k = 1$, find gains k_p and k_v for a position-regulation control law that results in the system's being critically damped with a closed-loop stiffness of 16.0.

If we wish k' to be 16.0, then, for critical damping, we require that $b' = 2\sqrt{mk'} = 8.0$. Now, $k = 1$ and $b = 1$, so we need

$$k_p = 15.0,$$

$$k_v = 7.0. \tag{9.39}$$

9.5 CONTROL-LAW PARTITIONING

In preparation for designing control laws for more complicated systems, let us consider a slightly different controller structure for the sample problem of Fig. 9.6. In this method, we will partition the controller into a **model-based portion** and a **servo portion**. The result is that the system's parameters (i.e., m, b, and k, in this case) appear only in the model-based portion and that the servo portion is independent of these parameters. At the moment, this distinction might not seem important, but it will become more obviously important as we consider nonlinear systems in Chapter 10. We will adopt this **control-law partitioning** approach throughout the book.

The open-loop equation of motion for the system is

$$m\ddot{x} + b\dot{x} + kx = f. \tag{9.40}$$

We wish to decompose the controller for this system into two parts. In this case, the model-based portion of the control law will make use of supposed knowledge of m, b, and k. This portion of the control law is set up such that it *reduces the system so that it appears to be a unit mass*. This will become clear when we do Example 9.5. The second part of the control law makes use of feedback to modify the behavior of the system. The model-based portion of the control law has the effect of making the system appear as a unit mass, so the design of the servo portion is very simple—gains are chosen to control a system composed of a single unit mass (i.e., no friction, no stiffness).

The model-based portion of the control appears in a control law of the form

$$f = \alpha f' + \beta, \tag{9.41}$$

where α and β are functions or constants and are chosen so that, if f' is taken as the *new input* to the system, *the system appears to be a unit mass*. With this structure of the control law, the system equation (the result of combining (9.40) and (9.41)) is

$$m\ddot{x} + b\dot{x} + kx = \alpha f' + \beta. \tag{9.42}$$

Clearly, in order to make the system appear as a unit mass from the f' input, for this particular system we should choose α and β as follows:

$$\alpha = m,$$

$$\beta = b\dot{x} + kx. \tag{9.43}$$

Making these assignments and plugging them into (9.42), we have the system equation

$$\ddot{x} = f'. \tag{9.44}$$

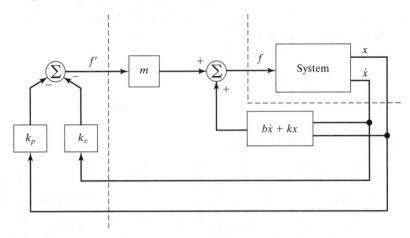

FIGURE 9.8: A closed-loop control system employing the partitioned control method.

This is the equation of motion for a unit mass. We now proceed as if (9.44) were the open-loop dynamics of a system to be controlled. We design a control law to compute f', just as we did before:

$$f' = -k_v \dot{x} - k_p x. \tag{9.45}$$

Combining this control law with (9.44) yields

$$\ddot{x} + k_v \dot{x} + k_p x = 0. \tag{9.46}$$

Under this methodology, the setting of the control gains is simple and is independent of the system parameters; that is,

$$k_v = 2\sqrt{k_p} \tag{9.47}$$

must hold for critical damping. Figure 9.8 shows a block diagram of the partitioned controller used to control the system of Fig. 9.6.

EXAMPLE 9.5

If the parameters of the system in Fig. 9.6 are $m = 1$, $b = 1$, and $k = 1$, find α, β, and the gains k_p and k_v for a position-regulation control law that results in the system's being critically damped with a closed-loop stiffness of 16.0.

We choose

$$\alpha = 1,$$
$$\beta = \dot{x} + x, \tag{9.48}$$

so that the system appears as a unit mass from the fictitious f' input. We then set gain k_p to the desired closed-loop stiffness and set $k_v = 2\sqrt{k_p}$ for critical damping.

This gives

$$k_p = 16.0,$$

$$k_v = 8.0. \tag{9.49}$$

9.6 TRAJECTORY-FOLLOWING CONTROL

Rather than just maintaining the block at a desired location, let us enhance our controller so that the block can be made to follow a trajectory. The trajectory is given by a function of time, $x_d(t)$, that specifies the desired position of the block. We assume that the trajectory is smooth (i.e., the first two derivatives exist) and that our trajectory generator provides x_d, $\dot{x}_d$, and $\ddot{x}_d$ at all times t. We define the servo error between the desired and actual trajectory as $e = x_d - x$. A servo-control law that will cause trajectory following is

$$f' = \ddot{x}_d + k_v \dot{e} + k_p e. \tag{9.50}$$

We see that (9.50) is a good choice if we combine it with the equation of motion of a unit mass (9.44), which leads to

$$\ddot{x} = \ddot{x}_d + k_v \dot{e} + k_p e, \tag{9.51}$$

or

$$\ddot{e} + k_v \dot{e} + k_p e = 0. \tag{9.52}$$

This is a second-order differential equation for which we can choose the coefficients, so we can design any response we wish. (Often, critical damping is the choice made.) Such an equation is sometimes said to be written in **error space**, because it describes the evolution of errors relative to the desired trajectory. Figure 9.9 shows a block diagram of our trajectory-following controller.

If our model is perfect (i.e., our knowledge of m, b, and k), and if there is no noise and no initial error, the block will follow the desired trajectory exactly. If there is an initial error, it will be suppressed according to (9.52), and thereafter the system will follow the trajectory exactly.

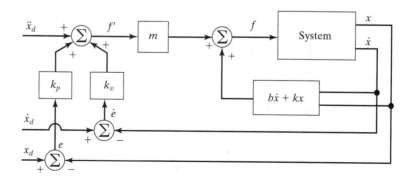

FIGURE 9.9: A trajectory-following controller for the system in Fig. 9.6.

9.7 DISTURBANCE REJECTION

One of the purposes of a control system is to provide **disturbance rejection**, that is, to maintain good performance (i.e., minimize errors) even in the presence of some external disturbances or **noise**. In Fig. 9.10, we show the trajectory-following controller with an additional input: a disturbance force f_{dist}. An analysis of our closed-loop system leads to the error equation

$$\ddot{e} + k_v \dot{e} + k_p e = f_{\text{dist}}. \tag{9.53}$$

Equation (9.53) is that of a differential equation driven by a forcing function. If it is known that f_{dist} is **bounded**—that is, that a constant a exists such that

$$\max_t f_{\text{dist}}(t) < a, \tag{9.54}$$

then the solution of the differential equation, $e(t)$, is also bounded. This result is due to a property of stable linear systems known as **bounded-input, bounded-output** or **BIBO** stability [3, 4]. This very basic result ensures that, for a large class of possible disturbances, we can at least be assured that the system remains stable.

Steady-state error

Let's consider the simplest kind of disturbance—namely, that f_{dist} is a constant. In this case, we can perform a **steady-state analysis** by analyzing the system at rest (i.e., the derivatives of all system variables are zero). Setting derivatives to zero in (9.53) yields the steady-state equation

$$k_p e = f_{\text{dist}}, \tag{9.55}$$

or

$$e = f_{\text{dist}}/k_p. \tag{9.56}$$

The value of e given by (9.56) represents a **steady-state error**. Thus, it is clear that the higher the position gain k_p, the smaller will be the steady-state error.

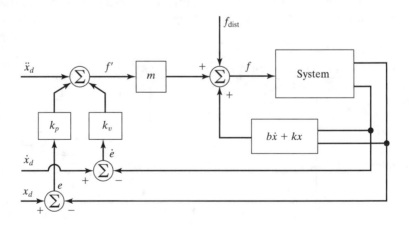

FIGURE 9.10: A trajectory-following control system with a disturbance acting.

Addition of an integral term

In order to eliminate steady-state error, a modified control law is sometimes used. The modification involves the addition of an *integral* term to the control law. The control law becomes

$$f' = \ddot{x}_d + k_v\dot{e} + k_pe + k_i \int edt,$$ (9.57)

which results in the error equation

$$\ddot{e} + k_v\dot{e} + k_pe + k_i \int edt = f_{\text{dist}}.$$ (9.58)

The term is added so that the system will have no steady-state error in the presence of constant disturbances. If $e(t) = 0$ for $t < 0$, we can write (9.58) for $t > 0$ as

$$\dddot{e} + k_v\ddot{e} + k_p\dot{e} + k_ie = \dot{f}_{\text{dist}},$$ (9.59)

which, in the steady state (for a constant disturbance), becomes

$$k_ie = 0,$$ (9.60)

so

$$e = 0.$$ (9.61)

With this control law, the system becomes a third-order system, and one can solve the corresponding third-order differential equation to work out the response of the system to initial conditions. Often, k_i is kept quite small so that the third-order system is "close" to the second-order system without this term (i.e., a dominant-pole analysis can be performed). The form of control law (9.57) is called a **PID control law**, or "proportional, integral, derivative" control law [4]. For simplicity, the displayed equations generally do not show an integral term in the control laws that we develop in this book.

9.8   CONTINUOUS VS. DISCRETE TIME CONTROL

In the control systems we have discussed, we implicitly assumed that the control computer performs the computation of the control law in zero time (i.e., infinitely fast), so that the value of the actuator force f is a continuous function of time. Of course, in reality, the computation requires some time, and the resulting commanded force is therefore a discrete "staircase" function. We shall employ this approximation of a very fast control computer throughout the book. This approximation is good if the rate at which new values of f are computed is much faster than the natural frequency of the system being controlled. In the field of **discrete time control** or **digital control**, one does not make this approximation but rather takes the **servo rate** of the control system into account when analyzing the system [3].

We will generally assume that the computations can be performed quickly enough that our continuous time assumption is valid. This raises a question: How quick is quick enough? There are several points that need to be considered in choosing a sufficiently fast servo (or sample) rate:

Tracking reference inputs: The frequency content of the desired or reference input places an absolute lower bound on the sample rate. The sample rate must be at least twice the bandwidth of reference inputs. This is usually not the limiting factor.

Disturbance rejection: In disturbance rejection, an upper bound on performance is given by a continuous-time system. If the sample period is longer than the correlation time of the disturbance effects (assuming a statistical model for random disturbances), then these disturbances will not be suppressed. Perhaps a good rule of thumb is that the sample period should be 10 times shorter than the correlation time of the noise [3].

Antialiasing: Any time an analog sensor is used in a digital control scheme, there will be a problem with aliasing unless the sensor's output is strictly band limited. In most cases, sensors do not have a band limited output, and so sample rate should be chosen such that the amount of energy that appears in the aliased signal is small.

Structural resonances: We have not included bending modes in our characterization of a manipulator's dynamics. All real mechanisms have finite stiffness and so will be subject to various kinds of vibrations. If it is important to suppress these vibrations (and it often is), we must choose a sample rate at least twice the natural frequency of these resonances. We will return to the topic of resonance later in this chapter.

9.9   MODELING AND CONTROL OF A SINGLE JOINT

In this section, we will develop a simplified model of a single rotary joint of a manipulator. A few assumptions will be made that will allow us to model the resulting system as a second-order linear system. For a more complete model of an actuated joint, see [5].

A common actuator found in many industrial robots is the direct current (DC) torque motor (as in Fig. 8.18). The nonturning part of the motor (the **stator**) consists of a housing, bearings, and either permanent magnets or electromagnets. These stator magnets establish a magnetic field across the turning part of the motor (the **rotor**). The rotor consists of a shaft and windings through which current moves to power the motor. The current is conducted to the windings via brushes, which make contact with the commutator. The commutator is wired to the various windings (also called the **armature**) in such a way that torque is always produced in the desired direction. The underlying physical phenomenon [6] that causes a motor to generate a torque when current passes through the windings can be expressed as

$$F = qV \times B, \tag{9.62}$$

where charge q, moving with velocity V through a magnetic field B, experiences a force F. The charges are those of electrons moving through the windings, and the magnetic field is that set up by the stator magnets. Generally, the torque-producing ability of a motor is stated by means of a single **motor torque constant**, which relates armature current to the output torque as

$$\tau_m = k_m i_a. \tag{9.63}$$

When a motor is rotating, it acts as a generator, and a voltage develops across the armature. A second motor constant, the **back emf constant**,[3] describes the voltage generated for a given rotational velocity:

$$v = k_e \dot{\theta}_m. \tag{9.64}$$

Generally, the fact that the commutator is switching the current through various sets of windings causes the torque produced to contain some **torque ripple**. Although sometimes important, this effect can usually be ignored. (In any case, it is quite hard to model—and quite hard to compensate for, even if it is modeled.)

Motor-armature inductance

Figure 9.11 shows the electric circuit of the armature. The major components are a voltage source, v_a, the inductance of the armature windings, l_a, the resistance of the armature windings, r_a, and the generated back emf, v. The circuit is described by a first-order differential equation:

$$l_a \dot{i}_a + r_a i_a = v_a - k_e \dot{\theta}_m. \tag{9.65}$$

It is generally desirable to control the torque generated by the motor (rather than the velocity) with electronic motor driver circuitry. These drive circuits sense the current through the armature and continuously adjust the voltage source v_a so that a desired current i_a flows through the armature. Such a circuit is called a **current amplifier** motor driver [7]. In these current-drive systems, the rate at which the armature current can be commanded to change is limited by the motor inductance l_a and by an upper limit on the voltage capability of the voltage source v_a. The net effect is that of a **low-pass filter** between the requested current and output torque.

Our first simplifying assumption is that the inductance of the motor can be neglected. This is a reasonable assumption when the natural frequency of the closed-loop control system is quite low compared to the cut-off frequency of the implicit low-pass filter in the current-drive circuitry due to the inductance. This assumption, along with the assumption that torque ripple is a negligible effect, means that we can essentially command torque directly. Although there might be a scale factor (such as k_m) to contend with, we will assume that the actuator acts as a pure torque source that we can command directly.

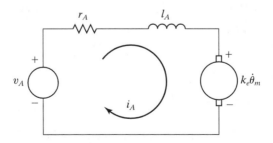

FIGURE 9.11: The armature circuit of a DC torque motor.

---

[3]"emf" stands for electromotive force.

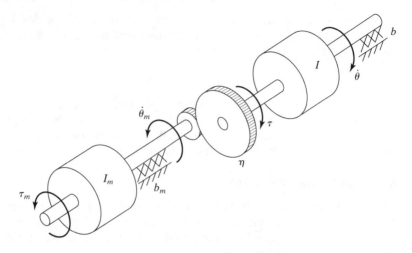

FIGURE 9.12: Mechanical model of a DC torque motor connected through gearing to an inertial load.

Effective inertia

Figure 9.12 shows the mechanical model of the rotor of a DC torque motor connected through a gear reduction to an inertial load. The torque applied to the rotor, τ_m, is given by (9.63) as a function of the current i_a flowing in the armature circuit. The gear ratio (η) causes an increase in the torque seen at the load and a reduction in the speed of the load, given by

$$\tau = \eta \tau_m,$$
$$\dot{\theta} = (1/\eta)\dot{\theta}_m, \tag{9.66}$$

where $\eta > 1$. Writing a torque balance for this system in terms of torque at the rotor yields

$$\tau_m = I_m \ddot{\theta}_m + b_m \dot{\theta}_m + (1/\eta)\left(I\ddot{\theta} + b\dot{\theta}\right), \tag{9.67}$$

where I_m and I are the inertias of the motor rotor and of the load, respectively, and b_m and b are viscous friction coefficients for the rotor and load bearings, respectively. Using the relations (9.66), we can write (9.67) in terms of motor variables as

$$\tau_m = \left(I_m + \frac{I}{\eta^2}\right)\ddot{\theta}_m + \left(b_m + \frac{b}{\eta^2}\right)\dot{\theta}_m \tag{9.68}$$

or in terms of load variables as

$$\tau = (I + \eta^2 I_m)\ddot{\theta} + (b + \eta^2 b_m)\dot{\theta}. \tag{9.69}$$

The term $I + \eta^2 I_m$ is sometimes called the **effective inertia** "seen" at the output (link side) of the gearing. Likewise, the term $b + \eta^2 b_m$ can be called the **effective damping**. Note that, in a highly geared joint (i.e., $\eta \gg 1$), the inertia of the motor rotor can be a significant portion of the combined effective inertia. It is this effect that allows us to make the assumption that the effective inertia is a constant. We know from Chapter 6 that the inertia, I, of a joint of the mechanism actually varies with configuration and load. However, in highly geared robots, the variations represent a smaller percentage than they would in a **direct-drive** manipulator (i.e., $\eta = 1$). To ensure that the motion of the robot link is never underdamped, the value used for I should be the maximum of the range of values that I takes on; we'll call this value I_{max}. This choice results in a system that is critically damped or overdamped in all situations. In Chapter 10, we will deal with varying inertia directly and will not have to make this assumption.

EXAMPLE 9.6

If the apparent link inertia, I, varies between 2 and 6 Kg-m^2, the rotor inertia is $I_m = 0.01$, and the gear ratio is $\eta = 30$, what are the minimum and maximum of the effective inertia?

The minimum effective inertia is

$$I_{min} + \eta^2 I_m = 2.0 + (900)(0.01) = 11.0; \tag{9.70}$$

the maximum is

$$I_{max} + \eta^2 I_m = 6.0 + (900)(0.01) = 15.0. \tag{9.71}$$

Hence, we see that, as a percentage of the total effective inertia, the variation of inertia is reduced by the gearing.

Unmodeled flexibility

The other major assumption we have made in our model is that the gearing, the shafts, the bearings, and the driven link are not flexible. In reality, all of these elements have finite stiffness, and their flexibility, if modeled, would increase the order of the system. The argument for ignoring flexibility effects is that, if the system is sufficiently stiff, the natural frequencies of these **unmodeled resonances** are very high and can be neglected compared to the influence of the dominant second-order poles that we have modeled.[4] The term "unmodeled" refers to the fact that, for purposes of control-system analysis and design, we neglect these effects and use a simpler dynamic model, such as (9.69).

Because we have chosen not to model structural flexibilities in the system, we must be careful not to excite these resonances. A rule of thumb [8] is that, if the lowest structural resonance is ω_{res}, then we must limit our closed-loop natural frequency according to

$$\omega_n \leq \frac{1}{2}\omega_{res}. \tag{9.72}$$

This provides some guidance on how to choose gains in our controller. We have seen that increasing gains leads to faster response and lower steady-state error, but we now see that unmodeled structural resonances limit the magnitude of gains. Typical industrial manipulators have structural resonances in the range from 5 Hz to 25 Hz [8]. Recent designs using direct-drive arrangements that do not contain flexibility

---

[4]This is basically the same argument we used to neglect the pole due to the motor inductance. Including it would also have raised the order of the overall system.

introduced by reduction and transmission systems have their lowest structural resonances as high as 70 Hz [9].

EXAMPLE 9.7

Consider the system of Fig. 9.7 with the parameter values $m = 1$, $b = 1$, and $k = 1$. Additionally, it is known that the lowest unmodeled resonance of the system is at 8 radians/second. Find α, β, and gains k_p and k_v for a position-control law so the system is critically damped, doesn't excite unmodeled dynamics, and has as high a closed-loop stiffness as possible.

We choose

$$\alpha = 1,$$

$$\beta = \dot{x} + x, \tag{9.73}$$

so that the system appears as a unit mass from the fictitious f' input. Using our rule of thumb (9.72), we choose the closed-loop natural frequency to be $\omega_n = 4$ radians/second. From (9.18) and (9.46), we have $k_p = \omega_n^2$, so

$$k_p = 16.0,$$

$$k_v = 8.0. \tag{9.74}$$

Estimating resonant frequency

The same sources of structural flexibility discussed in Chapter 8 give rise to resonances. In each case where a structural flexibility can be identified, an approximate analysis of the resulting vibration is possible if we can describe the effective mass or inertia of the flexible member. This is done by approximating the situation by a simple spring–mass system, which, as given in (9.20), exhibits the natural frequency

$$\omega_n = \sqrt{k/m}, \tag{9.75}$$

where k is the stiffness of the flexible member and m is the equivalent mass displaced in vibrations.

EXAMPLE 9.8

A shaft (assumed massless) with a stiffness of 400 Nt-m/radian drives a rotational inertia of 1 Kg-m^2. If the shaft stiffness was neglected in the modeling of the dynamics, what is the frequency of this unmodeled resonance?

Using (9.75), we have

$$\omega_{res} = \sqrt{400/1} = 20 \text{ rad/second} = 20/(2\pi)\text{Hz} \cong 3.2 \text{ Hz}. \tag{9.76}$$

For the purposes of a rough estimate of the lowest resonant frequency of beams and shafts, [10] suggests using a **lumped model** of the mass. We already

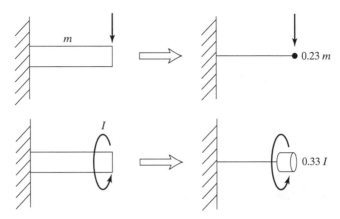

FIGURE 9.13: Lumped models of beams for estimation of lowest lateral and torsional resonance.

have formulas for estimating stiffness at the ends of beams and shafts; these lumped models provide the effective mass or inertia needed for our estimation of resonant frequency. Figure 9.13 shows the results of an energy analysis [10] which suggests that a beam of mass m be replaced by a point mass at the end of 0.23 m and, likewise, that a distributed inertia of I be replaced by a lumped 0.33 I at the end of the shaft.

EXAMPLE 9.9

A link of mass 4.347 Kg has an end-point lateral stiffness of 3600 Nt/m. Assuming the drive system is completely rigid, the resonance due to the flexibility of the link will limit control gains. What is ω_{res}?

The 4.347 Kg mass is distributed along the link. Using the method of Fig. 9.13, the effective mass is $(0.23)(4.347) \cong 1.0$ Kg. Hence, the vibration frequency is

$$\omega_{res} = \sqrt{3600/1.0} = 60 \text{ radians/second} = 60/(2\pi)\text{Hz} \cong 9.6 \text{ Hz.} \qquad (9.77)$$

The inclusion of structural flexibilities in the model of the system used for control-law synthesis is required if we wish to achieve closed-loop bandwidths higher than that given by (9.75). The resulting system models are of high order, and the control techniques applicable to this situation become quite sophisticated. Such control schemes are currently beyond the state of the art of industrial practice but are an active area of research [11, 12].

Control of a single joint

In summary, we make the following three major assumptions:

1. The motor inductance l_a can be neglected.
2. Taking into account high gearing, we model the effective inertia as a constant equal to $I_{max} + \eta^2 I_m$.
3. Structural flexibilities are neglected, except that the lowest structural resonance ω_{res} is used in setting the servo gains.

With these assumptions, a single joint of a manipulator can be controlled with the partitioned controller given by

$$\alpha = I_{\max} + \eta^2 I_m,$$

$$\beta = (b + \eta^2 b_m)\dot{\theta}, \tag{9.78}$$

$$\tau' = \ddot{\theta}_d + k_v \dot{e} + k_p e. \tag{9.79}$$

The resulting system closed-loop dynamics are

$$\ddot{e} + k_v \dot{e} + k_p e = \tau_{\text{dist}}, \tag{9.80}$$

where the gains are chosen as

$$k_p = \omega_n^2 = \frac{1}{4}\omega_{\text{res}}^2,$$

$$k_v = 2\sqrt{k_p} = \omega_{\text{res}}. \tag{9.81}$$

9.10 ARCHITECTURE OF AN INDUSTRIAL-ROBOT CONTROLLER

In this section, we briefly look at the architecture of the control system of the Unimation PUMA 560 industrial robot. As shown in Fig. 9.14, the hardware architecture is that of a two-level hierarchy, with a DEC LSI-11 computer serving as the top-level "master" control computer passing commands to six Rockwell 6503 microprocessors.[5] Each of these microprocessors controls an individual joint with a PID control law not unlike that presented in this chapter. Each joint of the PUMA 560 is instrumented with an incremental optical encoder. The encoders are interfaced to an up/down counter, which the microprocessor can read to obtain the current joint position. There are no tachometers in the PUMA 560; rather, joint positions are differenced on subsequent servo cycles to obtain an estimate of joint velocity. In order to command torques to the DC torque motors, the microprocessor

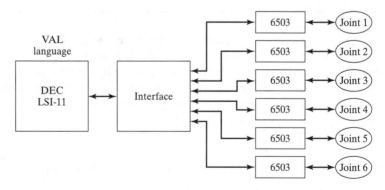

FIGURE 9.14: Hierarchical computer architecture of the PUMA 560 robot-control system.

---

[5]These simple 8-bit computers are already old technology. It is common these days for robot controllers to be based on 32-bit microprocessors.

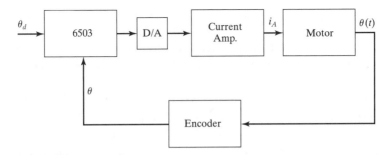

FIGURE 9.15: Functional blocks of the joint-control system of the PUMA 560.

is interfaced to a digital-to-analog converter (DAC) so that motor currents can be commanded to the current-driver circuits. The current flowing through the motor is controlled in analog circuitry by adjusting the voltage across the armature as needed to maintain the desired armature current. A block diagram is shown in Fig. 9.15.

Each 28 milliseconds, the LSI-11 computer sends a new position command (**set-point**) to the joint microprocessors. The joint microprocessors are running on a 0.875 millisecond cycle. In this time, they interpolate the desired position set-point, compute the servo error, compute the PID control law, and command a new value of torque to the motors.

The LSI-11 computer carries out all the "high-level" operations of the overall control system. First of all, it takes care of interpreting the VAL (Unimation's robot programming language) program commands one by one. When a motion command is interpreted, the LSI-11 must perform any needed inverse kinematic computations, plan a desired trajectory, and begin generating trajectory via points every 28 milliseconds for the joint controllers.

The LSI-11 is also interfaced to such standard peripherals as the terminal and a floppy disk drive. In addition, it is interfaced to a **teach pendant**. A teach pendant is a handheld button box that allows the operator to move the robot around in a variety of modes. For example, the PUMA 560 system allows the user to move the robot incrementally in joint coordinates or in Cartesian coordinates from the teach pendant. In this mode, teach-pendant buttons cause a trajectory to be computed "on the fly" and passed down to the joint-control microprocessors.

BIBLIOGRAPHY

[1] W. Boyce and R. DiPrima, *Elementary Differential Equations*, 3rd edition, John Wiley and Sons, New York, 1977.

[2] E. Purcell, *Calculus with Analytic Geometry*, Meredith Corporation, New York, 1972.

[3] G. Franklin and J.D. Powell, *Digital Control of Dynamic Systems*, Addison-Wesley, Reading, MA, 1980.

[4] G. Franklin, J.D. Powell, and A. Emami-Naeini, *Feedback Control of Dynamic Systems*, Addison-Wesley, Reading, MA, 1986.

[5] J. Luh, "Conventional Controller Design for Industrial Robots—a Tutorial," *IEEE Transactions on Systems, Man, and Cybernetics*, Vol. SMC-13, No. 3, June 1983.

[6] D. Halliday and R. Resnik, *Fundamentals of Physics*, Wiley, New York 1970.

[7] Y. Koren and A. Ulsoy, "Control of DC Servo-Motor Driven Robots," *Proceedings of Robots 6 Conference*, SME, Detroit, March 1982.

[8] R.P. Paul, *Robot Manipulators*, MIT Press, Cambridge, MA, 1981.

[9] H. Asada and K. Youcef-Toumi, *Direct-Drive Robots—Theory and Practice*, MIT Press, Cambridge, MA, 1987.

[10] J. Shigley, *Mechanical Engineering Design*, 3rd edition, McGraw-Hill, New York, 1977.

[11] W. Book, "Recursive Lagrangian Dynamics of Flexible Manipulator Arms," *The International Journal of Robotics Research*, Vol. 3, No. 3, 1984.

[12] R. Cannon and E. Schmitz, "Initial Experiments on the End-Point Control of a Flexible One Link Robot," *The International Journal of Robotics Research*, Vol. 3, No. 3, 1984.

[13] R.J. Nyzen, "*Analysis and Control of an Eight-Degree-of-Freedom Manipulator*," Ohio University Master's Thesis, Mechanical Engineering, Dr. Robert L. Williams II, Advisor, August 1999.

[14] R.L. Williams II, "*Local Performance Optimization for a Class of Redundant Eight-Degree-of-Freedom Manipulators*," **NASA Technical Paper 3417**, NASA Langley Research Center, Hampton, VA, March 1994.

EXERCISES

9.1 [20] For a second-order differential equation with complex roots

$$s_1 = \lambda + \mu i,$$
$$s_2 = \lambda - \mu i,$$

show that the general solution

$$x(t) = c_1 e^{s_1 t} + c_2 e^{s_2 t},$$

can be written

$$x(t) = c_1 e^{\lambda t} \cos(\mu t) + c_2 e^{\lambda t} \sin(\mu t).$$

9.2 [13] Compute the motion of the system in Fig. 9.2 if parameter values are $m = 2$, $b = 6$, and $k = 4$ and the block (initially at rest) is released from the position $x = 1$.

9.3 [13] Compute the motion of the system in Fig. 9.2 if parameter values are $m = 1$, $b = 2$, and $k = 1$ and the block (initially at rest) is released from the position $x = 4$.

9.4 [13] Compute the motion of the system in Fig. 9.2 if parameter values are $m = 1$, $b = 4$, and $k = 5$ and the block (initially at rest) is released from the position $x = 2$.

9.5 [15] Compute the motion of the system in Fig. 9.2 if parameter values are $m = 1$, $b = 7$, and $k = 10$ and the block is released from the position $x = 1$ with an initial velocity of $\dot{x} = 2$.

9.6 [15] Use the (1, 1) element of (6.60) to compute the variation (as a percentage of the maximum) of the inertia "seen" by joint 1 of this robot as it changes configuration. Use the numerical values

$$l_1 = l_2 = 0.5 \text{ m},$$
$$m_1 = 4.0 \text{ Kg},$$
$$m_2 = 2.0 \text{ Kg}.$$

Consider that the robot is direct drive and that the rotor inertia is negligible.

9.7 [17] Repeat Exercise 9.6 for the case of a geared robot (use $\eta = 20$) and a rotor inertia of $I_m = 0.01$ Kg m^2.

9.8 [18] Consider the system of Fig. 9.6 with the parameter values $m = 1$, $b = 4$, and $k = 5$. The system is also known to possess an unmodeled resonance at $\omega_{res} = 6.0$ radians/second. Determine the gains k_v and k_p that will critically damp the system with as high a stiffness as is reasonable.

9.9 [25] In a system like that of Fig. 9.12, the inertial load, I, varies between 4 and 5 Kg-m^2. The rotor inertia is $I_m = 0.01$ Kg-m^2, and the gear ratio is $\eta = 10$. The system possesses unmodeled resonances at 8.0, 12.0, and 20.0 radians/second. Design α and β of the partitioned controller and give the values of k_p and k_v such that the system is never underdamped and never excites resonances, but is as stiff as possible.

9.10 [18] A designer of a direct-drive robot suspects that the resonance due to beam flexibility of the link itself will be the cause of the lowest unmodeled resonance. If the link is approximately a square-cross-section beam of dimensions $5 \times 5 \times 50$ cm with a 1-cm wall thickness and a total mass of 5 Kg, estimate ω_{res}.

9.11 [15] A direct-drive robot link is driven through a shaft of stiffness 1000 Nt-m/radian. The link inertia is 1 Kg-m^2. Assuming the shaft is massless, what is ω_{res}?

9.12 [18] A shaft of stiffness 500 Nt-m/radian drives the input of a rigid gear pair with $\eta = 8$. The output of the gears drives a rigid link of inertia 1 Kg-m^2. What is the ω_{res} caused by flexibility of the shaft?

9.13 [25] A shaft of stiffness 500 Nt-m/radian drives the input of a rigid gear pair with $\eta = 8$. The shaft has an inertia of 0.1 Kg-m^2. The output of the gears drives a rigid link of inertia 1 Kg-m^2. What is the ω_{res} caused by flexibility of the shaft?

9.14 [28] In a system like that of Fig. 9.12, the inertial load, I, varies between 4 and 5 Kg-m^2. The rotor inertia is $I_m = 0.01$ Kg-m^2, and the gear ratio is $\eta = 10$. The system possesses an unmodeled resonance due to an end-point stiffness of the link of 2400 Nt-m/radian. Design α and β of the partitioned controller, and give the values of k_p and k_v such that the system is never underdamped and never excites resonances, but is as stiff as possible.

9.15 [25] A steel shaft of length 30 cm and diameter 0.2 cm drives the input gear of a reduction of $\eta = 8$. The rigid output gear drives a steel shaft of length 30 cm and diameter 0.3 cm. What is the range of resonant frequencies observed if the load inertia varies between 1 and 4 Kg-m^2?

PROGRAMMING EXERCISE (PART 9)

We wish to simulate a simple trajectory-following control system for the three-link planar arm. This control system will be implemented as an independent-joint PD (proportional plus derivative) control law. Set the servo gains to achieve closed-loop stiffnesses of 175.0, 110.0, and 20.0 for joints 1 through 3 respectively. Try to achieve approximate critical damping.

Use the simulation routine **UPDATE** to simulate a discrete-time servo running at 100 Hz—that is, calculate the control law at 100 Hz, not at the frequency of the numerical integration process. Test the control scheme on the following tests:

1. Start the arm at $\Theta = (60, -110, 20)$ and command it to stay there until $time = 3.0$, when the set-points should instantly change to $\Theta = (60, -50, 20)$. That is, give a step input of 60 degrees to joint 2. Record the error–time history for each joint.

2. Control the arm to follow the cubic-spline trajectory from Programming Exercise Part 7. Record the error–time history for each joint.

MATLAB EXERCISE 9

This exercise focuses on linearized independent joint-control simulation for the shoulder joint (joint 2) of the NASA eight-axis AAI ARMII (Advanced Research Manipulator II) manipulator arm—see [14]. Familiarity with linear classical feedback-control systems, including block diagrams and Laplace transforms, is assumed. We will use Simulink, the graphical user interface of MATLAB.

Figure 9.16 shows a linearized open-loop system-dynamics model for the ARMII electromechanical shoulder joint/link, actuated by an armature-controller DC servomotor. The open-loop input is reference voltage V_{ref} (boosted to armature voltage via an amplifier), and the output of interest is the load shaft angle ThetaL. The figure also shows the feedback-control diagram, where the load-shaft angle is sensed via an optical encoder and provided as feedback to the PID controller. The table describes all system parameters and variables.

If we reflect the load shaft inertia and damping to the motor shaft, the effective polar inertia and damping coefficient are $J = J_M + J_L(t)/n^2$ and $C = C_M + C_L/n^2$. By virtue of the large gear ratio n, these effective values are not much different from the motor-shaft values. Thus, the gear ratio allows us to ignore variations in the configuration-dependent load-shaft inertia $J_L(t)$ and just set a reasonable average value.

The ARMII shoulder joint constant parameters are given in the accompanying table [13]. Note that we can use the English units directly, because their effect cancels out inside the control diagram. Also, we can directly use deg units for the angle. Develop a Simulink model to simulate the single-joint control model from the model and feedback-control diagram shown; use the specific parameters from the table. For the nominal case, determine the PID gains by trial and error for "good" performance (reasonable percent overshoot, rise time, peak time, and settling time). Simulate the resulting motion for moving this shoulder joint for a step input of 0 to 60 deg. Plot the simulated load-angle value over time, plus the load-shaft angular velocity over time. In addition, plot the

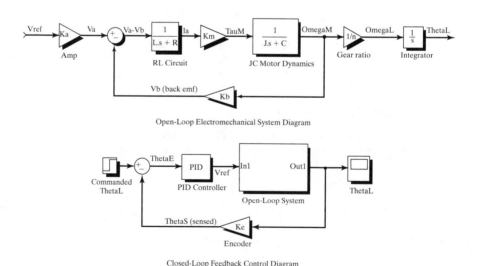

FIGURE 9.16: Linearized open-loop system-dynamics model for the ARMII electromechanical shoulder joint/link, actuated by an armature-controller DC servomotor.

TABLE 9.1: ARMII shoulder joint constant parameters.

$V_a(t)$	armature voltage	$\tau_M(t)$	generated motor torque	$\tau_L(t)$	load torque
$L = 0.0006H$	armature inductance	$\theta_M(t)$	motor shaft angle	$\theta_L(t)$	load shaft angle
$R = 1.40\Omega$	armature resistance	$\omega_M(t)$	motor shaft velocity	$\omega_L(t)$	load shaft velocity
$i_a(t)$	armature current	$J_M = 0.00844$ $lb_f\text{-in-s}^2$	lumped motor polar inertia	$J_L(t) = 1$ $lb_f\text{-in-s}^2$	lumped load polar inertia
$V_b(t)$	back emf voltage	$C_M = 0.00013$ $lb_f\text{-in/deg/s}$	motor shaft viscous damping coefficient	$C_L = 0.5$ $lb_f\text{-}$ $in/deg/s$	load shaft viscous damping coefficient
$K_a = 12$	amplifier gain	$n = 200$	gear ratio	$g = 0$ in/s^2	gravity (ignore gravity at first)
$K_b = 0.00867$ $V/deg/s$	back emf constant	$K_M = 4.375$ $lb_f\text{-in/A}$	torque constant	$K_e = 1$	encoder transfer function

control effort—that is, the armature voltage V_a over time. (On the same graph, also give the back emf V_b.)

Now, try some changes—Simulink is so easy and enjoyable to change:

1) The step input is frustrating for controller design, so try a ramped step input instead: Ramp from 0 to 60 deg in 1.5 sec, then hold the 60-deg command for all time greater than 1.5 sec. Redesign PID gains and restimulate.

2) Investigate whether the inductor L is significant in this system. (The electrical system rises much faster than the mechanical system—this effect can be represented by time constants.)

3) We don't have a good estimate for the load inertia and damping (J_L and C_L). With your best PID gains from before, investigate how big these values can grow (scale the nominal parameters up equally) before they affect the system.

4) Now, include the effect of gravity as a disturbance to the motor torque T_M. Assume that the moving robot mass is 200 lb and the moving length beyond joint 2 is 6.4 feet. Test for the nominal "good" PID gains you found; redesign if necessary. The shoulder load angle θ_2 zero configuration is straight up.

C H A P T E R 10

Nonlinear control
of manipulators

10.1 INTRODUCTION
10.2 NONLINEAR AND TIME-VARYING SYSTEMS
10.3 MULTI-INPUT, MULTI-OUTPUT CONTROL SYSTEMS
10.4 THE CONTROL PROBLEM FOR MANIPULATORS
10.5 PRACTICAL CONSIDERATIONS
10.6 CURRENT INDUSTRIAL-ROBOT CONTROL SYSTEMS
10.7 LYAPUNOV STABILITY ANALYSIS
10.8 CARTESIAN-BASED CONTROL SYSTEMS
10.9 ADAPTIVE CONTROL

10.1 INTRODUCTION

In the previous chapter, we made several approximations to allow a linear analysis of the manipulator-control problem. Most important among these approximations was that each joint could be considered independent and that the inertia "seen" by each joint actuator was constant. In implementations of linear controllers as introduced in the previous chapter, this approximation results in nonuniform damping throughout the workspace and other undesirable effects. In this chapter, we will introduce a more advanced control technique for which this assumption will not have to be made.

In Chapter 9, we modeled the manipulator by n independent second-order differential equations and based our controller on that model. In this chapter, we will base our controller design directly on the $n \times 1$-nonlinear vector differential equation of motion, derived in Chapter 6 for a general manipulator.

The field of nonlinear control theory is large; we must therefore restrict our attention to one or two methods that seem well suited to mechanical manipulators. Consequently, the major focus of the chapter will be one particular method, apparently first proposed in [1] and named the **computed-torque method** in [2, 3]. We will also introduce one method of stability analysis of nonlinear systems, known as **Lyapunov's** method [4].

To begin our discussion of nonlinear techniques for controlling a manipulator, we return again to a very simple single-degree-of-freedom mass–spring friction system.

10.2 NONLINEAR AND TIME-VARYING SYSTEMS

In the preceding development, we dealt with a linear constant-coefficient differential equation. This mathematical form arose because the mass–spring friction system of Fig. 9.6 was modeled as a linear time-invariant system. For systems whose parameters vary in time or systems that by nature are nonlinear, solutions are more difficult.

When nonlinearities are not severe, **local linearization** can be used to derive linear models that are approximations of the nonlinear equations in the neighborhood of an **operating point**. Unfortunately, the manipulator-control problem is not well suited to this approach, because manipulators constantly move among regions of their workspaces so widely separated that no linearization valid for all regions can be found.

Another approach is to move the operating point with the manipulator as it moves, always linearizing about the desired position of the manipulator. The result of this sort of *moving linearization* is a linear, but time-varying, system. Although this quasi-static linearization of the original system is useful in some analysis and design techniques, we will not make use of it in our control-law synthesis procedure. Rather, we will deal with the nonlinear equations of motion directly and will not resort to linearizations in deriving a controller.

If the spring in Fig. 9.6 were not linear but instead contained a nonlinear element, we could consider the system quasi-statically and, at each instant, figure out where the poles of the system are located. We would find that the poles "move" around in the real–imaginary plane as a function of the position of the block. Hence, we could not select fixed gains that would keep the poles in a desirable location (for example, at critical damping). So we may be tempted to consider a more complicated control law, in which the gains are time-varying (actually, varying as a function of the block's position) in such a manner that the system is always critically damped. Essentially, this would be done by computing k_p such that the combination of the nonlinear effect of the spring would be exactly cancelled by a nonlinear term in the control law so that the overall stiffness would stay a constant at all times. Such a control scheme might be called a **linearizing** control law, because it uses a nonlinear control term to "cancel" a nonlinearity in the controlled system, so that the overall closed loop system is linear.

We will now return to our partitioned control law and see that it can perform this linearizing function. In our partitioned control-law scheme, the servo law remains the same as always, but the model-based portion now will contain a model of the nonlinearity. Thus, the model-based portion of the control performs a linearization function. This is best shown in an example.

EXAMPLE 10.1

Consider the nonlinear spring characteristic shown in Fig. 10.1. Rather than the usual linear spring relationship, $f = kx$, this spring is described by $f = qx^3$. If this spring is part of the physical system shown in Fig. 9.6, construct a control law to keep the system critically damped with a stiffness of k_{CL}.

The open-loop equation is

$$m\ddot{x} + b\dot{x} + qx^3 = f. \tag{10.1}$$

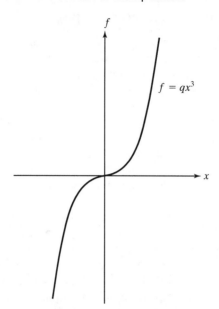

FIGURE 10.1: The force-vs.-distance characteristic of a nonlinear spring.

The model-based portion of the control is $f = \alpha f' + \beta$, where now we use

$$\alpha = m,$$
$$\beta = b\dot{x} + qx^3; \tag{10.2}$$

the servo portion is, as always

$$f' = \ddot{x}_d + k_v\dot{e} + k_p e, \tag{10.3}$$

where the values of the gains are calculated from some desired performance specification. Figure 10.2 shows a block diagram of this control system. The resulting closed-loop system maintains poles in fixed locations.

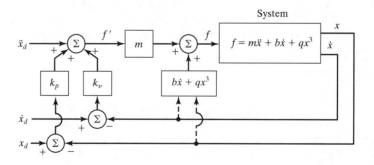

FIGURE 10.2: A nonlinear control system for a system with a nonlinear spring.

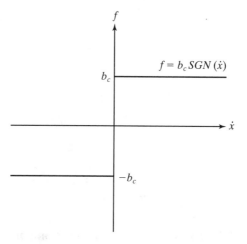

FIGURE 10.3: The force-vs.-velocity characteristic of Coulomb friction.

EXAMPLE 10.2

Consider the nonlinear friction characteristic shown in Fig. 10.3. Whereas linear friction is described by $f = b\dot{x}$, this **Coulomb friction** is described by $f = b_c sgn(\dot{x})$. For most of today's manipulators, the friction of the joint in its bearing (be it rotational or linear) is modeled more accurately by this nonlinear characteristic than by the simpler, linear model. If this type of friction is present in the system of Fig. 9.6, design a control system that uses a nonlinear model-based portion to damp the system critically at all times.

The open-loop equation is

$$m\ddot{x} + b_c sgn(\dot{x}) + kx = f. \tag{10.4}$$

The partitioned control law is $f = \alpha f' + \beta$, where

$$\alpha = m,$$
$$\beta = b_c sgn(\dot{x}) + kx, \tag{10.5}$$
$$f' = \ddot{x}_d + k_v \dot{e} + k_p e,$$

where the values of the gains are calculated from some desired performance specification.

EXAMPLE 10.3

Consider the single-link manipulator shown in Fig. 10.4. It has one rotational joint. The mass is considered to be located at a point at the distal end of the link, and so the moment of inertia is ml^2. There is Coulomb and viscous friction acting at the joint, and there is a load due to gravity.

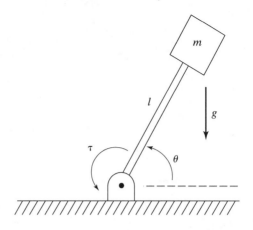

FIGURE 10.4: An inverted pendulum or a one-link manipulator.

The model of the manipulator is

$$\tau = ml^2\ddot{\theta} + v\dot{\theta} + csgn(\dot{\theta}) + mlg\cos(\theta). \tag{10.6}$$

As always, the control system has two parts, the linearizing model-based portion and the servo-law portion.

The model-based portion of the control is $f = \alpha f' + \beta$, where

$$\alpha = ml^2,$$
$$\beta = v\dot{\theta} + csgn(\dot{\theta}) + mlg\cos(\theta); \tag{10.7}$$

the servo portion is, as always,

$$f' = \ddot{\theta}_d + k_v\dot{e} + k_p e, \tag{10.8}$$

where the values of the gains are calculated from some desired performance specification.

We have seen that, in certain simple cases, it is not difficult to design a nonlinear controller. The general method used in the foregoing simple examples is the same method we will use for the problem of manipulator control:

1. Compute a nonlinear model-based control law that "cancels" the nonlinearities of the system to be controlled.
2. Reduce the system to a linear system that can be controlled with the simple linear servo law developed for the unit mass.

In some sense, the linearizing control law implements an *inverse model* of the system being controlled. The nonlinearities in the system cancel those in the inverse model; this, together with the servo law, results in a linear closed-loop system. Obviously, to do this cancelling, we must know the parameters and the structure of the nonlinear system. This is often a problem in practical application of this method.

10.3   MULTI-INPUT, MULTI-OUTPUT CONTROL SYSTEMS

Unlike the simple examples we have discussed in this chapter so far, the problem of controlling a manipulator is a multi-input, multi-output (MIMO) problem. That is, we have a *vector* of desired joint positions, velocities, and accelerations, and the control law must compute a *vector* of joint-actuator signals. Our basic scheme, partitioning the control law into a model-based portion and a servo portion, is still applicable, but it now appears in a matrix–vector form. The control law takes the form

$$F = \alpha F' + \beta,\tag{10.9}$$

where, for a system of n degrees of freedom, F, F', and β are $n \times 1$ vectors and α is an $n \times n$ matrix. Note that the matrix α is not necessarily diagonal, but rather is chosen to **decouple** the n equations of motion. If α and β are correctly chosen, then, from the F' input, the system appears to be n independent unit masses. For this reason, in the multidimensional case, the model-based portion of the control law is called a **linearizing and decoupling** control law. The servo law for a multidimensional system becomes

$$F' = \ddot{X}_d + K_v \dot{E} + K_p E,\tag{10.10}$$

where K_v and K_p are now $n \times n$ matrices, which are generally chosen to be diagonal with constant gains on the diagonal. E and $\dot{E}$ are $n \times 1$ vectors of the errors in position and velocity, respectively.

10.4   THE CONTROL PROBLEM FOR MANIPULATORS

In the case of manipulator control, we developed a model and the corresponding equations of motion in Chapter 6. As we saw, these equations are quite complicated. The rigid-body dynamics have the form

$$\tau = M(\Theta)\ddot{\Theta} + V(\Theta, \dot{\Theta}) + G(\Theta),\tag{10.11}$$

where $M(\Theta)$ is the $n \times n$ inertia matrix of the manipulator, $V(\Theta, \dot{\Theta})$ is an $n \times 1$ vector of centrifugal and Coriolis terms, and $G(\Theta)$ is an $n \times 1$ vector of gravity terms. Each element of $M(\Theta)$ and $G(\Theta)$ is a complicated function that depends on Θ, the position of all the joints of the manipulator. Each element of $V(\Theta, \dot{\Theta})$ is a complicated function of both Θ and $\dot{\Theta}$.

Additionally, we could incorporate a model of friction (or other non-rigid-body effects). Assuming that our model of friction is a function of joint positions and velocities, we add the term $F(\Theta, \dot{\Theta})$ to (10.11), to yield the model

$$\tau = M(\Theta)\ddot{\Theta} + V(\Theta, \dot{\Theta}) + G(\Theta) + F(\Theta, \dot{\Theta}).\tag{10.12}$$

The problem of controlling a complicated system like (10.12) can be handled by the partitioned controller scheme we have introduced in this chapter. In this case, we have

$$\tau = \alpha\tau' + \beta,\tag{10.13}$$

where τ is the $n \times 1$ vector of joint torques. We choose

$$\alpha = M(\Theta),$$
$$\beta = V(\Theta, \dot{\Theta}) + G(\Theta) + F(\Theta, \dot{\Theta}),\tag{10.14}$$

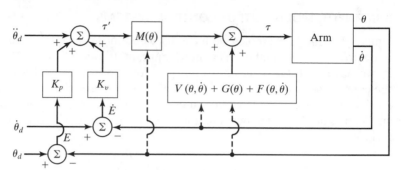

FIGURE 10.5: A model-based manipulator-control system.

with the servo law

$$\tau' = \ddot{\Theta}_d + K_v \dot{E} + K_p E,$$ (10.15)

where

$$E = \Theta_d - \Theta.$$ (10.16)

The resulting control system is shown in Fig. 10.5.

Using (10.12) through (10.15), it is quite easy to show that the closed-loop system is characterized by the error equation

$$\ddot{E} + K_v \dot{E} + K_p E = 0.$$ (10.17)

Note that this vector equation is decoupled: The matrices K_v and K_p are diagonal, so that (10.17) could just as well be written on a joint-by-joint basis as

$$\ddot{e}_i + k_{vi} \dot{e} + k_{pi} e = 0.$$ (10.18)

The ideal performance represented by (10.17) is unattainable in practice, for many reasons, the most important two being

1. The discrete nature of a digital-computer implementation, as opposed to the ideal continuous-time control law implied by (10.14) and (10.15).
2. Inaccuracy in the manipulator model (needed to compute (10.14)).

In the next section, we will (at least partially) address these two issues.

10.5 PRACTICAL CONSIDERATIONS

In developing the decoupling and linearizing control in the last few sections, we have implicitly made a few assumptions that rarely are true in practice.

Time required to compute the model

In all our considerations of the partitioned-control-law strategy, we have implicitly assumed that the entire system was running in continuous time and that the computations in the control law require zero time for their computation. Given any amount of computation, with a large enough computer we can do the computations sufficiently fast that this is a reasonable approximation; however, the expense of the computer could make the scheme economically unfeasible. In the manipulator-control case, the entire dynamic equation of the manipulator, (10.14), must be computed in the control law. These computations are quite involved; consequently, as was discussed in Chapter 6, there has been a great deal of interest in developing fast computational schemes to compute them in an efficient way. As computer power becomes more and more affordable, control laws that require a great deal of computation will become more practical. Several experimental implementations of nonlinear-model-based control laws have been reported [5–9], and partial implementations are beginning to appear in industrial controllers.

As was discussed in Chapter 9, almost all manipulator-control systems are now performed in digital circuitry and are run at a certain **sampling rate**. This means that the position (and possibly other) sensors are read at discrete points in time. From the values read, an actuator command is computed and sent to the actuator. Thus, reading sensors and sending actuator commands are not done continuously, but rather at a finite sampling rate. To analyze the effect of delay due to computation and finite sample rate, we must use tools from the field of **discrete-time control**. In discrete time, differential equations turn into difference equations, and a complete set of tools has been developed to answer questions about stability and pole placement for these systems. Discrete-time control theory is beyond the scope of this book, although, for researchers working in the area of manipulator control, many of the concepts from discrete-time systems are essential. (See [10].)

Although important, ideas and methods from discrete-time control theory are often difficult to apply to the case of nonlinear systems. Whereas we have managed to write a complicated differential equation of motion for the manipulator dynamic equation, a discrete-time equivalent is impossible to obtain in general because, for a general manipulator, the only way to solve for the motion of the manipulator for a given set of initial conditions, an input, and a finite interval is by numerical integration (as we saw in Chapter 6). Discrete-time models are possible if we are willing to use series solutions to the differential equations, or if we make approximations. However, if we need to make approximations to develop a discrete model, then it is not clear whether we have a better model than we have when just using the continuous model and making the continuous-time approximation. Suffice it to say that analysis of the discrete-time manipulator-control problem is difficult, and usually simulation is resorted to in order to judge the effect that a certain sample rate will have on performance.

We will generally assume that the computations can be performed quickly enough and often enough that the continuous-time approximation is valid.

Feedforward nonlinear control

The use of **feedforward control** has been proposed as a method of using a nonlinear dynamic model in a control law without the need for complex and time-consuming computations to be performed at servo rates [11]. In Fig. 10.5, the model-based control portion of the control law is "in the servo loop" in that signals "flow" through that black box with each tick of the servo clock. If we wish to select a sample

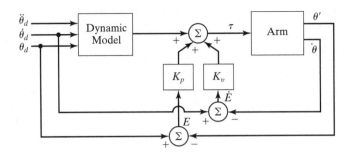

FIGURE 10.6: Control scheme with the model-based portion "outside" the servo loop.

rate of 200 Hz, then the dynamic model of the manipulator must be computed at this rate. Another possible control system is shown in Fig. 10.6. Here, the model-based control is "outside" the servo loop. Hence, it is possible to have a fast inner servo loop, consisting simply of multiplying errors by gains, with the model-based torques added at a slower rate.

Unfortunately, the feedforward scheme of Fig. 10.6 does not provide complete decoupling. If we write the system equations,[1] we will find that the error equation of this system is

$$\ddot{E} + M^{-1}(\Theta)K_v\dot{E} + M^{-1}(\Theta)K_pE = 0. \tag{10.19}$$

Clearly, as the configuration of the arm changes, the effective closed-loop gain changes, and the quasi-static poles move around in the real–imaginary plane. However, equation (10.19) could be used as a starting point for designing a **robust controller**—one that finds a good set of constant gains such that, despite the "motion" of the poles, they are guaranteed to remain in reasonably favorable locations. Alternatively, one might consider schemes in which variable gains are precomputed which change with configuration of the robot, so that the system's quasi-static poles remain in fixed positions.

Note that, in the system of Fig. 10.6, the dynamic model is computed as a function of the desired path only, so when the desired path is known in advance, values could be computed "off-line" before motion begins. At run time, the precomputed torque histories would then be read out of memory. Likewise, if time-varying gains are computed, they too could be computed beforehand and stored. Hence, such a scheme could be quite inexpensive computationally at run time and thus achieve a high servo rate.

Dual-rate computed-torque implementation

Figure 10.7 shows the block diagram of a possible practical implementation of the decoupling and linearizing position-control system. The dynamic model is expressed in its *configuration space* form so that the dynamic parameters of the manipulator will appear as functions of manipulator position only. These functions might then

---

[1]We have used the simplifying assumptions $M(\Theta_d) \cong M(\Theta)$, $V(\Theta_d, \dot{\Theta}_d) \cong (V(\Theta, \dot{\Theta}))$, $G(\Theta_d) \cong G(\Theta)$, and $F(\Theta_d, \dot{\Theta}_d) \cong F(\Theta, \dot{\Theta})$.

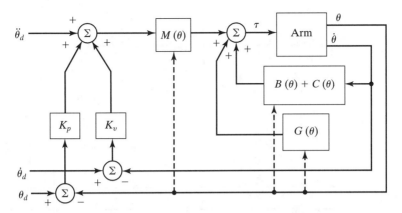

FIGURE 10.7: An implementation of the model-based manipulator-control system.

be computed by a *background* process or by a second control computer [8] or be looked up in a precomputed table [12]. In this architecture, the dynamic parameters can be updated at a rate slower than the rate of the closed-loop servo. For example, the background computation might proceed at 60 Hz while the closed-loop servo was running at 250 Hz.

Lack of knowledge of parameters

The second potential difficulty encountered in employing the computed-torque control algorithm is that the manipulator dynamic model is often not known accurately. This is particularly true of certain components of the dynamics, such as friction effects. In fact, it is usually extremely difficult to know the structure of the friction model, let alone the parameter values [13]. Finally, if the manipulator has some portion of its dynamics that is not repeatable—because, for example, it changes as the robot ages—it is difficult to have good parameter values in the model at all times.

By nature, most robots will be picking up various parts and tools. When a robot is holding a tool, the inertia and the weight of the tool change the dynamics of the manipulator. In an industrial situation, the mass properties of the tools might be known—in this case, they can be accounted for in the modeled portion of the control law. When a tool is grasped, the inertia matrix, total mass, and center of mass of the last link of the manipulator can be updated to new values that represent the combined effect of the last link plus tool. However, in many applications, the mass properties of objects that the manipulator picks up are not generally known, so maintenance of an accurate dynamic model is difficult.

The simplest possible nonideal situation is one in which we still assume a perfect model implemented in continuous time, but with external noise acting to disturb the system. In Fig. 10.8, we indicate a vector of disturbance torques acting at the joints. Writing the system error equation with inclusion of these unknown disturbances, we arrive at

$$\ddot{E} + K_v\dot{E} + K_pE = M^{-1}(\Theta)\tau_d, \tag{10.20}$$

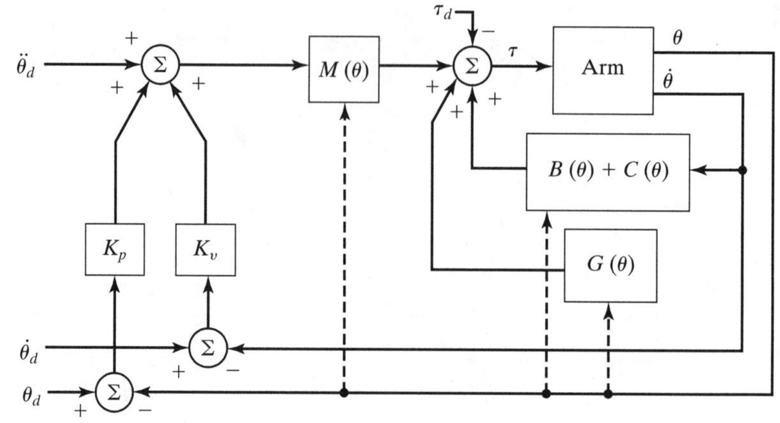

FIGURE 10.8: The model-based controller with an external disturbance acting.

where τ_d is the vector of disturbance torques at the joints. The left-hand side of (10.20) is uncoupled, but, from the right-hand side, we see that a disturbance on any particular joint will introduce errors at all the other joints, because $M\,(\Theta)$ is not, in general, diagonal.

Some simple analyses might be performed on the basis of (10.20). For example, it is easy to compute the steady-state servo error due to a constant disturbance as

$$E = K_p^{-1} M^{-1}(\Theta)\tau_d. \tag{10.21}$$

When our model of the manipulator dynamics is not perfect, analysis of the resulting closed-loop system becomes more difficult. We define the following notation: $\hat{M}(\Theta)$ is our model of the manipulator inertia matrix, $M\,(\Theta)$. Likewise, $\hat{V}(\Theta, \dot{\Theta})$, $\hat{G}(\Theta)$, and $\hat{F}(\Theta, \dot{\Theta})$ are our models of the velocity terms, gravity terms, and friction terms of the actual mechanism. Perfect knowledge of the model would mean that

$$\hat{M}(\Theta) = M(\Theta),$$

$$\hat{V}(\Theta, \dot{\Theta}) = V(\Theta, \dot{\Theta}), \tag{10.22}$$

$$\hat{G}(\Theta) = G(\Theta),$$

$$\hat{F}(\Theta, \dot{\Theta}) = F(\Theta, \dot{\Theta}).$$

Therefore, although the manipulator dynamics are given by

$$\tau = M(\Theta)\ddot{\Theta} + V(\Theta, \dot{\Theta}) + G(\Theta) + F(\Theta, \dot{\Theta}), \tag{10.23}$$

our control law computes

$$\tau = \alpha\tau' + \beta,$$

$$\alpha = \hat{M}(\Theta), \tag{10.24}$$

$$\beta = \hat{V}(\Theta, \dot{\Theta}) + \hat{G}(\Theta) + \hat{F}(\Theta, \dot{\Theta}).$$

Decoupling and linearizing will not, therefore, be perfectly accomplished when parameters are not known exactly. Writing the closed-loop equation for the system, we have

$$\ddot{E} + K_v \dot{E} + K_p E$$
$$= \hat{M}^{-1}[(M - \hat{M})\ddot{\Theta} + (V - \hat{V}) + (G - \hat{G}) + (F - \hat{F})], \qquad (10.25)$$

where the arguments of the dynamic functions are not shown for brevity. Note that, if the model were exact, so that (10.22) were true, then the right-hand side of (10.25) would be zero and the errors would disappear. When the parameters are not known exactly, the mismatch between actual and modeled parameters will cause servo errors to be excited (possibly even resulting in an unstable system [21]) according to the rather complicated equation (10.25).

Discussion of stability analysis of a nonlinear closed-loop system is deferred until Section 10.7.

10.6 CURRENT INDUSTRIAL-ROBOT CONTROL SYSTEMS

Because of the problems with having good knowledge of parameters, it is not clear whether it makes sense to go to the trouble of computing a complicated model-based control law for manipulator control. The expense of the computer power needed to compute the model of the manipulator at a sufficient rate might not be worthwhile, especially when lack of knowledge of parameters could nullify the benefits of such an approach. Manufacturers of industrial robots have decided, probably for economic reasons, that attempting to use a complete manipulator model in the controller is not worthwhile. Instead, present-day manipulators are controlled with very simple control laws that generally are completely error driven and are implemented in architectures such as those studied in Section 9.10. An industrial robot with a high-performance servo system is shown in Fig. 10.9.

Individual-joint PID control

Most industrial robots nowadays have a control scheme that, in our notation, would be described by

$$\alpha = I,$$
$$\beta = 0, \qquad (10.26)$$

where I is the $n \times n$ identity matrix. The servo portion is

$$\tau' = \ddot{\Theta}_d + K_v \dot{E} + K_p E + K_i \int E \, dt, \qquad (10.27)$$

where K_v, K_p, and K_i are constant diagonal matrices. In many cases, $\ddot{\Theta}_d$ is not available, and this term is simply set to zero. That is, most simple robot controllers do not use a model-based component *at all* in their control law. This type of PID control scheme is simple because each joint is controlled as a separate control system. Often, one microprocessor per joint is used to implement (10.27), as was discussed in Section 9.10.

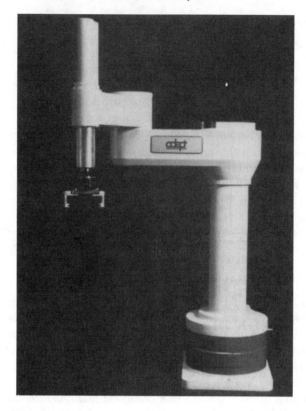

FIGURE 10.9: The Adept One, a direct-drive robot by Adept Technology, Inc.

The performance of a manipulator controlled in this way is not simple to describe. No decoupling is being done, so the motion of each joint affects the other joints. These interactions cause errors, which are suppressed by the error-driven control law. It is impossible to select fixed gains that will critically damp the response to disturbances for all configurations. Therefore, "average" gains are chosen, which approximate critical damping in the center of the robot's workspace. In various extreme configurations of the arm, the system becomes either underdamped or overdamped. Depending on the details of the mechanical design of the robot, these effects could be fairly small; then control would be good. In such systems, it is important to keep the gains as high as possible, so that the inevitable disturbances will be suppressed quickly.

Addition of gravity compensation

The gravity terms will tend to cause static positioning errors, so some robot manufacturers include a gravity model, $G(\theta)$, in the control law (that is, $\beta = \hat{G}(\Theta)$ in our notation). The complete control law takes the form

$$\tau' = \ddot{\Theta}_d + K_v \dot{E} + K_p E + K_i \int E dt + \hat{G}(\Theta). \tag{10.28}$$

Such a control law is perhaps the simplest example of a model-based controller. Because (10.28) can no longer be implemented on a strict joint-by-joint basis, the controller architecture must allow communication between the joint controllers or must make use of a central processor rather than individual-joint processors.

Various approximations of decoupling control

There are various ways to simplify the dynamic equations of a particular manipulator [3,14]. After the simplification, an approximate decoupling and linearizing law can be derived. A usual simplification might be to disregard components of torque due to the velocity terms—that is, to model only the inertial and gravity terms. Often, friction models are not included in the controller, because friction is so hard to model correctly. Sometimes, the inertia matrix is simplified so that it accounts for the major coupling between axes but not for minor cross-coupling effects. For example, [14] presents a simplified version of the PUMA 560's mass matrix that requires only about 10% of the calculations needed to compute the complete mass matrix, yet is accurate to within 1%.

10.7   LYAPUNOV STABILITY ANALYSIS

In Chapter 9, we examined linear control systems analytically to evaluate stability and also performance of the dynamic response in terms of damping and closed-loop bandwidth. The same analyses are valid for a nonlinear system that has been decoupled and linearized by means of a perfect model-based nonlinear controller, because the overall resulting system is again linear. However, when decoupling and linearizing are not performed by the controller, or are incomplete or inaccurate, the overall closed-loop system remains nonlinear. For nonlinear systems, stability and performance analysis is much more difficult. In this section, we introduce one method of stability analysis that is applicable to both linear and nonlinear systems.

Consider the simple mass–spring friction system originally introduced in Chapter 9, whose equation of motion is

$$m\ddot{x} + b\dot{x} + kx = 0. \tag{10.29}$$

The total energy of the system is given by

$$v = \frac{1}{2}m\dot{x}^2 + \frac{1}{2}kx^2, \tag{10.30}$$

where the first term gives the kinetic energy of the mass and the second term gives the potential energy stored in the spring. Note that the value, v, of the system energy is always nonnegative (i.e., it is positive or zero). Let's find out the *rate* of change of the total energy by differentiating (10.30) with respect to time, to obtain

$$\dot{v} = m\dot{x}\ddot{x} + kx\dot{x}. \tag{10.31}$$

Substituting (10.29) for $m\ddot{x}$ in (10.31) yields

$$\dot{v} = -b\dot{x}^2, \tag{10.32}$$

which we note is always nonpositive (because $b > 0$). Thus, energy is always leaving the system, unless $\dot{x} = 0$. This implies that, however initially perturbed, the system will lose energy until it comes to rest. Investigating possible resting positions by means of a steady-state analysis of (10.29) yields

$$kx = 0,$$ (10.33)

or

$$x = 0.$$ (10.34)

Hence, by means of an energy analysis, we have shown that the system of (10.29) with any initial conditions (i.e., any initial energy) will eventually come to rest at the equilibrium point. This stability proof by means of an energy analysis is a simple example of a more general technique called **Lyapunov stability analysis** or **Lyapunov's second** (or **direct**) **method**, after a Russian mathematician of the 19th century [15].

An interesting feature of this method of stability analysis is that we can conclude stability without solving for the solution of the differential equation governing the system. However, while Lyapunov's method is useful for examining *stability*, it generally does not provide any information about the transient response or *performance* of the system. Note that our energy analysis yielded no information on whether the system was overdamped or underdamped or on how long it would take the system to suppress a disturbance. It is important to distinguish between stability and performance: A stable system might nonetheless exhibit control performance unsatisfactory for its intended use.

Lyapunov's method is somewhat more general than our example indicated. It is one of the few techniques that can be applied directly to nonlinear systems to investigate their stability. As a means of quickly getting an idea of Lyapunov's method (in sufficient detail for our needs), we will look at an extremely brief introduction to the theory and then proceed directly to several examples. A more complete treatment of Lyapunov theory can be found in [16, 17].

Lyapunov's method is concerned with determining the stability of a differential equation

$$\dot{X} = f(X),$$ (10.35)

where X is $m \times 1$ and $f(\cdot)$ could be nonlinear. Note that higher order differential equations can always be written as a set of first-order equations in the form (10.35). To prove a system stable by Lyapunov's method, one is required to propose a generalized energy function $v(X)$ that has the following properties:

1. $v(X)$ has continuous first partial derivatives, and $v(X) > 0$ for all X except $v(0) = 0$.
2. $\dot{v}(X) \leq 0$. Here, $\dot{v}(X)$ means the change in $v(X)$ along all system trajectories.

These properties might hold only in a certain region, or they might be global, with correspondingly weaker or stronger stability results. The intuitive idea is that a positive definite "energy-like" function of state is shown to always decrease or remain constant—hence, the system is stable in the sense that the size of the state vector is bounded.

When $\dot{v}(X)$ is strictly less than zero, asymptotic convergence of the state to the zero vector can be concluded. Lyapunov's original work was extended in an important way by LaSalle and Lefschetz [4], who showed that, in certain situations, even when $\dot{v}(X) \leq 0$ (note equality included), asymptotic stability can be shown. For our purposes, we can deal with the case $\dot{v}(X) = 0$ by performing a steady-state analysis in order to learn whether the stability is asymptotic or the system under study can "get stuck" somewhere other than $v(X) = 0$.

A system described by (10.35) is said to be **autonomous** because the function $f(\cdot)$ is not an explicit function of time. Lyapunov's method also extends to **nonautonomous** systems, in which time is an argument of the nonlinear function. See [4, 17] for details.

EXAMPLE 10.4

Consider the linear system

$$\dot{X} = -AX, \tag{10.36}$$

where A is $m \times m$ and positive definite. Propose the **candidate Lyapunov function**

$$v(X) = \frac{1}{2}X^T X, \tag{10.37}$$

which is continuous and everywhere nonnegative. Differentiating yields

$$\begin{aligned} \dot{v}(X) &= X^T \dot{X} \\ &= X^T(-AX) \\ &= -X^T AX, \end{aligned} \tag{10.38}$$

which is everywhere nonpositive because A is a positive definite matrix. Hence, (10.37) is indeed a Lyapunov function for the system of (10.36). The system is asymptotically stable because $\dot{v}(X)$ can be zero only at $X = 0$; everywhere else, X must decrease.

EXAMPLE 10.5

Consider a mechanical spring–damper system in which both the spring and damper are nonlinear:

$$\ddot{x} + b(\dot{x}) + k(x) = 0. \tag{10.39}$$

The functions $b(\cdot)$ and $k(\cdot)$ are first- and third-quadrant continuous functions such that

$$\dot{x}b(\dot{x}) > 0 \ for \ x \neq 0,$$

$$xk(x) > 0 \ for \ x \neq 0. \tag{10.40}$$

Once having proposed the Lyapunov function

$$v(x, \dot{x}) = \frac{1}{2}\dot{x}^2 + \int_0^x k(\lambda)d\lambda, \tag{10.41}$$

we are led to

$$\dot{v}(x, \dot{x}) = \dot{x}\ddot{x} + k(x)\dot{x},$$

$$= -\dot{x}b(\dot{x}) - k(x)\dot{x} + k(x)\dot{x}, \qquad (10.42)$$

$$= -\dot{x}b(\dot{x}).$$

Hence, $\dot{v}(\cdot)$ is nonpositive but is only semidefinite, because it is not a function of x but only of $\dot{x}$. In order to conclude asymptotic stability, we have to ensure that it is not possible for the system to "get stuck" with nonzero x. To study all trajectories for which $\dot{x} = 0$, we must consider

$$\ddot{x} = -k(x), \qquad (10.43)$$

for which $x = 0$ is the only solution. Hence, the system will come to rest only if $x = \dot{x} = \ddot{x} = 0$.

EXAMPLE 10.6

Consider a manipulator with dynamics given by

$$\tau = M(\Theta)\ddot{\Theta} + V(\Theta, \dot{\Theta}) + G(\Theta) \qquad (10.44)$$

and controlled with the control law

$$\tau = K_p E - K_d \dot{\Theta} + G(\Theta), \qquad (10.45)$$

where K_p and K_d are diagonal gain matrices. Note that this controller does not force the manipulator to follow a trajectory, but moves the manipulator to a goal point along a path specified by the manipulator dynamics and then regulates the position there. The resulting closed-loop system obtained by equating (10.44) and (10.45) is

$$M(\Theta)\ddot{\Theta} + V(\Theta, \dot{\Theta}) + K_d \dot{\Theta} + K_p \Theta = K_p \Theta_d; \qquad (10.46)$$

it can be proven globally asymptotically stable by Lyapunov's method [18, 19].
    Consider the candidate Lyapunov function

$$v = \frac{1}{2}\dot{\Theta}^T M(\Theta)\dot{\Theta} + \frac{1}{2}E^T K_p E. \qquad (10.47)$$

The function (10.47) is always positive or zero, because the manipulator mass matrix, $M(\Theta)$, and the position gain matrix, K_p, are positive definite matrices. Differentiating (10.47) yields

$$\dot{v} = \frac{1}{2}\dot{\Theta}^T \dot{M}(\Theta)\dot{\theta} + \dot{\theta}^T M(\theta)\ddot{\Theta} - E^T K_p \dot{\Theta}$$

$$= \frac{1}{2}\dot{\Theta}^T \dot{M}(\Theta)\dot{\Theta} - \dot{\Theta}^T K_d \dot{\Theta} - \dot{\Theta}^T V(\Theta, \dot{\Theta}) \qquad (10.48)$$

$$= -\dot{\Theta}^T K_d \dot{\Theta},$$

which is nonpositive as long as K_d is positive definite. In taking the last step in (10.48), we have made use of the interesting identity

$$\frac{1}{2}\dot{\Theta}^T \dot{M}(\Theta)\dot{\Theta} = \dot{\Theta}^T V(\Theta, \dot{\Theta}), \qquad (10.49)$$

which can be shown by investigation of the structure of Lagrange's equations of motion [18–20]. (See also Exercise 6.17.)

Next, we investigate whether the system can get "stuck" with nonzero error. Because $\dot{v}$ can remain zero only along trajectories that have $\dot{\Theta} = 0$ and $\ddot{\Theta} = 0$, we see from (10.46) that, in this case,

$$K_p E = 0, \qquad (10.50)$$

and because K_p is nonsingular, we have

$$E = 0. \qquad (10.51)$$

Hence, control law (10.45) applied to the system (10.44) achieves global asymptotic stability.

This proof is important in that it explains, to some extent, why today's industrial robots work. Most industrial robots use a simple error-driven servo, occasionally with gravity models, and so are quite similar to (10.45).

See Exercises 10.11 through 10.16 for more examples of nonlinear manipulator-control laws that can be proven stable by Lyapunov's method. Recently, Lyapunov theory has become increasingly prevalent in robotics research publications [18–25].

10.8   CARTESIAN-BASED CONTROL SYSTEMS

In this section, we introduce the notion of **Cartesian-based control**. Although such approaches are not currently used in industrial robots, there is activity at several research institutions on such schemes.

Comparison with joint-based schemes

In all the control schemes for manipulators we have discussed so far, we assumed that the desired trajectory was available in terms of time histories of joint position, velocity, and acceleration. Given that these desired inputs were available, we designed **joint-based control** schemes, that is, schemes in which we develop trajectory errors by finding the difference between desired and actual quantities expressed in joint space. Very often, we wish the manipulator end-effector to follow straight lines or other path shapes described in Cartesian coordinates. As we saw in Chapter 7, it is possible to compute the time histories of the joint-space trajectory that correspond to Cartesian straight-line paths. Figure 10.10 shows this approach to manipulator-trajectory control. A basic feature of the approach is the **trajectory-conversion** process, which is used to compute the joint trajectories. This is then followed by some kind of joint-based servo scheme such as we have been studying.

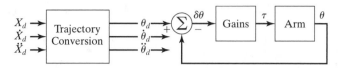

FIGURE 10.10: A joint-based control scheme with Cartesian-path input.

The trajectory-conversion process is quite difficult (in terms of computational expense) if it is to be done analytically. The computations that would be required are

$$\Theta_d = INVKIN(\chi_d),$$
$$\dot{\Theta}_d = J^{-1}(\Theta)\dot{\chi}_d, \tag{10.52}$$
$$\ddot{\Theta}_d = \dot{J}^{-1}(\Theta)\dot{\chi}_d + J^{-1}(\Theta)\ddot{\chi}_d.$$

To the extent that such a computation is done at all in present-day systems, usually just the solution for Θ_d is performed, by using the inverse kinematics, and then the joint velocities and accelerations are computed numerically by first and second differences. However, such numerical differentiation tends to amplify noise and introduces a lag unless it can be done with a noncausal filter.[2] Therefore, we are interested in either finding a less computationally expensive way of computing (10.52) or suggesting a control scheme in which this information is not needed.

An alternative approach is shown in Fig. 10.11. Here, the sensed position of the manipulator is immediately transformed by means of the kinematic equations into a Cartesian description of position. This Cartesian description is then compared to the desired Cartesian position in order to form errors in Cartesian space. Control schemes based on forming errors in Cartesian space are called **Cartesian-based control** schemes. For simplicity, velocity feedback is not shown in Fig. 10.11, but it would be present in any implementation.

The trajectory-conversion process is replaced by some kind of coordinate conversion inside the servo loop. Note that Cartesian-based controllers must perform many computations in the loop; the kinematics and other transformations are now "inside the loop." This can be a drawback of the Cartesian-based methods; the resulting system could run at a lower sampling frequency compared to joint-based

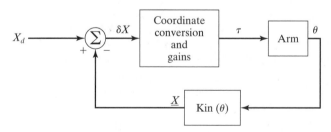

FIGURE 10.11: The concept of a Cartesian-based control scheme.

---

[2]Numerical differentiation introduces a lag unless it can be based on past, present, and future values. When the entire path is preplanned, this kind of noncausal numerical differentiation can be done.

systems (given the same size of computer). This would, in general, degrade the stability and disturbance-rejection capabilities of the system.

Intuitive schemes of Cartesian control

One possible control scheme that comes to mind rather intuitively is shown in Fig. 10.12. Here, Cartesian position is compared to the desired position to form an error, δX, in Cartesian space. This error, which may be presumed small if the control system is doing its job, may be mapped into a small displacement in joint space by means of the inverse Jacobian. The resulting errors in joint space, $\delta \theta$, are then multiplied by gains to compute torques that will tend to reduce these errors. Note that Fig. 10.12 shows a simplified controller in which, for clarity, the velocity feedback has not been shown. It could be added in a straightforward manner. We will call this scheme the **inverse-Jacobian controller**.

Another scheme which could come to mind is shown in Fig. 10.13. Here, the Cartesian error vector is multiplied by a gain to compute a Cartesian force vector. This can be thought of as a Cartesian force which, if applied to the end-effector of the robot, would push the end-effector in a direction that would tend to reduce the Cartesian error. This Cartesian force vector (actually a force–moment vector) is then mapped through the Jacobian transpose in order to compute the equivalent joint torques that would tend to reduce the observed errors. We will call this scheme the **transpose-Jacobian controller**.

The inverse-Jacobian controller and the transpose-Jacobian controller have both been arrived at intuitively. We cannot be sure that such arrangements would be stable, let alone perform well. It is also curious that the schemes are extremely similar, except that the one contains the Jacobian's inverse, the other its transpose. Remember, the inverse is not equal to the transpose in general (only in the case of a strictly Cartesian manipulator does $J^T = J^{-1}$). The exact dynamic performance

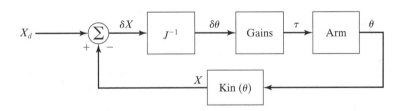

FIGURE 10.12: The inverse-Jacobian Cartesian-control scheme.

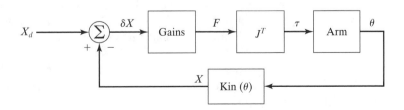

FIGURE 10.13: The transpose-Jacobian Cartesian-control scheme.

of such systems (if expressed in a second-order error-space equation, for example) is very complicated. It turns out that both schemes will work (i.e., can be made stable), but not well (i.e., performance is not good over the entire workspace). Both can be made stable by appropriate gain selection, including some form of velocity feedback (which was not shown in Figs. 10.12 and 10.13). While both will work, neither is *correct*, in the sense that we cannot choose fixed gains that will result in fixed closed-loop poles. The dynamic response of such controllers will vary with arm configuration.

Cartesian decoupling scheme

For Cartesian-based controllers, like joint-based controllers, good performance would be characterized by constant error dynamics over all configurations of the manipulator. Errors are expressed in Cartesian space in Cartesian-based schemes, so this means that we would like to design a system which, over all possible configurations, would suppress Cartesian errors in a critically damped fashion.

Just as we achieved good control with a joint-based controller that was based on a linearizing and decoupling model of the arm, we can do the same for the Cartesian case. However, we must now write the dynamic equations of motion of the manipulator in terms of Cartesian variables. This can be done, as was discussed in Chapter 6. The resulting form of the equations of motion is quite analogous to the joint-space version. The rigid-body dynamics can be written as

$$\mathcal{F} = M_x(\Theta)\ddot{\chi} + V_x(\Theta, \dot{\Theta}) + G_x(\Theta), \tag{10.53}$$

where $\mathcal{F}$ is a fictitious force–moment vector acting on the end-effector of the robot and χ is an appropriate Cartesian vector representing position and orientation of the end-effector [8]. Analogous to the joint-space quantities, $M_x(\Theta)$ is the mass matrix in Cartesian space, $V_x(\Theta, \dot{\Theta})$ is a vector of velocity terms in Cartesian space, and $G_x(\Theta)$ is a vector of gravity terms in Cartesian space.

Just as we did in the joint-based case, we can use the dynamic equations in a decoupling and linearizing controller. Because (10.53) computes $\mathcal{F}$, a fictitious Cartesian force vector which should be applied to the hand, we will also need to use the transpose of the Jacobian in order to implement the control—that is, after $\mathcal{F}$ is calculated by (10.53), we cannot actually cause a Cartesian force to be applied to the end-effector; we instead compute the joint torques needed to effectively balance the system if we were to apply this force:

$$\tau = J^T(\Theta)\mathcal{F}. \tag{10.54}$$

Figure 10.14 shows a Cartesian arm-control system using complete dynamic decoupling. Note that the arm is preceded by the Jacobian transpose. Notice that the controller of Fig. 10.14 allows Cartesian paths to be described directly, with no need for trajectory conversion.

As in the joint-space case, a practical implementation might best be achieved through use of a dual-rate control system. Figure 10.15 shows a block diagram of a Cartesian-based decoupling and linearizing controller in which the dynamic parameters are written as functions of manipulator position only. These dynamic parameters are updated at a rate slower than the servo rate by a background

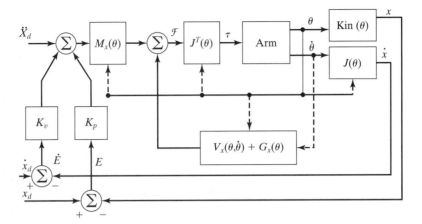

FIGURE 10.14: The Cartesian model-based control scheme.

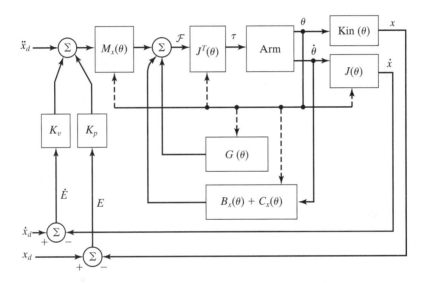

FIGURE 10.15: An implementation of the Cartesian model-based control scheme.

process or a second control computer. This is appropriate, because we desire a fast servo (perhaps running at 500 Hz or even higher) to maximize disturbance rejection and stability. The dynamic parameters are functions of manipulator position only, so they need be updated at a rate related only to how fast the manipulator is changing configuration. The parameter-update rate probably need not be higher than 100 Hz [8].

10.9   ADAPTIVE CONTROL

In the discussion of model-based control, it was noted that, often, parameters of the manipulator are not known exactly. When the parameters in the model do not

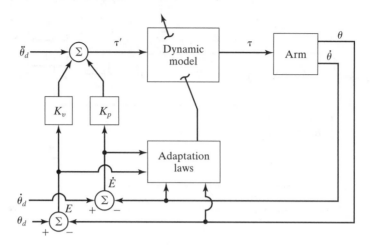

FIGURE 10.16: The concept of an adaptive manipulator controller.

match the parameters of the real device, servo errors will result, as is made explicit in (10.25). These servo errors could be used to drive some adaptation scheme that attempts to update the values of the model parameters until the errors disappear. Several such adaptive schemes have been proposed.

An ideal adaptive scheme might be like the one in Fig. 10.16. Here, we are using a model-based control law as developed in this chapter. There is an adaptation process that, given observations of manipulator state and servo errors, readjusts the parameters in the nonlinear model until the errors disappear. Such a system would *learn* its own dynamic properties. The design and analysis of adaptive schemes are beyond the scope of this book. A method that possesses exactly the structure shown in Fig. 10.16 and has been proven globally stable is presented in [20, 21]. A related technique is that of [22].

BIBLIOGRAPHY

[1] R.P. Paul, "Modeling, Trajectory Calculation, and Servoing of a Computer Controlled Arm," Technical Report AIM-177, Stanford University Artificial Intelligence Laboratory, 1972.

[2] B. Markiewicz, "Analysis of the Computed Torque Drive Method and Comparison with Conventional Position Servo for a Computer-Controlled Manipulator," Jet Propulsion Laboratory Technical Memo 33–601, March 1973.

[3] A. Bejczy, "Robot Arm Dynamics and Control," Jet Propulsion Laboratory Technical Memo 33–669, February 1974.

[4] J. LaSalle and S. Lefschetz, *Stability by Liapunov's Direct Method with Applications*, Academic Press, New York, 1961.

[5] P.K. Khosla, "Some Experimental Results on Model-Based Control Schemes," IEEE Conference on Robotics and Automation, Philadelphia, April 1988.

[6] M. Leahy, K. Valavanis, and G. Saridis, "The Effects of Dynamic Models on Robot Control," IEEE Conference on Robotics and Automation, San Francisco, April 1986.

[7] L. Sciavicco and B. Siciliano, *Modelling and Control of Robot Manipulators*, 2nd Edition, Springer-Verlag, London, 2000.

[8] O. Khatib, "A Unified Approach for Motion and Force Control of Robot Manipulators: The Operational Space Formulation," *IEEE Journal of Robotics and Automation*, Vol. RA-3, No. 1, 1987.

[9] C. An, C. Atkeson, and J. Hollerbach, "Model-Based Control of a Direct Drive Arm, Part II: Control," IEEE Conference on Robotics and Automation, Philadelphia, April 1988.

[10] G. Franklin, J. Powell, and M. Workman, *Digital Control of Dynamic Systems*, 2nd edition, Addison-Wesley, Reading, MA, 1989.

[11] A. Liegeois, A. Fournier, and M. Aldon, "Model Reference Control of High Velocity Industrial Robots," *Proceedings of the Joint Automatic Control Conference*, San Francisco, 1980.

[12] M. Raibert, "Mechanical Arm Control Using a State Space Memory," SME paper MS77-750, 1977.

[13] B. Armstrong, "Friction: Experimental Determination, Modeling and Compensation," IEEE Conference on Robotics and Automation, Philadelphia, April 1988.

[14] B. Armstrong, O. Khatib, and J. Burdick, "The Explicit Dynamic Model and Inertial Parameters of the PUMA 560 Arm," IEEE Conference on Robotics and Automation, San Francisco, April 1986.

[15] A.M. Lyapunov, "On the General Problem of Stability of Motion," (in Russian), Kharkov Mathematical Society, Soviet Union, 1892.

[16] C. Desoer and M. Vidyasagar, *Feedback Systems: Input–Output Properties*, Academic Press, New York, 1975.

[17] M. Vidyasagar, *Nonlinear Systems Analysis*, Prentice-Hall, Englewood Cliffs, NJ, 1978.

[18] S. Arimoto and F. Miyazaki, "Stability and Robustness of PID Feedback Control for Robot Manipulators of Sensory Capability," Third International Symposium of Robotics Research, Gouvieux, France, July 1985.

[19] D. Koditschek, "Adaptive Strategies for the Control of Natural Motion," *Proceedings of the 24th Conference on Decision and Control*, Ft. Lauderdale, FL, December 1985.

[20] J. Craig, P. Hsu, and S. Sastry, "Adaptive Control of Mechanical Manipulators," IEEE Conference on Robotics and Automation, San Francisco, April 1986.

[21] J. Craig, *Adaptive Control of Mechanical Manipulators*, Addison-Wesley, Reading, MA, 1988.

[22] J.J. Slotine and W. Li, "On the Adaptive Control of Mechanical Manipulators," *The International Journal of Robotics Research*, Vol. 6, No. 3, 1987.

[23] R. Kelly and R. Ortega, "Adaptive Control of Robot Manipulators: An Input–Output Approach," IEEE Conference on Robotics and Automation, Philadelphia, 1988.

[24] H. Das, J.J. Slotine, and T. Sheridan, "Inverse Kinematic Algorithms for Redundant Systems," IEEE Conference on Robotics and Automation, Philadelphia, 1988.

[25] T. Yabuta, A. Chona, and G. Beni, "On the Asymptotic Stability of the Hybrid Position/Force Control Scheme for Robot Manipulators," IEEE Conference on Robotics and Automation, Philadelphia, 1988.

EXERCISES

10.1 [15] Give the nonlinear control equations for an α,β-partitioned controller for the system

$$\tau = (2\sqrt{\theta} + 1)\ddot{\theta} + 3\dot{\theta}^2 - \sin(\theta).$$

Choose gains so that this system is always critically damped with $k_{CL} = 10$.

10.2 [15] Give the nonlinear control equations for an α,β-partitioned controller for the system

$$\tau = 5\theta\dot{\theta} + 2\ddot{\theta} - 13\dot{\theta}^3 + 5.$$

Choose gains so that this system is always critically damped with $k_{CL} = 10$.

10.3 [19] Draw a block diagram showing a joint-space controller for the two-link arm from Section 6.7, such that the arm is critically damped over its entire workspace. Show the equations inside the blocks of a block diagram.

10.4 [20] Draw a block diagram showing a Cartesian-space controller for the two-link arm from Section 6.7, such that the arm is critically damped over its entire workspace. (See Example 6.6.) Show the equations inside the blocks of a block diagram.

10.5 [18] Design a trajectory-following control system for the system whose dynamics are given by

$$\tau_1 = m_1 l_1^2 \ddot{\theta}_1 + m_1 l_1 l_2 \dot{\theta}_1 \dot{\theta}_2,$$

$$\tau_2 = m_2 l_2^2 (\ddot{\theta}_1 + \ddot{\theta}_2) + v_2 \dot{\theta}_2.$$

Do you think these equations could represent a real system?

10.6 [17] For the control system designed for the one-link manipulator in Example 10.3, give an expression for the steady-state position error as a function of error in the mass parameter. Let $\psi_m = m - \hat{m}$. The result should be a function of $l, g, \theta, \psi_m, \hat{m}$, and k_p. For what position of the manipulator is this at a maximum?

10.7 [26] For the two-degree-of-freedom mechanical system of Fig. 10.17, design a controller that can cause x_1 and x_2 to follow trajectories and suppress disturbances in a critically damped fashion.

10.8 [30] Consider the dynamic equations of the two-link manipulator from Section 6.7 in configuration-space form. Derive expressions for the sensitivity of the computed

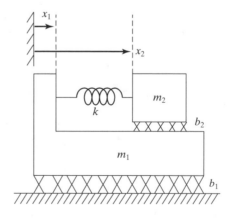

FIGURE 10.17: Mechanical system with two degrees of freedom.

torque value versus small deviations in Θ. Can you say something about how often the dynamics should be recomputed in a controller like that of Fig. 10.7 as a function of average joint velocities expected during normal operations?

10.9 [32] Consider the dynamic equations of the two-link manipulator from Example 6.6 in Cartesian configuration-space form. Derive expressions for the sensitivity of the computed torque value versus small deviations in Θ. Can you say something about how often the dynamics should be recomputed in a controller like that of Fig. 10.15 as a function of average joint velocities expected during normal operations?

10.10 [15] Design a control system for the system

$$f = 5x\dot{x} + 2\ddot{x} - 12.$$

Choose gains so that this system is always critically damped with a closed-loop stiffness of 20.

10.11 [20] Consider a position-regulation system that (without loss of generality) attempts to maintain $\Theta_d = 0$. Prove that the control law

$$\tau = -K_p\Theta - M(\Theta)K_v\dot{\Theta} + G(\Theta)$$

yields an asymptotically stable nonlinear system. You may take K_v to be of the form $K_v = k_v I_n$ where k_v is a scalar and I_n is the $n \times n$ identity matrix. *Hint*: This is similar to example 10.6.

10.12 [20] Consider a position-regulation system that (without loss of generality) attempts to maintain $\Theta_d = 0$. Prove that the control law

$$\tau = -K_p\Theta - \hat{M}(\Theta)K_v\dot{\Theta} + G(\Theta)$$

yields an asymptotically stable nonlinear system. You may take K_v to be of the form $K_v = k_v I_n$ where k_v is a scalar and I_n is the $n \times n$ identity matrix. The matrix $\hat{M}(\Theta)$ is a positive definite estimate of the manipulator mass matrix. *Hint*: This is similar to example 10.6.

10.13 [25] Consider a position-regulation system that (without loss of generality) attempts to maintain $\Theta_d = 0$. Prove that the control law

$$\tau = -M(\Theta)[K_p\Theta + K_v\dot{\Theta}] + G(\Theta)$$

yields an asymptotically stable nonlinear system. You may take K_v to be of the form $K_v = k_v I_n$ where k_v is a scalar and I_n is the $n \times n$ identity matrix. *Hint*: This is similar to example 10.6.

10.14 [25] Consider a position-regulation system that (without loss of generality) attempts to maintain $\Theta_d = 0$. Prove that the control law

$$\tau = -\hat{M}(\Theta)[K_p\Theta + K_v\dot{\Theta}] + G(\Theta)$$

yields an asymptotically stable nonlinear system. You may take K_v to be of the form $K_v = k_v I_n$, where k_v is a scalar and I_n is the $n \times n$ identity matrix. The matrix $\hat{M}(\Theta)$ is a positive definite estimate of the manipulator mass matrix. *Hint*: This is similar to example 10.6.

10.15 [28] Consider a position-regulation system that (without loss of generality) attempts to maintain $\Theta_d = 0$. Prove that the control law

$$\tau = -K_p\Theta - K_v\dot{\Theta}$$

yields a stable nonlinear system. Show that stability is not asymptotic and give an expression for the steady-state error. *Hint*: This is similar to Example 10.6.

10.16 [30] Prove the global stability of the Jacobian-transpose Cartesian controller introduced in Section 10.8. Use an appropriate form of velocity feedback to stabilize the system. *Hint*: See [18].

10.17 [15] Design a trajectory-following controller for a system with dynamics given by

$$f = ax^2 \dot{x} \ddot{x} + b\dot{x}^2 + c\sin(x),$$

such that errors are suppressed in a critically damped fashion over all configurations.

10.18 [15] A system with open-loop dynamics given by

$$\tau = m\ddot{\theta} + b\dot{\theta}^2 + c\dot{\theta}$$

is controlled with the control law

$$\tau = m[\ddot{\theta}_d + k_v \dot{e} + k_p e] + \sin(\theta).$$

Give the differential equation that characterizes the closed-loop action of the system.

PROGRAMMING EXERCISE (PART 10)

Repeat Programming Exercise Part 9, and use the same tests, but with a new controller that uses a complete dynamic model of the 3-link to decouple and linearize the system. For this case, use

$$K_p = \begin{bmatrix} 100.0 & 0.0 & 0.0 \\ 0.0 & 100.0 & 0.0 \\ 0.0 & 0.0 & 100.0 \end{bmatrix}.$$

Choose a diagonal K_v that guarantees critical damping over all configurations of the arm. Compare the results with those obtained with the simpler controller used in Programming Exercise Part 9.

C H A P T E R   11

Force control of manipulators

11.1 INTRODUCTION
11.2 APPLICATION OF INDUSTRIAL ROBOTS TO ASSEMBLY TASKS
11.3 A FRAMEWORK FOR CONTROL IN PARTIALLY CONSTRAINED TASKS
11.4 THE HYBRID POSITION/FORCE CONTROL PROBLEM
11.5 FORCE CONTROL OF A MASS–SPRING SYSTEM
11.6 THE HYBRID POSITION/FORCE CONTROL SCHEME
11.7 CURRENT INDUSTRIAL-ROBOT CONTROL SCHEMES

11.1 INTRODUCTION

Position control is appropriate when a manipulator is following a trajectory through space, but when any contact is made between the end-effector and the manipulator's environment, mere position control might not suffice. Consider a manipulator washing a window with a sponge. The compliance of the sponge might make it possible to regulate the force applied to the window by controlling the position of the end-effector relative to the glass. If the sponge is very compliant or the position of the glass is known very accurately, this technique could work quite well.

If, however, the stiffness of the end-effector, tool, or environment is high, it becomes increasingly difficult to perform operations in which the manipulator presses against a surface. Instead of washing with a sponge, imagine that the manipulator is scraping paint off a glass surface, using a rigid scraping tool. If there is any uncertainty in the position of the glass surface or any error in the position of the manipulator, this task would become impossible. Either the glass would be broken, or the manipulator would wave the scraping tool over the glass with no contact taking place.

In both the washing and scraping tasks, it would be more reasonable not to specify the position of the plane of the glass, but rather *to specify a force that is to be maintained normal to the surface.*

More so than in previous chapters, in this chapter we present methods that are not yet employed by industrial robots, except in an extremely simplified way. The major thrust of the chapter is to introduce the **hybrid position/force controller**, which is one formalism through which industrial robots might someday be controlled in order to perform tasks requiring force control. However, regardless of which method(s) emerge as practical for industrial application, many of the concepts introduced in this chapter will certainly remain valid.

317

11.2 APPLICATION OF INDUSTRIAL ROBOTS TO ASSEMBLY TASKS

The majority of the industrial robot population is employed in relatively **simple applications**, such as spot welding, spray painting, and pick-and-place operations. Force control has already appeared in a few applications; for example, some robots are already capable of simple force control that allows them to do such tasks as grinding and deburring. Apparently, the next big area of application will be to assembly-line tasks in which one or more parts are mated. In such **parts-mating** tasks, monitoring and control of the forces of contact are extremely important.

Precise control of manipulators in the face of uncertainties and variations in their work environments is a prerequisite to application of robot manipulators to assembly operations in industry. It seems that, by providing manipulator hands with sensors that can give information about the state of manipulation tasks, important progress can be made toward using robots for assembly tasks. Currently, the dexterity of manipulators remains quite low and continues to limit their application in the automated assembly area.

The use of manipulators for assembly tasks requires that the precision with which parts are positioned with respect to one another be quite high. Current industrial robots are often not accurate enough for these tasks, and building robots that are might not make sense. Manipulators of greater precision can be achieved only at the expense of size, weight, and cost. The ability to measure and control contact forces generated at the hand, however, offers a possible alternative for extending the effective precision of a manipulation. Because relative measurements are used, absolute errors in the position of the manipulator and the manipulated objects are not as important as they would be in a purely position-controlled system. Small variations in relative position generate large contact forces when parts of moderate stiffness interact, so knowledge and control of these forces can lead to a tremendous increase in effective positional accuracy.

11.3 A FRAMEWORK FOR CONTROL IN PARTIALLY CONSTRAINED TASKS

The approach presented in this chapter is based on a framework for control in situations in which motion of the manipulator is partially constrained by contact with one or more surfaces [1–3]. This framework for understanding partially constrained tasks is based on a simplified model of interaction between the manipulator's end-effector and the environment: We are interested in describing contact and freedoms, so we consider only the forces due to contact. This is equivalent to doing a quasi-static analysis and ignoring other static forces, such as certain friction components and gravity. The analysis is reasonable where forces due to contact between relatively stiff objects are the dominant source of forces acting on the system. Note that the methodology presented here is somewhat simplistic and has some limitations, but it is a good way to introduce the basic concepts involved and do so at a level appropriate for this text. For a related, but more general and rigorous methodology, see [19].

Every manipulation task can be broken down into subtasks that are defined by a particular contact situation occurring between the manipulator end-effector (or tool) and the work environment. With each such subtask, we can associate a set of constraints, called the **natural constraints**, that result from the particular mechanical and geometric characteristics of the task configuration. For instance, a hand in contact with a stationary, rigid surface is not free to move through that surface; hence, a *natural* position constraint exists. If the surface is frictionless, the hand is not free to apply arbitrary forces tangent to the surface; thus, a *natural* force constraint exists.

In our model of contact with the environment, for each subtask configuration, a **generalized surface** can be defined with position constraints along the normals to this surface and force constraints along the tangents. These two types of constraint, force and position, partition the degrees of freedom of possible end-effector motions into two orthogonal sets that must be controlled according to different criteria. Note that this model of contact does not include all possible contacting situations. (See [19] for a more general scheme.)

Figure 11.1 shows two representative tasks along with their associated natural constraints. Notice that, in each case, the task is described in terms of a frame {*C*}, the so-called **constraint frame**, which is located in a task-relevant location. According to the task, {*C*} could be fixed in the environment or could move with the end-effector of the manipulator. In Fig. 11.1(a), the constraint frame is attached to the crank as shown and moves with the crank, with the $\hat{X}$ direction always directed toward the pivot point of the crank. Friction acting at the fingertips ensures a secure grip on the handle, which is on a spindle so that it can rotate relative to the crank arm. In Fig. 11.1(b), the constraint frame is attached to the tip of the screwdriver and moves with it as the task proceeds. Notice that, in the $\hat{Y}$ direction, the force is constrained to be zero, because the slot of the screw would allow the screwdriver to slip out in that direction. In these examples, a given set of constraints remains true throughout the task. In more complex situations, the task is broken into subtasks for which a constant set of natural constraints can be identified.

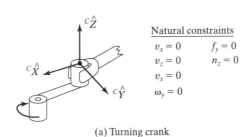

Natural constraints

$$v_x = 0 \qquad f_y = 0$$
$$v_z = 0 \qquad n_z = 0$$
$$v_x = 0$$
$$\omega_y = 0$$

(a) Turning crank

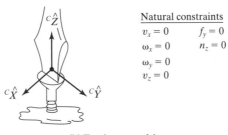

Natural constraints

$$v_x = 0 \qquad f_y = 0$$
$$\omega_x = 0 \qquad n_z = 0$$
$$\omega_y = 0$$
$$v_z = 0$$

(b) Turning screwdriver

FIGURE 11.1: The natural constraints for two different tasks.

In Fig. 11.1, position constraints have been indicated by giving values for components of velocity of the end-effector, V, described in frame $\{C\}$. We could just as well have indicated position constraints by giving expressions for position, rather than velocities; however, in many cases, it is simpler to specify a position constraint as a "velocity equals zero" constraint. Likewise, force constraints have been specified by giving values to components of the force-moment vector, $\mathcal{F}$, acting on the end-effector described in frame $\{C\}$. Note that when we say *position constraints*, we mean position or orientation constraints, and when we say *force constraints*, we mean force or moment constraints. The term *natural constraints* is used to indicate that these constraints arise naturally from the particular contacting situation. They have nothing to do with the desired or intended motion of the manipulator.

Additional constraints, called **artificial constraints**, are introduced in accordance with the natural constraints to specify desired motions or force application. That is, each time the user specifies a desired trajectory in either position or force, an artificial constraint is defined. These constraints also occur along the tangents and normals of the generalized constraint surface, but, unlike natural constraints, artificial force constraints are specified along surface normals, and artificial position constraints along tangents—hence, consistency with the natural constraints is preserved.

Figure 11.2 shows the natural and artificial constraints for two tasks. Note that when a natural position constraint is given for a particular degree of freedom in $\{C\}$,

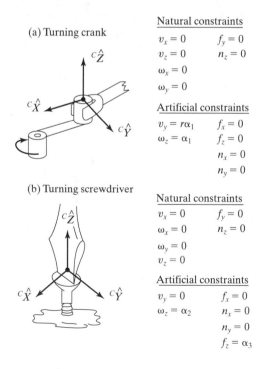

(a) Turning crank

Natural constraints

$$v_x = 0 \qquad f_y = 0$$
$$v_z = 0 \qquad n_z = 0$$
$$\omega_x = 0$$
$$\omega_y = 0$$

Artificial constraints

$$v_y = r\alpha_1 \qquad f_x = 0$$
$$\omega_z = \alpha_1 \qquad f_z = 0$$
$$\qquad\qquad n_x = 0$$
$$\qquad\qquad n_y = 0$$

(b) Turning screwdriver

Natural constraints

$$v_x = 0 \qquad f_y = 0$$
$$\omega_x = 0 \qquad n_z = 0$$
$$\omega_y = 0$$
$$v_z = 0$$

Artificial constraints

$$v_y = 0 \qquad f_x = 0$$
$$\omega_z = \alpha_2 \qquad n_x = 0$$
$$\qquad\qquad n_y = 0$$
$$\qquad\qquad f_z = \alpha_3$$

FIGURE 11.2: The natural and artificial constraints for two tasks.

an artificial force constraint should be specified, and vice versa. At any instant, any given degree of freedom in the constraint frame is controlled to meet either a position or a force constraint.

Assembly strategy is a term that refers to a sequence of planned artificial constraints that will cause the task to proceed in a desirable manner. Such strategies must include methods by which the system can detect a change in the contacting situation so that transitions in the natural constraints can be tracked. With each such change in natural constraints, a new set of artificial constraints is recalled from the set of assembly strategies and enforced by the control system. Methods for automatically choosing the constraints for a given assembly task await further research. In this chapter, we will assume that a task has been analyzed in order to determine the natural constraints and that a human planner has determined an **assembly strategy** with which to control the manipulator.

Note that we will usually ignore friction forces between contacting surfaces in our analysis of tasks. This will suffice for our introduction to the problem and in fact will yield strategies that work in many cases. Usually friction forces of sliding are acting in directions chosen to be position controlled, and so these forces appear as disturbances to the position servo and are overcome by the control system.

EXAMPLE 11.1

Figure 11.3(a)–(d) shows an assembly sequence used to put a round peg into a round hole. The peg is brought down onto the surface to the left of the hole and then slid along the surface until it drops into the hole. It is then inserted until the peg reaches the bottom of the hole, at which time the assembly is complete. Each of the four indicated contacting situations defines a subtask. For each of the subtasks shown, give the natural and artificial constraints. Also, indicate how the system senses the change in the natural constraints as the operation proceeds.

First, we will attach the constraint frame to the peg as shown in Fig. 11.3(a). In Fig. 11.3(a), the peg is in free space, and so the natural constraints are

$$^C\mathcal{F} = 0. \tag{11.1}$$

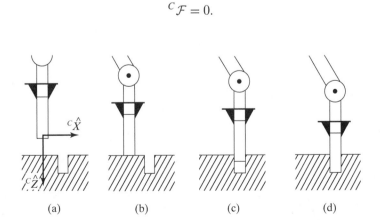

$$(a) \qquad (b) \qquad (c) \qquad (d)$$

FIGURE 11.3: The sequence of four contacting situations for peg insertion.

Therefore, the artificial constraints in this case constitute an entire position trajectory, which moves the peg in the $^C\hat{Z}$ direction toward the surface. For example,

$$
^Cv = \begin{bmatrix} 0 \\ 0 \\ v_{\text{approach}} \\ 0 \\ 0 \\ 0 \end{bmatrix},
\tag{11.2}
$$

where v_{approach} is the speed with which to approach the surface.

In Fig. 11.3(b), the peg has reached the surface. To detect that this has happened, we observe the force in the $^C\hat{Z}$ direction. When this sensed force exceeds a threshold, we sense contact, which implies a new contacting situation with a new set of natural constraints. Assuming that the contacting situation is as shown in Fig. 11.3(b), the peg is not free to move in $^C\hat{Z}$, or to rotate about $^C\hat{X}$ or $^C\hat{Y}$. In the other three degrees of freedom, it is not free to apply forces; hence, the natural constraints are

$$
^Cv_z = 0,
$$
$$
^C\omega_x = 0,
$$
$$
^C\omega_y = 0,
\tag{11.3}
$$
$$
^Cf_x = 0,
$$
$$
^Cf_y = 0,
$$
$$
^Cn_z = 0.
$$

The artificial constraints describe the strategy of sliding along the surface in the $^C\hat{X}$ direction while applying small forces to ensure that contact is maintained. Thus, we have

$$
^Cv_x = v_{\text{slide}},
$$
$$
^Cv_y = 0,
$$
$$
^C\omega_z = 0,
$$
$$
^Cf_z = f_{\text{contact}},
\tag{11.4}
$$
$$
^Cn_x = 0,
$$
$$
^Cn_y = 0.
$$

where f_{contact} is the force applied normal to the surface as the peg is slid, and v_{slide} is the velocity with which to slide across the surface.

In Fig. 11.3(c), the peg has fallen slightly into the hole. This situation is sensed by observing the velocity in the $^C\hat{Z}$ direction and waiting for it to cross a threshold (to become nonzero, in the ideal case). When this is observed, it signals that once again the natural constraints have changed, and thus our strategy (as embodied in the artificial constraints) must change. The new natural constraints are

$$^C v_x = 0,$$
$$^C v_y = 0,$$
$$^C \omega_x = 0,$$
$$^C \omega_y = 0, \tag{11.5}$$
$$^C f_x = 0,$$
$$^C n_z = 0.$$

We choose the artificial constraints to be

$$^C v_z = v_{\text{insert}},$$
$$^C \omega_z = 0,$$
$$^C f_x = 0, \tag{11.6}$$
$$^C f_y = 0,$$
$$^C n_x = 0,$$
$$^C n_y = 0,$$

where v_{insert} is the velocity at which the peg is inserted into the hole. Finally, the situation shown in Fig. 11.3(d) is detected when the force in the $^C\hat{Z}$ direction increases above a threshold.

It is interesting to note that changes in the natural constraints are always detected by observing the position or force variable that is *not* being controlled. For example, to detect the transition from Fig. 11.3(b) to Fig. 11.3(c), we monitor the velocity in $^C\hat{Z}$ while we are controlling force in $^C\hat{Z}$. To discover when the peg has hit the bottom of the hole, we monitor $^C f_z$, although we are controlling $^C v_z$.

The framework we have introduced is somewhat simplistic. A more general and rigorous method of "splitting" tasks into position-controlled and force-controlled components can be found in [19].

Determining assembly strategies for fitting more complicated parts together is quite complex. We have also neglected the effects of uncertainty in our simple analysis of this task. The development of automatic planning systems that include the effects of uncertainty and can be applied to practical situations has been a research topic [4–8]. For a good review of these methods, see [9].

11.4   THE HYBRID POSITION/FORCE CONTROL PROBLEM

Figure 11.4 shows two extreme examples of contacting situations. In Fig. 11.4(a), the manipulator is moving through free space. In this case, the natural constraints are all force constraints—there is nothing to react against, so all forces are constrained

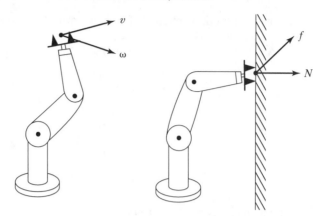

FIGURE 11.4: The two extremes of contacting situations. The manipulator on the left is moving in free space where no reaction surface exits. The manipulator on the right is glued to the wall so that no free motion is possible.

to be zero.[1] With an arm having six degrees of freedom, we are free to move in six degrees of freedom in position, but we are unable to exert forces in any direction. Figure 11.4(b) shows the extreme situation of a manipulator with its end-effector glued to a wall. In this case, the manipulator is subject to six natural position constraints, because it is not free to be repositioned. However, the manipulator is free to exert forces and torques to the object with six degrees of freedom.

In Chapters 9 and 10, we studied the position-control problem that applies to the situation of Fig. 11.4(a). The situation of Fig. 11.4(b) does not occur very often in practice; we usually must consider force control in the context of partially constrained tasks, in which some degrees of freedom of the system are subject to position control and others are subject to force control. Thus, in this chapter, we are interested in considering **hybrid position/force control** schemes.

The hybrid position/force controller must solve three problems:

1. Position control of a manipulator along directions in which a natural force constraint exists.
2. Force control of a manipulator along directions in which a natural position constraint exists.
3. A scheme to implement the arbitrary mixing of these modes along orthogonal degrees of freedom of an arbitrary frame, $\{C\}$.

11.5 FORCE CONTROL OF A MASS–SPRING SYSTEM

In Chapter 9, we began our study of the complete position-control problem with the study of the very simple problem of controlling a single block of mass. We were then able, in Chapter 10, to use a model of the manipulator in such a way that the problem of controlling the entire manipulator became equivalent to controlling n independent

---

[1]It is important to remember that we are concerned here with *forces of contact* between end-effector and environment, not inertial forces.

masses (for a manipulator with n joints). In a similar way, we begin our look at force control by controlling the force applied by a simple single-degree-of-freedom system.

In considering forces of contact, we must make some model of the environment upon which we are acting. For the purposes of conceptual development, we will use a very simple model of interaction between a controlled body and the environment. We model contact with an environment as a spring—that is, we assume our system is rigid and the environment has some stiffness, k_e.

Let us consider the control of a mass attached to a spring, as in Fig. 11.5. We will also include an unknown disturbance force, f_{dist}, which might be thought of as modeling unknown friction or cogging in the manipulator's gearing. The variable we wish to control is the force acting on the environment, f_e, which is the force acting in the spring:

$$f_e = k_e x. \tag{11.7}$$

The equation describing this physical system is

$$f = m\ddot{x} + k_e x + f_{\text{dist}}, \tag{11.8}$$

or, written in terms of the variable we wish to control, f_e,

$$f = mk_e^{-1}\ddot{f}_e + f_e + f_{\text{dist}}. \tag{11.9}$$

Using the partitioned-controller concept, as well as

$$\alpha = mk_e^{-1}$$

and

$$\beta = f_e + f_{\text{dist}}$$

we arrive at the control law,

$$f = mk_e^{-1}\left[\ddot{f}_d + k_{vf}\dot{e}_f + k_{pf}e_f\right] + f_e + f_{\text{dist}}, \tag{11.10}$$

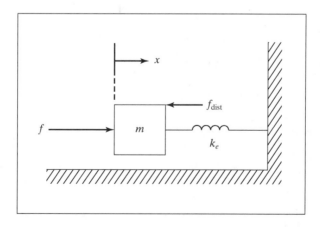

FIGURE 11.5: A spring–mass system.

where $e_f = f_d - f_e$ is the force error between the desired force, f_d, and the sensed force on the environment, f_e. If we could compute (11.10), we would have the closed-loop system

$$\ddot{e}_f + k_{vf}\dot{e}_f + k_{pf}e_f = 0. \tag{11.11}$$

However, we cannot use knowledge of f_{dist} in our control law, and so (11.10) is not feasible. We might leave that term out of the control law, but a steady-state analysis shows that there is a better choice, especially when the stiffness of the environment, k_e, is high (the usual situation).

If we choose to leave the f_{dist} term out of our control law, equate (11.9) and (11.10), and do a steady-state analysis by setting all time derivatives to zero, we find that

$$e_f = \frac{f_{\text{dist}}}{\alpha}, \tag{11.12}$$

where $\alpha = mk_e^{-1}k_{pf}$, the effective force-feedback gain; however, if we choose to use f_d in the control law (11.10) in place of the term $f_e + f_{\text{dist}}$, we find the steady-state error to be

$$e_f = \frac{f_{\text{dist}}}{1 + \alpha}. \tag{11.13}$$

When the environment is stiff, as is often the case, α might be small, and so the steady-state error calculated in (11.13) is quite an improvement over that of (11.12). Therefore, we suggest the control law

$$f = mk_e^{-1}\left[\ddot{f}_d + k_{vf}\dot{e}_f + k_{pf}e_f\right] + f_d. \tag{11.14}$$

Figure 11.6 is a block diagram of the closed-loop system using the control law (11.14).

Generally, practical considerations change the implementation of a force-control servo quite a bit from the ideal shown in Fig. 11.6. First, force trajectories are

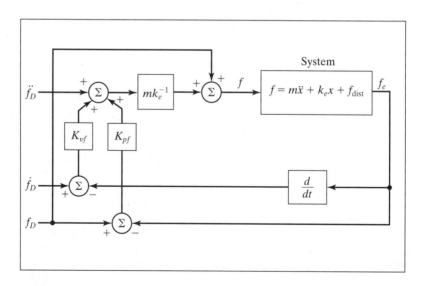

FIGURE 11.6: A force control system for the spring–mass system.

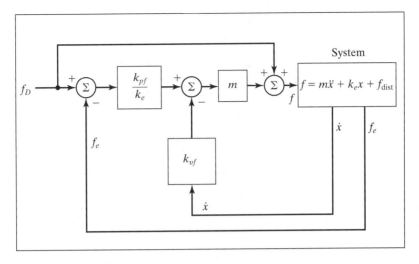

FIGURE 11.7: A practical force-control system for the spring–mass system.

usually constants—that is, we are usually interested in controlling the contact force to be at some constant level. Applications in which contact forces should follow some arbitrary function of time are rare. Therefore, the $\dot{f}_d$ and $\ddot{f}_d$ inputs of the control system are very often permanently set to zero. Another reality is that sensed forces are quite "noisy," and numerical differentiation to compute $\dot{f}_e$ is ill-advised. However, $f_e = k_e x$, so we can obtain the derivative of the force on the environment as $\dot{f}_e = k_e \dot{x}$. This is much more realistic, in that most manipulators have means of obtaining good measures of velocity. Having made these two pragmatic choices, we write the control law as

$$f = m \left[k_{pf} k_e^{-1} e_f - k_{vf} \dot{x} \right] + f_d, \tag{11.15}$$

with the corresponding block diagram shown in Fig. 11.7.

Note that an interpretation of the system of Fig. 11.7 is that force errors generate a set-point for an inner velocity-control loop with gain k_{vf}. Some force-control laws also include an integral term to improve steady-state performance.

An important remaining problem is that the stiffness of the environment, k_e, appears in our control law, but is often unknown and perhaps changes from time to time. However, often an assembly robot is dealing with rigid parts, and k_e could be guessed to be quite high. Generally this assumption is made, and gains are chosen such that the system is somewhat robust with respect to variations in k_e.

The purpose in constructing a control law to control the force of contact has been to show one suggested structure and to expose a few issues. For the remainder of this chapter, we will simply assume that such a force-controlling servo could be built and abstract it away into a black box, as shown in Fig. 11.8. In practice, it is not easy to build a high-performance force servo, and it is currently an area of active research [11–14]. For a good review of this area, see [15].

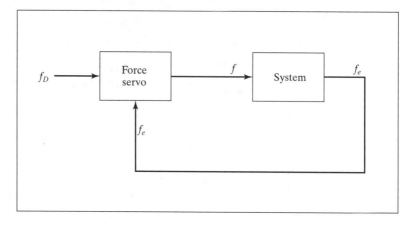

FIGURE 11.8: The force-control servo as a black box.

11.6   THE HYBRID POSITION/FORCE CONTROL SCHEME

In this section, we introduce an architecture for a control system that implements the hybrid position/force controller.

A Cartesian manipulator aligned with {C}

We will first consider the simple case of a manipulator having three degrees of freedom with prismatic joints acting in the $\hat{Z}$, $\hat{Y}$, and $\hat{X}$ directions. For simplicity, we will assume that each link has mass m and slides on frictionless bearings. Let us also assume that the joint motions are lined up exactly with the constraint frame, {C}. The end-effector is in contact with a surface of stiffness k_e that is oriented with its normal in the $-^C\hat{Y}$ direction. Hence, force control is required in that direction, but position control in the $^C\hat{X}$ and $^C\hat{Z}$ directions. (See Fig. 11.9.)

In this case, the solution to the hybrid position/force control problem is clear. We should control joints 1 and 3 with the position controller developed for a unit mass in Chapter 9. Joint 2 (operating in the $\hat{Y}$ direction) should be controlled with the force controller developed in Section 11.4. We could then supply a position trajectory in the $^C\hat{X}$ and $^C\hat{Z}$ directions, while independently supplying a force trajectory (perhaps just a constant) in the $^C\hat{Y}$ direction.

If we wish to be able to switch the nature of the constraint surface such that its normal might also be $\hat{X}$ or $\hat{Z}$, we can slightly generalize our Cartesian arm-control system as follows: We build the structure of the controller such that we may specify a complete position trajectory in all three degrees of freedom and also a force trajectory in all three degrees of freedom. Of course, we can't control so as to meet these six constraints at any one time—rather, we will set modes to indicate which components of which trajectory will be followed at any given time.

Consider the controller shown in Fig. 11.10. Here, we indicate the control of all three joints of our simple Cartesian arm in a single diagram by showing both the position controller and the force controller. The matrices S and S' have been introduced to control which mode—position or force—is used to control each joint of the Cartesian arm. The S matrix is diagonal, with ones and zeros on the diagonal.

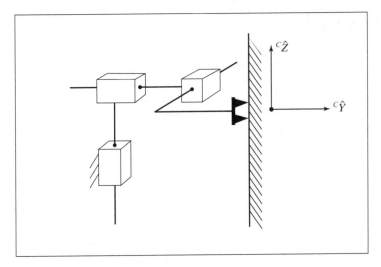

FIGURE 11.9: A Cartesian manipulator with three degrees of freedom in contact with a surface.

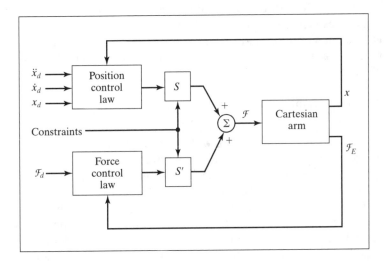

FIGURE 11.10: The hybrid controller for a 3-DOF Cartesian arm.

Where a one is present in S, a zero is present in S' and position control is in effect. Where a zero is present in S, a one is present in S' and force control is in effect. Hence the matrices S and S' are simply switches that set the mode of control to be used with each degree of freedom in $\{C\}$. In accordance with the setting of S, there are always three components of the trajectory being controlled, yet the mix between position control and force control is arbitrary. The other three components of desired trajectory and associated servo errors are being ignored. Hence, when a certain degree of freedom is under force control, position errors on that degree of freedom are ignored.

EXAMPLE 11.2

For the situation shown in Fig. 11.9, with motions in the $^C\hat{Y}$ direction constrained by the reaction surface, give the matrices S and S'.

Because the $\hat{X}$ and $\hat{Z}$ components are to be position controlled, we enter ones on the diagonal of S corresponding to these two components. This will cause the position servo to be active in these two directions, and the input trajectory will be followed. Any position trajectory input for the $\hat{Y}$ component will be ignored. The S' matrix has the ones and zeros on the diagonal inverted; hence, we have

$$S = \begin{bmatrix} 1 & 0 & 0 \\ 0 & 0 & 0 \\ 0 & 0 & 1 \end{bmatrix},$$

$$S' = \begin{bmatrix} 0 & 0 & 0 \\ 0 & 1 & 0 \\ 0 & 0 & 0 \end{bmatrix}. \tag{11.16}$$

Figure 11.10 shows the hybrid controller for the special case that the joints line up exactly with the constraint frame, $\{C\}$. In the following subsection, we use techniques studied in previous chapters to generalize the controller to work with general manipulators and for an arbitrary $\{C\}$; however, in the ideal case, the system performs as if the manipulator had an actuator "lined up" with each of the degrees of freedom in $\{C\}$.

A general manipulator

Generalizing the hybrid controller shown in Fig. 11.10 so that a general manipulator may be used is straightforward with the concept of Cartesian-based control. Chapter 6 discussed how the equations of motion of a manipulator could be written in terms of Cartesian motion of the end-effector, and Chapter 10 showed how such a formulation might be used to achieve decoupled Cartesian position control of a manipulator. The major idea is that, through use of a dynamic model written in Cartesian space, it is possible to control so that the combined system of the actual manipulator and computed model appear as a set of independent, uncoupled unit masses. Once this decoupling and linearizing are done, we can apply the simple servo already developed in Section 11.4.

Figure 11.11 shows the compensation based on the formulation of the manipulator dynamics in Cartesian space such that the manipulator appears as a set of uncoupled unit masses. For use in the hybrid control scheme, the Cartesian dynamics and the Jacobian are written in the constraint frame, $\{C\}$. Likewise, the kinematics are computed with respect to the constraint frame.

Because we have designed the hybrid controller for a Cartesian manipulator aligned with the constraint frame, and because the Cartesian decoupling scheme provides us with a system with the same input–output properties, we need only combine the two to generate the generalized hybrid position/force controller.

Figure 11.12 is a block diagram of the hybrid controller for a general manipulator. Note that the dynamics are written in the constraint frame, as is the Jacobian.

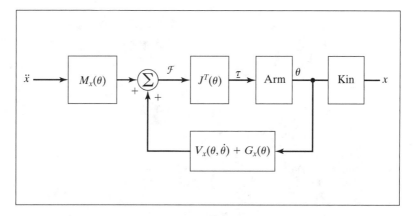

FIGURE 11.11: The Cartesian decoupling scheme introduced in Chapter 10.

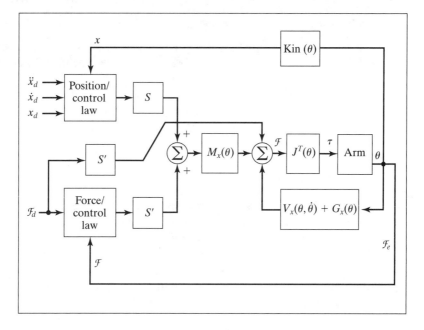

FIGURE 11.12: The hybrid position/force controller for a general manipulator. For simplicity, the velocity-feedback loop has not been shown.

The kinematics are written to include the transformation of coordinates into the constraint frame, and the sensed forces are likewise transformed into {C}. Servo errors are calculated in {C}, and control modes within {C} are set through proper choice of S.[2] Figure 11.13 shows a manipulator being controlled by such a system.

[2]The partitioning of control modes along certain task-related directions has been generalized in [10] from the more basic approach presented in this chapter.

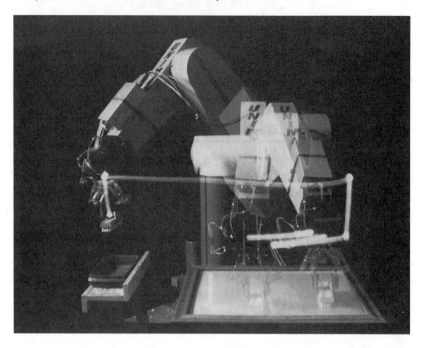

FIGURE 11.13: A PUMA 560 manipulator washes a window under control of the COSMOS system developed under O. Khatib at Stanford University. These experiments use force-sensing fingers and a control structure similar to that of Fig. 11.12 [10].

Adding variable stiffness

Controlling a degree of freedom in strict position or force control represents control at two ends of the spectrum of servo stiffness. An ideal position servo is infinitely stiff and rejects all force disturbances acting on the system. In contrast, an ideal force servo exhibits zero stiffness and maintains the desired force application regardless of position disturbances. It could be useful to be able to control the end-effector to exhibit stiffnesses other than zero or infinite. In general, we might wish to control the **mechanical impedance** of the end-effector [14, 16, 17].

In our analysis of contact, we have imagined that the environment is very stiff. When we contact a stiff environment, we use zero-stiffness force control. When we contact zero stiffness (moving in free space) we use high-stiffness position control. Hence, it appears that controlling the end-effector to exhibit a stiffness that is approximately the inverse of the local environment is perhaps a good strategy. Therefore, in dealing with plastic parts or springs, we could wish to set servo stiffness to other than zero or infinite.

Within the framework of the hybrid controller, this is done simply by using position control and lowering the position gain corresponding to the appropriate degree of freedom in {C}. Generally, if this is done, the corresponding velocity gain is lowered so that that degree of freedom remains critically damped. The ability to change both position and velocity gains of the position servo along the degrees of freedom of $\{C\}$ allows the hybrid position/force controller to implement a generalized impedance of the end-effector [17]. However, in many practical situations we are dealing with the interaction of stiff parts, so that pure position control or pure force control is desired.

11.7   CURRENT INDUSTRIAL-ROBOT CONTROL SCHEMES

True force control, such as the hybrid position/force controller introduced in this chapter, does not exist today in industrial robots. Among the problems of practical implementation are the rather large amount of computation required, lack of accurate parameters for the dynamic model, lack of rugged force sensors, and the burden of difficulty placed on the user in specifying a position/force strategy.

Passive compliance

Extremely rigid manipulators with very stiff position servos are ill-suited to tasks in which parts come into contact and contact forces are generated. In such situations, parts are often jammed or damaged. Ever since early experiments with manipulators attempting to do assembly, it was realized that, to the extent that the robots could perform such tasks, it was only thanks to the compliance of the parts, of the fixtures, or of the arm itself. This ability of one or more parts of the system to "give" a little was often enough to allow the successful mating of parts.

Once this was realized, devices were specially designed to introduce compliance into the system on purpose. The most successful such device is the RCC or *remote center compliance* device developed at Draper Labs [18]. The RCC was cleverly designed so that it introduced the "right" kind of compliance, which allowed certain tasks to proceed smoothly and rapidly with little or no chance of jamming. The RCC is essentially a spring with six degrees of freedom, which is inserted between the manipulator's wrist and the end-effector. By setting the stiffnesses of the six springs, various amounts of compliance can be introduced. Such schemes are called **passive-compliance** schemes and are used in industrial applications of manipulators in some tasks.

Compliance through softening position gains

Rather than achieving compliance in a passive, and therefore fixed, way, it is possible to devise schemes in which the apparent stiffness of the manipulator is altered through adjustment of the gains of a position-control system. A few industrial robots do something of this type for applications such as grinding, in which contact with a surface needs to be maintained but delicate force control is not required.

A particularly interesting approach has been suggested by Salisbury [16]. In this scheme, the position gains in a joint-based servo system are modified in such a way that the end-effector appears to have a certain stiffness along Cartesian degrees of freedom: Consider a general spring with six degrees of freedom. Its action could be described by

$$\mathcal{F} = K_{px}\delta\chi, \tag{11.17}$$

where K_{px} is a diagonal 6×6 matrix with three linear stiffnesses followed by three torsional stiffnesses on the diagonal. How could we make the end-effector of a manipulator exhibit this stiffness characteristic?

Recalling the definition of the manipulator Jacobian, we have

$$\delta \chi = J(\Theta) \delta \Theta. \tag{11.18}$$

Combining with (11.17) gives

$$\mathcal{F} = K_{px} J(\Theta) \delta \Theta. \tag{11.19}$$

From static-force considerations, we have

$$\tau = J^T(\Theta) \mathcal{F}, \tag{11.20}$$

which, combined with (11.19), yields

$$\tau = J^T(\Theta) K_{px} J(\Theta) \delta \Theta. \tag{11.21}$$

Here, the Jacobian is usually written in the tool frame. Equation (11.21) is an expression for how joint torques should be generated as a function of small changes in joint angles, $\delta \Theta$, in order to make the manipulator end-effector behave as a Cartesian spring with six degrees of freedom.

Whereas a simple joint-based position controller might use the control law

$$\tau = K_p E + K_v \dot{E}, \tag{11.22}$$

where K_p and K_v are constant diagonal gain matrices and E is servo error defined as $\Theta_d - \Theta$, Salisbury suggests using

$$\tau = J^T(\Theta) K_{px} J(\Theta) E + K_v \dot{E}, \tag{11.23}$$

where K_{px} is the desired stiffness of the end-effector in Cartesian space. For a manipulator with six degrees of freedom, K_{px} is diagonal with the six values on the diagonal representing the three translational and three rotational stiffnesses that the end-effector is to exhibit. Essentially, through use of the Jacobian, a Cartesian stiffness has been transformed to a joint-space stiffness.

Force sensing

Force sensing allows a manipulator to detect contact with a surface and, using this sensation, to take some action. For example, the term **guarded move** is sometimes used to mean the strategy of moving under position control until a force is felt, then halting motion. Additionally, force sensing can be used to weigh objects that the manipulator lifts. This can be used as a simple check during a parts-handling operation—to ensure that a part was acquired or that the appropriate part was acquired.

Some commercially available robots come equipped with force sensors in the end-effector. These robots can be programmed to stop motion or take other action when a force threshold is exceeded, and some can be programmed to weigh objects that are grasped in the end-effector.

BIBLIOGRAPHY

[1] M. Mason, "Compliance and Force Control for Computer Controlled Manipulators," M.S. Thesis, MIT AI Laboratory, May 1978.

[2] J. Craig and M. Raibert, "A Systematic Method for Hybrid Position/Force Control of a Manipulator," *Proceedings of the 1979 IEEE Computer Software Applications Conference*, Chicago, November 1979.

[3] M. Raibert and J. Craig, "Hybrid Position/Force Control of Manipulators," *ASME Journal of Dynamic Systems, Measurement, and Control*, June 1981.

[4] T. Lozano-Perez, M. Mason, and R. Taylor, "Automatic Synthesis of Fine-Motion Strategies for Robots," 1st International Symposium of Robotics Research, Bretton Woods, NH, August 1983.

[5] M. Mason, "Automatic Planning of Fine Motions: Correctness and Completeness," IEEE International Conference on Robotics, Atlanta, March 1984.

[6] M. Erdmann, "Using Backprojections for the Fine Motion Planning with Uncertainty," *The International Journal of Robotics Research*, Vol. 5, No. 1, 1986.

[7] S. Buckley, "Planning and Teaching Compliant Motion Strategies," Ph.D. Dissertation, Department of Electrical Engineering and Computer Science, MIT, January 1986.

[8] B. Donald, "Error Detection and Recovery for Robot Motion Planning with Uncertainty," Ph.D. Dissertation, Department of Electrical Engineering and Computer Science, MIT, July 1987.

[9] J.C. Latombe, "Motion Planning with Uncertainty: On the Preimage Backchaining Approach," in *The Robotics Review*, O. Khatib, J. Craig, and T. Lozano-Perez, Editors, MIT Press, Cambridge, MA, 1988.

[10] O. Khatib, "A Unified Approach for Motion and Force Control of Robot Manipulators: The Operational Space Formulation," *IEEE Journal of Robotics and Automation*, Vol. RA-3, No. 1, 1987.

[11] D. Whitney, "Force Feedback Control of Manipulator Fine Motions," Proceedings Joint Automatic Control Conference, San Francisco, 1976.

[12] S. Eppinger and W. Seering, "Understanding Bandwidth Limitations in Robot Force Control," *Proceedings of the IEEE Conference on Robotics and Automation*, Raleigh, NC, 1987.

[13] W. Townsend and J.K. Salisbury, "The Effect of Coulomb Friction and Stiction on Force Control," *Proceedings of the IEEE Conference on Robotics and Automation*, Raleigh, NC, 1987.

[14] N. Hogan, "Stable Execution of Contact Tasks Using Impedance Control," *Proceedings of the IEEE Conference on Robotics and Automation*, Raleigh, NC, 1987.

[15] N. Hogan and E. Colgate, "Stability Problems in Contact Tasks," in *The Robotics Review*, O. Khatib, J. Craig, and T. Lozano-Perez, Editors, MIT Press, Cambridge, MA, 1988.

[16] J.K. Salisbury, "Active Stiffness Control of a Manipulator in Cartesian Coordinates," 19th IEEE Conference on Decision and Control, December 1980.

[17] J.K. Salisbury and J. Craig, "Articulated Hands: Force Control and Kinematic Issues," *International Journal of Robotics Research*, Vol. 1, No. 1.

[18] S. Drake, "Using Compliance in Lieu of Sensory Feedback for Automatic Assembly," Ph.D. Thesis, Mechanical Engineering Department, MIT, September 1977.

[19] R. Featherstone, S.S. Thiebaut, and O. Khatib, "A General Contact Model for Dynamically-Decoupled Force/Motion Control," *Proceedings of the IEEE Conference on Robotics and Automation*, Detroit, 1999.

EXERCISES

11.1 [12] Give the natural constraints present for a peg of square cross section sliding into a hole of square cross section. Show your definition of $\{C\}$ in a sketch.

11.2 [10] Give the artificial constraints (i.e., the trajectory) you would suggest in order to cause the peg in Exercise 11.1 to slide further into the hole without jamming.

11.3 [20] Show that using the control law (11.14) with a system given by (11.9) results in the error-space equation

$$\ddot{e}_f + k_{v_f}\dot{e}_f + (k_{pf} + m^{-1}k_e)e_f = m^{-1}k_e f_{\text{dist}},$$

and, hence, that choosing gains to provide critical damping is possible only if the stiffness of the environment, k_e, is known.

11.4 [17] Given

$$
{}^A_B T = \begin{bmatrix} 0.866 & -0.500 & 0.000 & 10.0 \\ 0.500 & 0.866 & 0.000 & 0.0 \\ 0.000 & 0.000 & 1.000 & 5.0 \\ 0 & 0 & 0 & 1 \end{bmatrix},
$$

if the force–torque vector at the origin of $\{A\}$ is

$$
{}^A\nu = \begin{bmatrix} 0.0 \\ 2.0 \\ -3.0 \\ 0.0 \\ 0.0 \\ 4.0 \end{bmatrix},
$$

find the 6×1 force–torque vector with reference point at the origin of $\{B\}$.

11.5 [17] Given

$$
{}^A_B T = \begin{bmatrix} 0.866 & 0.500 & 0.000 & 10.0 \\ -0.500 & 0.866 & 0.000 & 0.0 \\ 0.000 & 0.000 & 1.000 & 5.0 \\ 0 & 0 & 0 & 1 \end{bmatrix},
$$

if the force–torque vector at the origin of $\{A\}$ is

$$
{}^A\nu = \begin{bmatrix} 6.0 \\ 6.0 \\ 0.0 \\ 5.0 \\ 0.0 \\ 0.0 \end{bmatrix},
$$

find the 6×1 force–torque vector with reference point at the origin of $\{B\}$.

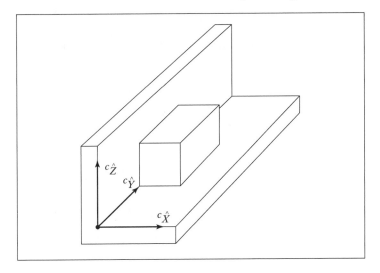

FIGURE 11.14: A block constrained by a floor below and a wall to the side.

11.6 [18] Describe in English how you accomplish the insertion of a book into a narrow crack between books on your crowded bookshelf.

11.7 [20] Give the natural and artificial constraints for the task of closing a hinged door with a manipulator. Make any reasonable assumptions needed. Show your definition of {C} in a sketch.

11.8 [20] Give the natural and artificial constraints for the task of uncorking a bottle of champagne with a manipulator. Make any reasonable assumptions needed. Show your definition of {C} in a sketch.

11.9 [41] For the stiffness servo system of Section 11.7, we have made no claim that the system is stable. Assume that (11.23) is used as the servo portion of a decoupled and linearized manipulator (so that the n joints appear as unit masses). Prove that the controller is stable for any K_v which is negative definite.

11.10 [48] For the stiffness servo system of Section 11.7, we have made no claim that the system is or can be critically damped. Assume that (11.23) is used as the servo portion of a decoupled and linearized manipulator (so that the n joints appear as unit masses). Is it possible to design a K_p that is a function of Θ and causes the system to be critically damped over all configurations?

11.11 [15] As shown in Fig. 11.14, a block is constrained below by a floor and to the side by a wall. Assuming this contacting situation is maintained over an interval of time, give the natural constraints that are present.

PROGRAMMING EXERCISE (PART 11)

Implement a Cartesian stiffness-control system for the three-link planar manipulator by using the control law (11.23) to control the simulated arm. Use the Jacobian written in frame {3}.

For the manipulator in position $\Theta = [60.0 - 90.030.0]$ and for K_{px} of the form

$$K_{px} = \begin{bmatrix} k_{\text{small}} & 0.0 & 0.0 \\ 0.0 & k_{\text{big}} & 0.0 \\ 0.0 & 0.0 & k_{\text{big}} \end{bmatrix},$$

simulate the application of the following static forces:

1. a 1-newton force acting at the origin of {3} in the $\hat{X}_3$ direction, and
2. a 1-newton force acting at the origin of {3} in the $\hat{Y}_3$ direction.

The values of k_{small} and k_{big} should be found experimentally. Use a large value of k_{big} for high stiffness in the $\hat{Y}_3$ direction and a low value of k_{small} for low stiffness in the $\hat{X}_3$ direction. What are the steady-state deflections in the two cases?

CHAPTER 12

Robot programming languages and systems

12.1 INTRODUCTION

In this chapter, we begin to consider the interface between the human user and an industrial robot. It is by means of this interface that a user takes advantage of all the underlying mechanics and control algorithms we have studied in previous chapters.

The sophistication of the user interface is becoming extremely important as manipulators and other programmable automation are applied to more and more demanding industrial applications. It turns out that the nature of the user interface is a very important concern. In fact, most of the challenge of the design and use of industrial robots focuses on this aspect of the problem.

Robot manipulators differentiate themselves from fixed automation by being "flexible," which means programmable. Not only are the movements of manipulators programmable, but, through the use of sensors and communications with other factory automation, manipulators can *adapt* to variations as the task proceeds.

In considering the programming of manipulators, it is important to remember that they are typically only a minor part of an automated process. The term **workcell** is used to describe a local collection of equipment, which may include one or more manipulators, conveyor systems, parts feeders, and fixtures. At the next higher level, workcells might be interconnected in factorywide networks so that a central control computer can control the overall factory flow. Hence, the programming of manipulators is often considered within the broader problem of programming a variety of interconnected machines in an automated factory workcell.

Unlike that in the previous 11 chapters, the material in this chapter (and the next chapter) is of a nature that constantly changes. It is therefore difficult to present this material in a detailed way. Rather, we attempt to point out the underlying fundamental concepts, and we leave it to the reader to seek out the latest examples, as industrial technology continues to move forward.

12.2   THE THREE LEVELS OF ROBOT PROGRAMMING

There have been many styles of user interface developed for programming robots. Before the rapid proliferation of microcomputers in industry, robot controllers resembled the simple sequencers often used to control fixed automation. Modern approaches focus on computer programming, and issues in programming robots include all the issues faced in general computer programming—and more.

Teach by showing

Early robots were all programmed by a method that we will call **teach by showing**, which involved moving the robot to a desired goal point and recording its position in a memory that the sequencer would read during playback. During the teach phase, the user would guide the robot either by hand or through interaction with a **teach pendant**. Teach pendants are handheld button boxes that allow control of each manipulator joint or of each Cartesian degree of freedom. Some such controllers allow testing and branching, so that simple programs involving logic can be entered. Some teach pendants have alphanumeric displays and are approaching hand-held terminals in complexity. Figure 12.1 shows an operator using a teach pendant to program a large industrial robot.

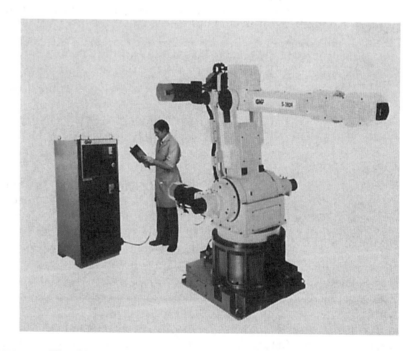

FIGURE 12.1: The GMF S380 is often used in automobile-body spot-welding applications. Here an operator uses a teach-pendant interface to program the manipulator. Photo courtesy of GMFanuc Corp.

Explicit robot programming languages

Ever since the arrival of inexpensive and powerful computers, the trend has been increasingly toward programming robots via programs written in computer programming languages. Usually, these computer programming languages have special features that apply to the problems of programming manipulators and so are called **robot programming languages** (RPLs). Most of the systems that come equipped with a robot programming language have nonetheless retained a teach-pendant-style interface also.

Robot programming languages have likewise taken on many forms. We will split them into three categories:

1. **Specialized manipulation languages**. These robot programming languages have been built by developing a completely new language that, although addressing robot-specific areas, might well be considered a general computer programming language. An example is the VAL language developed to control the industrial robots by Unimation, Inc [1]. VAL was developed especially as a manipulator control language; as a general computer language, it was quite weak. For example, it did not support floating-point numbers or character strings, and subroutines could not pass arguments. A more recent version, V-II, provided these features [2]. The current incarnation of this language, V+, includes many new features [13]. Another example of a specialized manipulation language is AL, developed at Stanford University [3]. Although the AL language is now a relic of the past, it nonetheless provides good examples of some features still not found in most modern languages (force control, parallelism). Also, because it was built in an academic environment, there are references available to describe it [3]. For these reasons, we continue to make reference to it.

2. **Robot library for an existing computer language**. These robot programming languages have been developed by starting with a popular computer language (e.g., Pascal) and adding a library of robot-specific subroutines. The user then writes a Pascal program making use of frequent calls to the predefined subroutine package for robot-specific needs. An examples is AR-BASIC from American Cimflex [4], essentially a subroutine library for a standard BASIC implementation. JARS, developed by NASA's Jet Propulsion Laboratory, is an example of such a robot programming language based on Pascal [5].

3. **Robot library for a new general-purpose language**. These robot programming languages have been developed by first creating a new general-purpose language as a programming base and then supplying a library of predefined robot-specific subroutines. Examples of such robot programming languages are RAPID developed by ABB Robotics [6], AML developed by IBM [7], and KAREL developed by GMF Robotics [8].

Studies of actual application programs for robotic workcells have shown that a large percentage of the language statements are not robot-specific [7]. Instead, a great deal of robot programming has to do with initialization, logic testing and branching, communication, and so on. For this reason, a trend might develop to move away from developing special languages for robot programming and move toward developing extensions to general languages, as in categories 2 and 3 above.

Task-level programming languages

The third level of robot programming methodology is embodied in **task-level programming languages**. These languages allow the user to command desired subgoals of the task directly, rather than specify the details of every action the robot is to take. In such a system, the user is able to include instructions in the application program at a significantly higher level than in an explicit robot programming language. A task-level robot programming system must have the ability to perform many planning tasks automatically. For example, if an instruction to "grasp the bolt" is issued, the system must plan a path of the manipulator that avoids collision with any surrounding obstacles, must automatically choose a good grasp location on the bolt, and must grasp it. In contrast, in an explicit robot programming language, all these choices must be made by the programmer.

The border between explicit robot programming languages and task-level programming languages is quite distinct. Incremental advances are being made to explicit robot programming languages to help to ease programming, but these enhancements cannot be counted as components of a task-level programming system. True task-level programming of manipulators does not yet exist, but it has been an active topic of research [9, 10] and continues as a research topic today.

12.3   A SAMPLE APPLICATION

Figure 12.2 shows an automated workcell that completes a small subassembly in a hypothetical manufacturing process. The workcell consists of a conveyor under computer control that delivers a workpiece; a camera connected to a vision system, used to locate the workpiece on the conveyor; an industrial robot (a PUMA 560 is pictured) equipped with a force-sensing wrist; a small feeder located on the work surface that supplies another part to the manipulator; a computer-controlled press that can be loaded and unloaded by the robot; and a pallet upon which the robot places finished assemblies.

The entire process is controlled by the manipulator's controller in a sequence, as follows:

1. The conveyor is signaled to start; it is stopped when the vision system reports that a bracket has been detected on the conveyor.
2. The vision system judges the bracket's position and orientation on the conveyor and inspects the bracket for defects, such as the wrong number of holes.
3. Using the output of the vision system, the manipulator grasps the bracket with a specified force. The distance between the fingertips is checked to ensure that the bracket has been properly grasped. If it has not, the robot moves out of the way and the vision task is repeated.
4. The bracket is placed in the fixture on the work surface. At this point, the conveyor can be signaled to start again for the next bracket—that is, steps 1 and 2 can begin in parallel with the following steps.

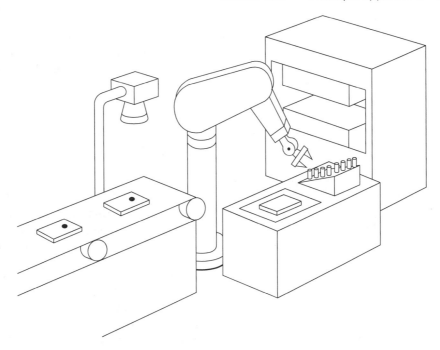

FIGURE 12.2: An automated workcell containing an industrial robot.

5. A pin is picked from the feeder and inserted partway into a tapered hole in the bracket. Force control is used to perform this insertion and to perform simple checks on its completion. (If the pin feeder is empty, an operator is notified and the manipulator waits until commanded to resume by the operator.)

6. The bracket–pin assembly is grasped by the robot and placed in the press.

7. The press is commanded to actuate, and it presses the pin the rest of the way into the bracket. The press signals that it has completed, and the bracket is placed back into the fixture for a final inspection.

8. By force sensing, the assembly is checked for proper insertion of the pin. The manipulator senses the reaction force when it presses sideways on the pin and can do several checks to discover how far the pin protrudes from the bracket.

9. If the assembly is judged to be good, the robot places the finished part into the next available pallet location. If the pallet is full, the operator is signaled. If the assembly is bad, it is dropped into the trash bin.

10. Once Step 2 (started earlier in parallel) is complete, go to Step 3.

This is an example of a task that is possible for today's industrial robots. It should be clear that the definition of such a process through "teach by showing" techniques is probably not feasible. For example, in dealing with pallets, it is laborious to have to teach all the pallet compartment locations; it is much preferable to teach only the corner location and then compute the others from knowledge of the dimensions of the pallet. Further, specifying interprocess signaling and setting up parallelism by using a typical teach pendant or a menu-style interface is usually not possible at all. This kind of application necessitates a robot programming-language approach to process description. (See Exercise 12.5.) On the other hand, this application is too complex for any existing task-level languages to deal with directly. It is typical of the great many applications that must be addressed with an explicit robot programming approach. We will keep this sample application in mind as we discuss features of robot programming languages.

12.4 REQUIREMENTS OF A ROBOT PROGRAMMING LANGUAGE

World modeling

Manipulation programs must, by definition, involve moving objects in three-dimensional space, so it is clear that any robot programming language needs a means of describing such actions. The most common element of robot programming languages is the existence of special **geometric types**. For example, *types* are introduced to represent joint-angle sets, Cartesian positions, orientations, and frames. Predefined operators that can manipulate these types often are available. The "standard frames" introduced in Chapter 3 might serve as a possible model of the world: All motions are described as tool frame relative to station frame, with goal frames being constructed from arbitrary expressions involving geometric types.

Given a robot programming environment that supports geometric types, the robot and other machines, parts, and fixtures can be modeled by defining named variables associated with each object of interest. Figure 12.3 shows part of our example workcell with frames attached in task-relevant locations. Each of these frames would be represented with a variable of type "frame" in the robot program.

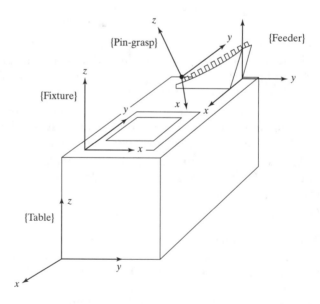

FIGURE 12.3: Often, a workcell is modeled simply, as a set of frames attached to relevant objects.

In many robot programming languages, this ability to define named variables of various geometric types and refer to them in the program forms the basis of the world model. Note that the physical shapes of the objects are not part of such a world model, and neither are surfaces, volumes, masses, or other properties. The extent to which objects in the world are modeled is one of the basic design decisions made when designing a robot programming system. Most present-day systems support only the style just described.

Some world-modeling systems allow the notion of **affixments** between named objects [3]—that is, the system can be notified that two or more named objects have become "affixed"; from then on, if one object is explicitly moved with a language statement, any objects affixed to it are moved with it. Thus, in our application, once the pin has been inserted into the hole in the bracket, the system would be notified (via a language statement) that these two objects have become affixed. Subsequent motions of the bracket (that is, changes to the value of the frame variable "bracket") would cause the value stored for variable "pin" to be updated along with it.

Ideally, a world-modeling system would include much more information about the objects with which the manipulator has to deal and about the manipulator itself. For example, consider a system in which objects are described by CAD-style models that represent the spatial shape of an object by giving definitions of its edges, surfaces, or volume. With such data available to the system, it begins to become possible to implement many of the features of a task-level programming system. These possibilities are discussed further in Chapter 13.

Motion specification

A very basic function of a robot programming language is to allow the description of desired motions of the robot. Through the use of motion statements in the language, the user interfaces to path planners and generators of the style described in Chapter 7. Motion statements allow the user to specify via points, the goal point, and whether to use joint-interpolated motion or Cartesian straight-line motion. Additionally, the user might have control over the speed or duration of a motion.

To illustrate various syntaxes for motion primitives, we will consider the following example manipulator motions: (1) move to position "goal1," then (2) move in a straight line to position "goal2," then (3) move without stopping through "via1" and come to rest at "goal3." Assuming all of these path points had already been taught or described textually, this program segment would be written as follows:

In VAL II,

```
move goal1
moves goal2
move via1
move goal3
```

In AL (here controlling the manipulator "garm"),

```
move garm to goal1;
move garm to goal2 linearly;
move garm to goal3 via via1;
```

Most languages have similar syntax for simple motion statements like these. Differences in the basic motion primitives from one robot programming language to another become more apparent if we consider features such as the following:

1. the ability to do math on such structured types as frames, vectors, and rotation matrices;

2. the ability to describe geometric entities like frames in several different convenient representations—along with the ability to convert between representations;

3. the ability to give constraints on the duration or velocity of a particular move—for example, many systems allow the user to set the speed to a fraction of maximum, but fewer allow the user to specify a desired duration or a desired maximum joint velocity directly;

4. the ability to specify goals relative to various frames, including frames defined by the user and frames in motion (on a conveyor, for example).

Flow of execution

As in more conventional computer programming languages, a robot programming system allows the user to specify the flow of execution—that is, concepts such as testing and branching, looping, calls to subroutines, and even interrupts are generally found in robot programming languages.

More so than in many computer applications, parallel processing is generally important in automated workcell applications. First of all, very often two or more robots are used in a single workcell and work simultaneously to reduce the cycle time of the process. Even in single-robot applications, such as the one shown in Fig. 12.2, other workcell equipment must be controlled by the robot controller in a parallel fashion. Hence, *signal* and *wait* primitives are often found in robot programming languages, and occasionally more sophisticated parallel-execution constructs are provided [3].

Another frequent occurrence is the need to monitor various processes with some kind of sensor. Then, either by interrupt or through polling, the robot system must be able to respond to certain events detected by the sensors. The ability to specify such **event monitors** easily is afforded by some robot programming languages [2, 3].

Programming environment

As with any computer languages, a good programming environment fosters programmer productivity. Manipulator programming is difficult and tends to be very interactive, with a lot of trial and error. If the user were forced to continually repeat the "edit-compile-run" cycle of compiled languages, productivity would be low. Therefore, most robot programming languages are now *interpreted*, so that individual language statements can be run one at a time during program development and debugging. Many of the language statements cause motion of a physical device, so the tiny amount of time required to interpret the language statements is insignificant. Typical programming support, such as text editors, debuggers, and a file system, are also required.

Sensor integration

An extremely important part of robot programming has to do with interaction with sensors. The system should have, at a minimum, the capability to query touch and force sensors and to use the response in if-then-else constructs. The ability to specify event monitors to watch for transitions on such sensors in a *background* mode is also very useful.

Integration with a vision system allows the vision system to send the manipulator system the coordinates of an object of interest. For example, in our sample application, a vision system locates the brackets on the conveyor belt and returns to the manipulator controller their position and orientation relative to the camera. The camera's frame is known relative to the station frame, so a desired goal frame for the manipulator can be computed from this information.

Some sensors could be part of other equipment in the workcell—for example, some robot controllers can use input from a sensor attached to a conveyor belt so that the manipulator can track the belt's motion and acquire objects from the belt as it moves [2].

The interface to force-control capabilities, as discussed in Chapter 9, comes through special language statements that allow the user to specify force strategies [3]. Such force-control strategies are by necessity an integrated part of the manipulator control system—the robot programming language simply serves as an interface to those capabilities. Programming robots that make use of active force control might require other special features, such as the ability to display force data collected during a constrained motion [3].

In systems that support active force control, the description of the desired force application could become part of the motion specification. The AL language describes active force control in the motion primitives by specifying six components of stiffness (three translational and three rotational) and a bias force. In this way, the manipulator's apparent stiffness is programmable. To apply a force, usually the stiffness is set to zero in that direction and a bias force is specified—for example,

```
move garm to goal
with stiffness=(80, 80, 0, 100, 100, 100)
with force=20*ounces along zhat;
```

12.5 PROBLEMS PECULIAR TO ROBOT PROGRAMMING LANGUAGES

Advances in recent years have helped, but programming robots is still difficult. Robot programming shares all the problems of conventional computer programming, plus some additional difficulties caused by effects of the physical world [12].

Internal world model versus external reality

A central feature of a robot programming system is the world model that is maintained internally in the computer. Even when this model is quite simple, there are ample difficulties in assuring that it matches the physical reality that it attempts to model. Discrepancies between internal model and external reality result in poor or failed grasping of objects, collisions, and a host of more subtle problems.

This correspondence between internal model and the external world must be established for the program's initial state and must be maintained throughout its execution. During initial programming or debugging, it is generally up to the user to suffer the burden of ensuring that the state represented in the program corresponds to the physical state of the workcell. Unlike more conventional programming, where only internal variables need to be saved and restored to reestablish a former situation, in robot programming, physical objects must usually be repositioned.

Besides the uncertainty inherent in each object's position, the manipulator itself is limited to a certain degree of accuracy. Very often, steps in an assembly will require the manipulator to make motions requiring greater precision than it is capable of. A common example of this is inserting a pin into a hole where the clearance is an order of magnitude less than the positional accuracy of the manipulator. To further complicate matters, the manipulator's accuracy usually varies over its workspace.

In dealing with those objects whose locations are not known exactly, it is essential to somehow refine the positional information. This can sometimes be done with sensors (e.g., vision, touch) or by using appropriate force strategies for constrained motions.

During debugging of manipulator programs, it is very useful to be able to modify the program and then back up and try a procedure again. Backing up entails restoring the manipulator and objects being manipulated to a former state. However, in working with physical objects, it is not always easy, or even possible, to undo an action. Some examples are the operations of painting, riveting, drilling, or welding, which cause a physical modification of the objects being manipulated. It might therefore be necessary for the user to get a new copy of the object to replace the old, modified one. Further, it is likely that some of the operations just prior to the one being retried will also need to be repeated to establish the proper state required before the desired operation can be successfully retried.

Context sensitivity

Bottom-up programming is a standard approach to writing a large computer program in which one develops small, low-level pieces of a program and then puts them together into larger pieces, eventually attaining a completed program. For this method to work, it is essential that the small pieces be relatively insensitive to the language statements that precede them and that there be no assumptions concerning the context in which these program pieces execute. For manipulator programming, this is often not the case; code that worked reliably when tested in isolation frequently fails when placed in the context of the larger program. These problems generally arise from dependencies on manipulator configuration and speed of motions.

Manipulator programs can be highly sensitive to initial conditions—for example, the initial manipulator position. In motion trajectories, the starting position will influence the trajectory that will be used for the motion. The initial manipulator position might also influence the velocity with which the arm will be moving during some critical part of the motion. For example, these statements are true for manipulators that follow the cubic-spline joint-space paths studied in Chapter 7. These effects can sometimes be dealt with by proper programming care, but often such problems do not arise until after the initial language statements have been debugged in isolation and are then joined with statements preceding them.

Because of insufficient manipulator accuracy, a program segment written to perform an operation at one location is likely to need to be tuned (i.e., positions retaught and the like) to make it work at a different location. Changes in location within the workcell result in changes in the manipulator's configuration in reaching goal locations. Such attempts at relocating manipulator motions within the workcell test the accuracy of the manipulator kinematics and servo system, and problems frequently arise. Such relocation could cause a change in the manipulator's kinematic configuration—for example, from left shoulder to right shoulder, or from elbow up to elbow down. Moreover, these changes in configuration could cause large arm motions during what had previously been a short, simple motion.

The nature of the spatial shape of trajectories is likely to change as paths are located in different portions of the manipulator's workspace. This is particularly true of joint-space trajectory methods, but use of Cartesian-path schemes can also lead to problems when singularities are nearby.

When testing a manipulator motion for the first time, it often is wise to have the manipulator move slowly. This allows the user a chance to stop the motion if it appears to be about to cause a collision. It also allows the user to inspect the motion closely. After the motion has undergone some initial debugging at a slower speed it is then desirable to increase motion speeds. Doing so might cause some aspects of the motion to change. Limitations in most manipulator control systems cause greater servo errors, which are to be expected if the quicker trajectory is followed. Also, in force-control situations involving contact with the environment, speed changes can completely change the force strategies required for success.

The manipulator's configuration also affects the delicacy and accuracy of the forces that can be applied with it. This is a function of how well conditioned the Jacobian of the manipulator is at a certain configuration, something generally difficult to consider when developing robot programs.

Error recovery

Another direct consequence of working with the physical world is that objects might not be exactly where they should be and, hence, motions that deal with them could fail. Part of manipulator programming involves attempting to take this into account and making assembly operations as robust as possible, but, even so, errors are likely, and an important aspect of manipulator programming is how to recover from these errors.

Almost any motion statement in the user's program can fail, sometimes for a variety of reasons. Some of the more common causes are objects shifting or dropping out of the hand, an object missing from where it should be, jamming during an insertion, and not being able to locate a hole.

The first problem that arises for error recovery is identifying that an error has indeed occurred. Because robots generally have quite limited sensing and reasoning capabilities, *error detection* is often difficult. In order to detect an error, a robot program must contain some type of explicit test. This test might involve checking the manipulator's position to see that it lies in the proper range; for example, when doing an insertion, lack of change in position might indicate jamming, or too much change might indicate that the hole was missed entirely or the object has slipped out of the hand. If the manipulator system has some type of visual capabilities, then it might take a picture and check for the presence or absence of an object and, if the object is present, report its location. Other checks might involve force, such as weighing the load being carried to check that the object is still there and has not been dropped, or checking that a contact force remains within certain bounds during a motion.

Every motion statement in the program might fail, so these explicit checks can be quite cumbersome and can take up more space than the rest of the program. Attempting to deal with all possible errors is extremely difficult; usually, just the few statements that seem most likely to fail are checked. The process of predicting which portions of a robot application program are likely to fail is one that requires a certain amount of interaction and partial testing with the robot during the program-development stage.

Once an error has been detected, an attempt can be made to recover from it. This can be done totally by the manipulator under program control, or it might involve manual intervention by the user, or some combination of the two. In any event, the recovery attempt could in turn result in new errors. It is easy to see how code to recover from errors can become the major part of the manipulator program.

The use of parallelism in manipulator programs can further complicate recovery from errors. When several processes are running concurrently and one causes an error to occur, it could affect other processes. In many cases, it will be possible to back up the offending process, while allowing the others to continue. At other times, it will be necessary to reset several or all of the running processes.

BIBLIOGRAPHY

[1] B. Shimano, "VAL: A Versatile Robot Programming and Control System," Proceedings of COMPSAC 1979, Chicago, November 1979.

[2] B. Shimano, C. Geschke, and C. Spalding, "VAL II: A Robot Programming Language and Control System," SME Robots VIII Conference, Detroit, June 1984.

[3] S. Mujtaba and R. Goldman, "AL Users' Manual," 3rd edition, Stanford Department of Computer Science, Report No. STAN-CS-81-889, December 1981.

[4] A. Gilbert et al., *AR-BASIC: An Advanced and User Friendly Programming System for Robots*, American Robot Corporation, June 1984.

[5] J. Craig, "JARS—JPL Autonomous Robot System: Documentation and Users Guide," JPL Interoffice memo, September 1980.

[6] ABB Robotics, "The RAPID Language," in the *SC4Plus Controller Manual*, ABB Robotics, 2002.

[7] R. Taylor, P. Summers, and J. Meyer, "AML: A Manufacturing Language," *International Journal of Robotics Research*, Vol. 1, No. 3, Fall 1982.

[8] FANUC Robotics, Inc., "KAREL Language Reference," FANUC Robotics North America, Inc, 2002.

[9] R. Taylor, "A Synthesis of Manipulator Control Programs from Task-Level Specifications," Stanford University AI Memo 282, July 1976.

[10] J.C. LaTombe, "Motion Planning with Uncertainty: On the Preimage Backchaining Approach," in *The Robotics Review*, O. Khatib, J. Craig, and T. Lozano-Perez, Editors, MIT Press, Cambridge, MA, 1989.

[11] W. Gruver and B. Soroka, "Programming, High Level Languages," in *The International Encyclopedia of Robotics*, R. Dorf and S. Nof, Editors, Wiley Interscience, New York, 1988.

[12] R. Goldman, *Design of an Interactive Manipulator Programming Environment*, UMI Research Press, Ann Arbor, MI, 1985.

[13] Adept Technology, *V+ Language Reference*, Adept Technology, Livermore, CA, 2002.

EXERCISES

12.1 [15] Write a robot program (in a language of your choice) to pick a block up from location *A* and place it in location *B*.

12.2 [20] Describe tying your shoelace in simple English commands that might form the basis of a robot program.

12.3 [32] Design the syntax of a new robot programming language. Include ways to give duration or speeds to motion trajectories, make I/O statements to peripherals, give commands to control the gripper, and produce force-sensing (i.e., guarded move) commands. You can skip force control and parallelism (to be covered in Exercise 12.4).

12.4 [28] Extend the specification of the new robot programming language that you started in Exercise 12.3 by adding force-control syntax and syntax for parallelism.

12.5 [38] Write a program in a commercially available robot programming language to perform the application outlined in Section 12.3. Make any reasonable assumptions concerning I/O connections and other details.

12.6 [28] Using any robot language, write a general routine for unloading an arbitrarily sized pallet. The routine should keep track of indexing through the pallet and signal a human operator when the pallet is empty. Assume the parts are unloaded onto a conveyor belt.

12.7 [35] Using any robot language, write a general routine for unloading an arbitrarily sized source pallet and loading an arbitrarily sized destination pallet. The routine should keep track of indexing through the pallets and signal a human operator when the source pallet is empty and when the destination pallet is full.

12.8 [35] Using any capable robot programming language, write a program that employs force control to fill a cigarette box with 20 cigarettes. Assume that the manipulator has an accuracy of about 0.25 inch, so force control should be used for many operations. The cigarettes are presented on a conveyor belt, and a vision system returns their coordinates.

12.9 [35] Using any capable robot programming language, write a program to assemble the hand-held portion of a standard telephone. The six components (handle, microphone, speaker, two caps, and cord) arrive in a *kit*, that is, a special pallet holding one of each kind of part. Assume there is a fixture into which the handle can be placed that holds it. Make any other reasonable assumptions needed.

12.10 [33] Write a robot program that uses two manipulators. One, called GARM, has a special end-effector designed to hold a wine bottle. The other arm, BARM, will hold a wineglass and is equipped with a force-sensing wrist that can be used to signal GARM to stop pouring when it senses the glass is full.

PROGRAMMING EXERCISE (PART 12)

Create a user interface to the other programs you have developed by writing a few subroutines in Pascal. Once these routines are defined, a "user" could write a Pascal program that contains calls to these routines to perform a 2-D robot application in simulation.

Define primitives that allow the user to set station and tool frames—namely,

```
setstation(Sre1B:vec3);

settool(Tre1W:vec3);
```

where "Sre1B" gives the station frame relative to the base frame of the robot and "Tre1W" defines the tool frame relative to the wrist frame of the manipulator. Define the motion primitives

```
moveto(goal:vec3);

moveby(increment:vec3);
```

where "goal" is a specification of the goal frame relative to the station frame and "increment" is a specification of a goal frame relative to the current tool frame. Allow multisegment paths to be described when the user first calls the "pathmode" function, then specifies motions to via points, and finally says "runpath"—for example,

```
pathmode; (* enter path mode *)

moveto(goal1);

moveto(goal2);

runpath; (* execute the path without stopping at goal1 *)
```

Write a simple "application" program, and have your system print the location of the arm every n seconds.

C H A P T E R   13

Off-line programming systems

13.1 INTRODUCTION
13.2 CENTRAL ISSUES IN OLP SYSTEMS
13.3 THE 'PILOT' SIMULATOR
13.4 AUTOMATING SUBTASKS IN OLP SYSTEMS

13.1   INTRODUCTION

We define an **off-line programming** (OLP) system as a robot programming language that has been sufficiently extended, generally by means of computer graphics, that the development of robot programs can take place without access to the robot itself.[1] Off-line programming systems are important both as aids in programming present-day industrial automation and as platforms for robotics research. Numerous issues must be considered in the design of such systems. In this chapter, first a discussion of these issues is presented [1] and then a closer look at one such system [2].

Over the past 20 years, the growth of the industrial robot market has not been as rapid as once was predicted. One primary reason for this is that robots are still too difficult to use. A great deal of time and expertise is required to install a robot in a particular application and bring the system to production readiness. For various reasons, this problem is more severe in some applications than in others; hence, we see certain application areas (e.g., spot welding and spray painting) being automated with robots much sooner than other application domains (e.g., assembly). It seems that lack of sufficiently trained robot-system implementors is limiting growth in some, if not all, areas of application. At some manufacturing companies, management encourages the use of robots to an extent greater than that realizable by applications engineers. Also, a large percentage of the robots delivered are being used in ways that do not take full advantage of their capabilities. These symptoms indicate that current industrial robots are not easy enough to use to allow successful installation and programming in a timely manner.

There are many factors that make robot programming a difficult task. First, it is intrinsically related to general computer programming and so shares in many of the problems encountered in that field; but the programming of robots, or of any programmable machine, has particular problems that make the development of production-ready software even more difficult. As we saw in the last chapter,

---

[1]Chapter 13 is an edited version of two papers: one reprinted with permission from *International Symposium of Robotics Research*, R. Bolles and B. Roth (editors), 1988 (ref [1]); the other from *Robotics: The Algorithmic Perspective*, P. Agarwal et al. (editors), 1998 (ref [2]).

most of these special problems arise from the fact that a robot manipulator interacts with its physical environment [3]. Even simple programming systems maintain a "world model" of this physical environment in the form of locations of objects and have "knowledge" about presence and absence of various objects encoded in the program strategies. During the development of a robot program (and especially later during production use), it is necessary to keep the internal model maintained by the programming system in correspondence with the actual state of the robot's environment. Interactive debugging of programs with a manipulator requires frequent manual resetting of the state of the robot's environment—parts, tools, and so forth must be moved back to their initial locations. Such state resetting becomes especially difficult (and sometimes costly) when the robot performs a irreversible operation on one or more parts (e.g., drilling or routing). The most spectacular effect of the presence of the physical environment is when a program bug manifests itself in some unintended irreversible operation on parts, on tools, or even on the manipulator itself.

Although difficulties exist in maintaining an accurate internal model of the manipulator's environment, there seems no question that great benefits result from doing so. Whole areas of sensor research, perhaps most notably computer vision, focus on developing techniques by which world models can be verified, corrected, or discovered. Clearly, in order to apply any computational algorithm to the robot command-generation problem, the algorithm needs access to a model of the robot and its surroundings.

In the development of programming systems for robots, advances in the power of programming techniques seem directly tied to the sophistication of the internal model referenced by the programming language. Early joint-space "teach by showing" robot systems employed a limited world model, and there were very limited ways in which the system could aid the programmer in accomplishing a task. Slightly more sophisticated robot controllers included kinematic models, so that the system could at least aid the user in moving the joints so as to accomplish Cartesian motions. Robot programming languages (RPLs) evolved to support many different data types and operations, which the programmer may use as needed to model attributes of the environment and compute actions for the robot. Some RPLs support such world-modeling primitives as affixments, data types for forces and moments, and other features [4].

The robot programming languages of today might be called "explicit program-ming languages," in that every action that the system takes must be programmed by the application engineer. At the other end of the spectrum are the so-called task-level-programming (TLP) systems, in which the programmer may state such high-level goals as "insert the bolt" or perhaps even "build the toaster oven." These systems use techniques from artificial-intelligence research to generate motion and strategy plans automatically. However, task-level languages this sophisticated do not yet exist; various pieces of such systems are currently under development by researchers [5]. Task-level-programming systems will require a very complete model of the robot and its environment to perform automated planning operations.

Although this chapter focuses to some extent on the particular problem of robot programming, the notion of an OLP system extends to any programmable device on the factory floor. An argument commonly raised in favor is that an OLP system will not tie up production equipment when it needs to be reprogrammed; hence, automated factories can stay in production mode a greater percentage of the time. They also serve as a natural vehicle to tie computer-aided design (CAD) data bases used in the design phase of a product's development to the actual manufacturing of the product. In some applications, this direct use of CAD design data can dramatically reduce the programming time required for the manufacturing machinery.

Off-line programming of robots offers other potential benefits, ones just beginning to be appreciated by industrial robot users. We have discussed some of the problems associated with robot programming, and most have to do with the fact that an external, physical workcell is being manipulated by the robot program. This makes backing up to try different strategies tedious. Programming of robots in simulation offers a way of keeping the bulk of the programming work strictly internal to a computer—until the application is nearly complete. Under this approach, many of the problems peculiar to robot programming tend to diminish.

Off-line programming systems should serve as the natural growth path from explicit programming systems to task-level-programming systems. The simplest OLP system is merely a graphical extension to a robot programming language, but from there it can be extended into a task-level-programming system. This gradual extension is accomplished by providing automated solutions to various subtasks (as these solutions become available) and letting the programmer use them to explore options in the simulated environment. Until we discover how to build task-level systems, the user must remain in the loop to evaluate automatically planned subtasks and guide the development of the application program. If we take this view, an OLP system serves as an important basis for research and development of task-level-planning systems, and, indeed, in support of their work, many researchers have developed various components of an OLP system (e.g., 3-D models and graphic display, language postprocessors). Hence, OLP systems should be a useful tool in research as well as an aid in current industrial practice.

13.2   CENTRAL ISSUES IN OLP SYSTEMS

This section raises many of the issues that must be considered in the design of an OLP system. The collection of topics discussed will help to set the scope of the definition of an OLP system.

User interface

A major motivation for developing an OLP system is to create an environment that makes programming manipulators easier, so the user interface is of crucial importance. However, another major motivation is to remove reliance on use of the physical equipment during programming. Upon initial consideration, these two goals seem to conflict—robots are hard enough to program when you can see them, so how can it be easier without the presence of physical device? This question touches upon the essence of the OLP design problem.

Manufacturers of industrial robots have learned that the RPLs they provide with their robots cannot be utilized successfully by a large percentage of manufacturing personnel. For this and other historical reasons, many industrial robots are provided with a two-level interface [6], one for programmers and one for nonprogrammers. Nonprogrammers utilize a teach pendant and interact directly with the robot to develop robot programs. Programmers write code in the RPL and interact with the robot in order to teach robot work points and to debug program flow. In general, these two approaches to program development trade off ease of use against flexibility.

When viewed as an extension of a RPL, an OLP system by nature contains an RPL as a subset of its user interface. This RPL should provide features that have already been discovered to be valuable in robot programming systems. For example, for use as an RPL, **interactive languages** are much more productive than compiled languages, which force the user to go through the "edit–compile–run" cycle for each program modification.

The language portion of the user interface inherits much from "traditional" RPLs; it is the lower-level (i.e., easier-to-use) interface that must be carefully considered in an OLP system. A central component of this interface is a computer-graphic view of the robot being programmed and of its environment. Using a pointing device such as a **mouse**, the user can indicate various locations or objects on the graphics screen. The design of the user interface addresses exactly how the user interacts with the screen to specify a robot program. The same pointing device can indicate items in a "menu" in order to specify modes or invoke various functions.

A central primitive is that for teaching a robot a work point or "frame" that has six degrees of freedom by means of interaction with the graphics screen. The availability of 3-D models of fixtures and workpieces in the OLP system often makes this task quite easy. The interface provides the user with the means to indicate locations on surfaces, allowing the orientation of the frame to take on a local surface normal, and then provides methods for offsetting, reorienting, and so on. Depending on the specifics of the application, such tasks are quite easily specified via the graphics window into the simulated world.

A well-designed user interface should enable nonprogrammers to accomplish many applications from start to finish. In addition, frames and motion sequences "taught" by nonprogrammers should be able to be translated by the OLP system into textual RPL statements. These simple programs can be maintained and embellished in RPL form by more experienced programmers. For programmers, the RPL availability allows arbitrary code development for more complex applications.

3-D modeling

A central element in OLP systems is the use of graphic depictions of the simulated robot and its workcell. This requires the robot and all fixtures, parts, and tools in the workcell to be modeled as three-dimensional objects. To speed up program development, it is desirable to use any CAD models of parts or tooling that are directly available from the CAD system on which the original design was done. As CAD systems become more and more prevalent in industry, it becomes more and more likely that this kind of geometric data will be readily available. Because of the strong desire for this kind of CAD integration from design to production, it makes sense for an OLP system either to contain a CAD modeling subsystem or to be, itself, a part of a CAD design system. If an OLP system is to be a stand-alone system, it must have appropriate interfaces to transfer models to and from external CAD systems; however, even a stand-alone OLP system should have at least a simple local CAD facility for quickly creating models of noncritical workcell items or for adding robot-specific data to imported CAD models.

OLP systems generally require multiple representations of spatial shapes. For many operations, an exact analytic description of the surface or volume is generally present; yet, in order to benefit from display technology, another representation is often needed. Current technology is well suited to systems in which the underlying display primitive is a planar polygon; hence, although an object shape might be well represented by a smooth surface, practical display (especially for animation) requires a faceted representation. User-interface graphical actions, such as pointing to a spot on a surface, should internally act so as to specify a point on the true surface, even if, graphically, the user sees a depiction of the faceted model.

An important use of the three-dimensional geometry of the object models is in **automatic collision detection**—that is, when any collisions occur between objects in the simulated environment, the OLP system should automatically warn the user and indicate exactly where the collision takes place. Applications such as assembly may involve many desired "collisions," so it is necessary to be able to inform the system that collisions between certain objects are acceptable. It is also valuable to be able to generate a collision warning when objects pass within a specified tolerance of a true collision. Currently, the exact collision-detection problem for general 3-D solids is difficult, but collision detection for faceted models is quite practical.

Kinematic emulation

A central component in maintaining the validity of the simulated world is the faithful emulation of the geometrical aspects of each simulated manipulator. With regard to inverse kinematics, the OLP system can interface to the robot controller in two distinct ways. First, the OLP system could replace the inverse kinematics of the robot controller and always communicate robot positions in mechanism joint space. The second choice is to communicate Cartesian locations to the robot controller and let the controller use the inverse kinematics supplied by the manufacturer to solve for robot configurations. The second choice is almost always preferable, especially as manufacturers begin to build *arm signature* style calibration into their robots. These calibration techniques customize the inverse kinematics for each individual robot. In this case, it becomes desirable to communicate information at the Cartesian level to robot controllers.

These considerations generally mean that the forward and inverse kinematic functions used by the simulator must reflect the nominal functions used in the robot controller supplied by the manufacturer of the robot. There are several details of the inverse-kinematic function specified by the manufacturer that must be emulated by the simulator software. Any inverse-kinematic algorithm must make arbitrary choices in order to resolve singularities. For example, when joint 5 of a PUMA 560 robot is at its zero location, axes 4 and 6 line up, and a singular condition exists. The inverse-kinematic function in the robot controller can solve for the sum of joint angles 4 and 6, but then must use an arbitrary rule to choose individual values for joints 4 and 6. The OLP system must emulate whatever algorithm is used. Choosing the nearest solution when many alternate solutions exist provides another example. The simulator must use the same algorithm as the controller in order to avoid potentially catastrophic errors in simulating the actual manipulator. A helpful feature occasionally found in robot controllers is the ability to command a Cartesian goal and specify which of the possible solutions the manipulator should use. The existence of this feature eliminates the requirement that the simulator emulate the solution-choice algorithm; the OLP system can simply force its choice on the controller.

Path-planning emulation

In addition to kinematic emulation for static positioning of the manipulator, an OLP system should accurately emulate the path taken by the manipulator in moving through space. Again, the central problem is that the OLP system needs to simulate the algorithms in the employed robot controller, and such path-planning and -execution algorithms vary considerably from one robot manufacturer to another. Simulation of the spatial shape of the path taken is important for detection of collisions between the robot and its environment. Simulation of the temporal aspects of the trajectory are important for predicting the cycle times of applications. When a robot is operating in a moving environment (e.g., near another robot), accurate simulation of the temporal attributes of motion is necessary to predict collisions accurately and, in some cases, to predict communication or synchronization problems, such as deadlock.

Dynamic emulation

Simulated motion of manipulators can neglect dynamic attributes if the OLP system does a good job of emulating the trajectory-planning algorithm of the controller and if the actual robot follows desired trajectories with negligible errors. However, at high speed or under heavy loading conditions, trajectory-tracking errors can become important. Simulation of these tracking errors necessitates both modeling the dynamics of the manipulator and of the objects that it moves and emulating the control algorithm used in the manipulator controller. Currently, practical problems exist in obtaining sufficient information from the robot vendors to make this kind of dynamic simulation of practical value, but, in some cases, dynamic simulation can be pursued fruitfully.

Multiprocess simulation

Some industrial applications involve two or more robots cooperating in the same environment. Even single-robot workcells often contain a conveyor belt, a transfer line, a vision system, or some other active device with which the robot must interact. For this reason, it is important that an OLP system be able to simulate multiple moving devices and other activities that involve **parallelism**. As a basis for this capability, the underlying language in which the system is implemented should be a multiprocessing language. Such an environment makes it possible to write independent robot-control programs for each of two or more robots in a single cell and then simulate the action of the cell with the programs running concurrently. Adding signal and wait primitives to the language enables the robots to interact with each other just as they might in the application being simulated.

Simulation of sensors

Studies have shown that a large component of robot programs consists not of motion statements, but rather of initialization, error-checking, I/O, and other kinds of statements [7]. Hence, the ability of the OLP system to provide an environment that allows simulation of complete applications, including interaction with sensors, various I/O, and communication with other devices, becomes important. An OLP system that supports simulation of sensors and multiprocessing not only can check robot motions for feasibility, but also can verify the communication and synchronization portion of the robot program.

Language translation to target system

An annoyance for current users of industrial robots (and of other programmable automation) is that almost every supplier of such systems has invented a unique language for programming its product. If an OLP system aspires to be universal in the equipment it can handle, it must deal with the problem of translating to and from several different languages. One choice for dealing with this problem is to choose a single language to be used by the OLP system and then postprocess the language in order to convert it into the format required by the target machine. An ability to upload programs that already exist on the target machines and bring them into the OLP system is also desirable.

Two potential benefits of OLP systems relate directly to the language-translation topic. Most proponents of OLP systems note that having a single, universal interface, one that enables users to program a variety of robots, solves the problem of learning and dealing with several automation languages. A second benefit stems from economic considerations in future scenarios in which hundreds or perhaps thousands of robots fill factories. The cost associated with a powerful programming environment (such as a language and graphical interface) might prohibit placing it at the site of each robot installation. Rather, it seems to make economic sense to place a very simple, "dumb," and cheap controller with each robot and have it downloaded from a powerful, "intelligent" OLP system that is located in an office environment. Hence, the general problem of translating an application program from a powerful universal language to a simple language designed to execute in a cheap processor becomes an important issue in OLP systems.

Workcell calibration

An inevitable reality of a computer model of any real-world situation is that of inaccuracy in the model. In order to make programs developed on an OLP system usable, methods for **workcell calibration** must be an integral part of the system. The magnitude of this problem varies greatly with the application; this variability makes off-line programming of some tasks much more feasible that of others. If the majority of the robot work points for an application must be retaught with the actual robot to solve inaccuracy problems, OLP systems lose their effectiveness.

Many applications involve the frequent performance of actions relative to a rigid object. Consider, for example, the task of drilling several hundred holes in a bulkhead. The actual location of the bulkhead relative to the robot can be taught by using the actual robot to take three measurements. From those data, the locations of all the holes can be updated automatically if they are available in part coordinates from a CAD system. In this situation, only these three points need be taught with the robot, rather than hundreds. Most tasks involve this sort of "many operations relative to a rigid object" paradigm—for example, PC-board component insertion, routing, spot welding, arc welding, palletizing, painting, and deburring.

13.3 THE 'PILOT' SIMULATOR

In this section, we consider one such off-line simulator system: the 'Pilot' system developed by Adept Technology [8]. The Pilot system is actually a suite of three closely related simulation systems; here, we look at the portion of Pilot (known as "Pilot/Cell") that is used to simulate an individual workcell in a factory. In particular, this system is unusual in that it attempts to model several aspects of the physical world, as a means of unburdening the programmer of the simulator. In this section, we will discuss the "geometric algorithms" that are used to empower the simulator to emulate certain aspects of physical reality.

The need for ease of use drives the need for the simulation system to behave like the actual physical world. The more the simulator acts like the real world, the simpler the user-interface paradigm becomes for the user, because the physical world is the one we are all familiar with. At the same time, trade-offs of ease against computational speed and other factors have driven a design in which a particular "slice" of reality is simulated while many details are not.

Pilot is well-suited as a host for a variety of geometric algorithms. The need to model various portions of the real world, together with the need to unburden the user by automating frequent geometric computations, drives the need for such algorithms. Pilot provides the environment in which some advanced algorithms can be brought to bear on real problems occurring in industry.

One decision made very early on in the design of the Pilot simulation system was that the *programming paradigm* should be as close as possible to the way the actual robot system would be programmed. Certain higher level planning and optimization tools are provided, but it was deemed important to have the basic programming interaction be similar to actual hardware systems. This decision has led the product's development down a path along which we find a genuine need for various geometric algorithms. The algorithms needed range widely from extremely simple to quite complex.

If a simulator is to be programmed as the physical system would be, then the actions and reactions of the physical world must be modeled "automatically" by the simulator. The goal is to free the user of the system from having to write any "simulation-specific code." As a simple example, if the robot gripper is commanded to open, a grasped part should fall in response to gravity and possibly should even bounce and settle into a certain stable state. Forcing the user of the system to specify these real-world actions would make the simulator fall short of its goal: being programmed just as the actual system is. Ultimate ease of use can be achieved only when the simulated world "knows how" to behave like the real world without burdening the user.

Most, if not all, commercial systems for simulating robots or other mechanisms do not attempt to deal directly with this problem. Rather, they typically "allow" the user (actually, *force* the user) to embed simulation-specific commands within the program written to control the simulated device. A simple example would be the following code sequence:

```
MOVE TO pick_part
CLOSE gripper
affix(gripper,part[i]);
MOVE TO place_part
OPEN gripper
unaffix(gripper,part[i]);
```

Here, the user has been forced to insert "affix" and "unaffix" commands, which (respectively) cause the part to move with the gripper when grasped and to stop moving with it when released. If the simulator allows the robot to be programmed in its native language, generally that language is not rich enough to support these required "simulation-specific" commands. Hence, there is a need for a second set of commands, possibly even with a different syntax, for dealing with interactions with the real world. Such a scheme is inherently *not* programmed "just as the physical system is" and must inherently cause an increased programming burden for the user.

From the preceding example, we see the first geometric algorithm that one finds a need for: From the geometry of the gripper and the relative placements of parts, figure out which part (if any) will be grasped when the gripper closes and possibly how the part will self-align within the gripper. In the case of Pilot, we solve the first part of this problem with a simple algorithm. In limited cases, the "alignment action" of the part in the gripper is computed, but, in general, such alignments need to be pretaught by the system's user. Hence, Pilot has not reached the ultimate goal yet, but has taken some steps in that direction.

Physical Modeling and Interactive Systems

In a simulation system, one always trades off complexity of the model in terms of computation time against accuracy of the simulation. In the case of Pilot and its intended goals, it is particularly important to keep the system fully interactive. This has led to designing Pilot so that it can use various approximate models—for example, the use of quasi-static approximations where a full dynamic model might be more accurate. Although there appears to be a possibility that "full dynamic" models might soon be applicable [9], given the current state of computer hardware, of dynamic algorithms, and of the complexity of the CAD models that industrial users wish to employ, we feel these trade-offs still need to be made.

Geometric Algorithms for Part Tumbling

In some feeding systems employed in industrial practice, parts tumble from some form of infeed conveyor onto a presentation surface; then computer vision is used to locate parts to be acquired by the robot. Designing such automation systems with the aid of a simulator means that the simulator must be able to predict how parts fall, bounce, and take on a stable orientation, or *stable state*.

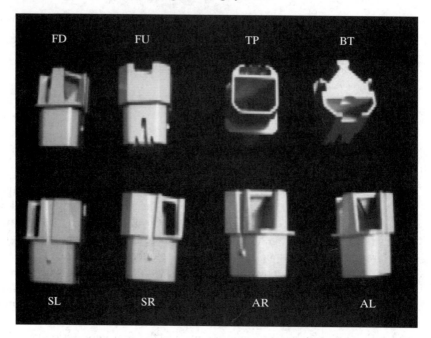

FIGURE 13.1: The eight stable states of the part.

Stable-state probabilities

As reported in [10], an algorithm has been implemented that takes as input any geometric shape (represented by a CAD model) and, for that shape, can compute the N possible ways that it can rest stably on a horizontal surface. These are called the *stable states* of the part. Further, the algorithm uses a perturbed quasi-static approach to estimate the probability associated with each of the N stable states. We have performed physical experiments with sample parts in order to assess the resulting accuracy of stable-state prediction.

Figure 13.1 shows the eight stable states of a particular test part. Using an Adept robot and vision system, we dropped this part more than 26,000 times and recorded the resulting stable state, in order to compare our stable-state prediction algorithm to reality. Table 13.1 shows the results for the test part. These results are characteristic of our current algorithm—stable-state likelihood prediction error typically ranges from 5% to 10%.

Adjusting probabilities as a function of drop height

Clearly, if a part is dropped from a gripper from a very small height (e.g., 1 mm) above a surface, the probabilities of the various stable states differ from those which occur when the part is dropped from higher than some *critical height*. In Pilot, we use probabilities from the *stable-state estimator* algorithm when parts are dropped from heights equal to or greater than the largest dimension of the part. For drop

TABLE 13.1: Predicted versus Actual Stable-State
Probabilities for the Test Part

Stable State	Actual #	% Actual	% Predicted
FU	1871	7.03%	8.91%
FD	10,600	39.80%	44.29%
TP	648	2.43%	7.42%
BT	33	0.12%	8.19%
SR	6467	24.28%	15.90%
SL	6583	24.72%	15.29%
AR/AL	428	1.61%	0.00%
Total	26,630	100%	100%

heights below that value, probabilities are adjusted to take into account the initial orientation of the part and the height of the drop. The adjustment is such that, as an infinitesimal drop height is approached, the part remains in its initial orientation (assuming it is a stable orientation). This is an important addition to the overall probability algorithm, because it is typical for parts to be released a small distance above a support surface.

Simulation of bounce

Parts in Pilot are tagged with their coefficient of restitution; so are all surfaces on which parts may be placed. The product of these two factors is used in a formula for predicting how far the part will bounce when dropped. These details are important, because they affect how parts scatter or clump in the simulation of some feeding systems. When bouncing, parts are scattered radially according to a uniform distribution. The distance of bounce (away from the initial contact point) is a certain distribution function out to a maximum distance, which is computed as a function of drop height (energy input) and the coefficients of restitution that apply.

Parts in Pilot can bounce recursively from surface to surface in certain arrangements. It is also possible to mark certain surfaces such that parts are not able to bounce *off* them, but can only bounce *within* them. Entities known as *bins* in Pilot have this property –parts can fall into them, but never bounce out.

Simulation of stacking and tangling

As a simplification, parts in Pilot always rest on planar support surfaces. If parts are tangled or stacked on one another, this is displayed as parts that are intersecting each other (that is, the boolean intersection of their volumes would be non-empty). This saves the enormous amount of computation that would be needed to compute the various ways a part might be stacked or tangled with another part's geometry.

Parts in Pilot are tagged with a *tangle factor*. For example, something like a marble would have a tangle factor of 0.0 because, when tumbled onto a support surface, marbles tend never to stack or tangle, but rather tend to spread out on the surface. On the other hand, parts like coiled springs might have a tangle factor near 1.0; they quite readily become entangled with one another. When a part falls and bounces, a *findspace* algorithm runs, in which the part tries to bounce into an open space on the surface. However, exactly "how hard it tries" to find an open space is a function of its tangle factor. By adjustment of this coefficient, Pilot can simulate parts that tumble and become entangled more or less. Currently, there is no algorithm for automatically computing the tangle factor from the part geometry—this is an interesting open problem. Through the user interface, the Pilot user can set the tangle factor to what seems appropriate.

Geometric Algorithms for Part Grasping

Much of the difficulty in programming and using actual robots has to do with the details of teaching grasp locations on parts and with the detailed design of grippers. This is an area in which additional planning algorithms in a simulator system could have a large impact. In this section, we discuss the algorithms currently in place in Pilot. The current approaches are quite simple, so this is an area of ongoing work.

Computing which part to grasp

When a tool closes, or a suction end-effector actuates, Pilot applies a simple algorithm to compute which part (if any) should become grasped by the robot. First, the system figures out which support surface is immediately beneath the gripper. Then, for all parts on that surface, it searches for each whose bounding box (for the current stable state) contains the TCP (tool center point) of the gripper. If more than one part satisfies this criterion, then it chooses the nearest among those which do.

Computation of default grasp location

Pilot automatically assigns a grasp location for each stable orientation predicted by the stable-state estimator previously described. The current algorithm is simplistic, so a graphical user interface is also provided so that the user can edit and redefine these grasp points. The current grasp algorithm is a function of the part's bounding box and the geometry of the gripper, which is assumed to be either a parallel-jaw gripper or a suction cup. Along with computing a default grasp location for each stable state, a default approach and depart height are also automatically computed.

Computation of alignment of the part during grasp

In some important cases in industrial practice, the system designer counts on the fact that, when the robot end-effector actuates, the captured part will align itself in some way with surfaces of the end-effector. This effect can be important in removing small misalignments in the presentation of parts to the robot.

A very real effect which needs to be simulated is that, with suction cup grippers, it can be the case that, when suction is applied, the part is "lifted" up against the suction cup in a way which significantly alters its orientation relative to the end-effector. Pilot simulates this effect by piercing the part geometry with a vertical line aligned with the center line of the suction cup. Whichever facet of the polygonal part model is pierced is used in computing the orientation at grasp—the normal of this facet becomes anti-aligned with the normal of the bottom of the suction cup. In altering the part orientation, rotation about this piercing line is minimized (the part does not spin about the axis of the suction cup when picked). Without simulation of this effect, the simulator would be unable to depict realistically some pick-and-place strategies employing suction grippers.

We have also implemented a planner that allows parts to rotate about the Z axis when a parallel jaw gripper closes on them. This case is automatic only for a simple case—in other situations, the user must teach the resulting alignment (i.e., we are still waiting for a more nearly complete algorithm).

Geometric Algorithms for Part Pushing

One style of part pushing occurs between the jaws of a gripper, as mentioned in the previous section. In current industrial practice, parts sometimes get pushed by simple mechanisms. For example, after a part is presented by a bowl feeder, it might be pushed by a linear actuator right into an assembly that has been brought into the cell by a tray-conveyor system.

Pilot has support for simulating the pushing of parts: an entity called a *push-bar*, which can be attached to a pneumatic cylinder or a leadscrew actuator in the simulator. When the actuator moves the push-bar along a linear path, the leading surface of the push-bar will move parts. In the future, it is planned, push-bars will also be able to be added as guides along conveyors or placed anywhere that requires that parts motion be affected by their presence. The current pushing is still very simple, but it suffices for many real-world tasks.

Geometric Algorithms for Tray Conveyors

Pilot supports the simulation of tray-conveyor systems in which trays move along tracks composed of straight-line and circular-section components. Placed along the tracks at key locations can be *gates*, which pop up temporarily to block a tray when so commanded. Additionally, *sensors* that detect a passing tray can be placed in the track at user-specified locations. Such conveyor systems are typical in many automation schemes.

Connecting tray conveyors and sources and sinks

Tray conveyors can be connected together to allow various styles of branching. Where two conveyors "flow together," a simple collision-avoidance scheme is provided to cause trays from the *spur* conveyor to be subordinate to trays on the *main* conveyor. Trays on the spur conveyor will wait whenever a collision would occur. At "flow apart" connections, a device called a *director* is added to the main conveyor, which can be used to control which direction a tray will take at the intersection. Digital I/O lines connected to the simulated robot controller are used to read sensors, activate gates, and activate directors.

At the ends of a tray conveyor are a *source* and a *sink*. Sources are set up by the user to generate trays at certain statistical intervals. The trays generated could either be empty or be preloaded with parts or fixtures. At the end of a tray conveyor, trays (and their contents) disappear into sinks. Each time a tray enters a sink, its arrival time and contents are recorded. These so-called *sink records* can then be replayed through a source elsewhere in the system. Hence, a line of cells can be studied in the simulator one cell at a time, by setting the source of cell $N + 1$ to the sink record from cell N.

Pushing of trays

Pushing is also implemented for trays: A push-bar can be used to push a tray off a tray conveyor system and into a particular work cell. Likewise, trays can be pushed onto a tray conveyor. The updating of various data structures when trays come off a conveyor or onto one is an automatic part of the pushing code.

Geometric Algorithms for Sensors

Simulation of various sensor systems is required, so that the user will not be burdened with the writing of code to emulate their behavior in the cell.

Proximity sensors

Pilot supports the simulation of proximity sensors and other sensors. In the case of proximity sensors, the user tags the device with its minimum and maximum range and with a threshold. If an object is within range and closer than the threshold, then the sensor will detect it. To perform this computation in the simulated world, a line segment is temporarily added to the world, one that stretches from minimum to maximum sensor range. Using a collision algorithm, the system computes the locations at which this line segment intersects other CAD geometry. The intersection point nearest the sensor corresponds to the real-world item that would have stopped the beam. A comparison of the distance to this point and the threshold gives the output of the sensor. At present, we do not make use of the angle of the encountered surface or of its reflectance properties, although those features might be added in the future.

2-D vision systems

Pilot simulates the performance of the Adept 2-D vision system. The way the simulated vision system works is closely related to the way the real vision system works, even to how it is programmed in the AIM language [11] used by Adept robots. The following elements of this vision system are simulated:

- The shape and extent of the field of view.
- The stand-off distance and a simple model of focus.
- The time required to perform vision processing (approximate).
- The spatial ordering of results in the queue in the case of many parts being found in one image.
- The ability to distinguish parts according to which stable state they are in.
- The inability to recognize parts that are touching or overlapping.
- Within the context of AIM, the ability to update robot goals based on vision results.

The use of a vision system is well integrated with the AIM robot programming system, so implementation of the AIM language in the simulator implies implementation of vision system emulation. AIM supports several constructs that make the use of vision easy for robot guidance. Picking parts that are identified visually from both indexing and tracking conveyors is easily accomplished.

A data structure keeps track of which support surface the vision system is looking at. For all parts supported on that surface, we compute which are within the vision system's field of view. We prune out any parts that are too near or too far from the camera (e.g., out of focus). We prune out any parts that are touching neighboring parts. From the remaining parts, we choose those which are in the sought-after stable state and put them in a list. Finally, this list is sorted to emulate the ordering the Adept vision system uses when multiple parts are found in one scene.

Inspector sensors

A special class of sensor is provided, called an *inspector*. The inspector is used to give a binary output for each part placed in front of it. Parts in Pilot can be tagged with a *defect rate*, and inspectors can ferret out the defective parts. Inspectors play the role of several real-world sensor systems.

Conclusion

As is mentioned throughout this section, although some simple geometric algorithms are currently in place in the simulator, there is a need for more and better algorithms. In particular, we would like to investigate the possibility of adding a quasi-static simulation capability for predicting the motion of objects in situations in which friction effects dominate any inertial effects. This could be used to simulate parts being pushed or tipped by various actions of end-effectors or other pushing mechanisms.

13.4   AUTOMATING SUBTASKS IN OLP SYSTEMS

In this section, we briefly mention some advanced features that could be integrated into the "baseline" OLP-system concept already presented. Most of these features accomplish automated planning of some small portion of an industrial application.

Automatic robot placement

One of the most basic tasks that can be accomplished by means of an OLP system is the determination of the workcell layout so that the manipulator(s) can reach all of the required workpoints. Determining correct robot or workpiece placement by trial and error is more quickly completed in a simulated world than in the physical cell. An advanced feature that automates the search for feasible robot or workpiece location(s) goes one step further in reducing burden on the user.

Automatic placement can be computed by direct search or (sometimes) by heuristic-guided search techniques. Most robots are mounted flat on the floor (or ceiling) and have the first rotary joint perpendicular to the floor, so no more is generally necessary than to search by tessellation of the three-dimensional space of robot-base placement. The search might optimize some criterion or might halt upon location of the first feasible robot or part placement. Feasibility can be defined as collision-free ability to reach all workpoints (or perhaps be given an even stronger definition). A reasonable criterion to maximize might be some form of a *measure of manipulability*, as was discussed in Chapter 8. An implementation using a similar measure of manipulability has been discussed in [12]. The result of such an automatic placement is a cell in which the robot can reach all of its workpoints in *well-conditioned* configurations.

Collision avoidance and path optimization

Research on the planning of collision-free paths [13,14] and the planning of time-optimal paths [15,16] generates natural candidates for inclusion in an OLP system. Some related problems that have a smaller scope and a smaller search space are also of interest. For example, consider the problem of using a six-degree-of-freedom robot for an arc-welding task whose geometry specifies only five degrees of freedom. Automatic planning of the redundant degree of freedom can be used to avoid collisions and singularities of the robot [17].

Automatic planning of coordinated motion

In many arc-welding situations, details of the process require that a certain relationship between the workpiece and the gravity vector be maintained during the weld. This results in a two- or three-degree-of-freedom-orienting system on which the part is mounted, operating simultaneously with the robot and in a coordinated fashion. In such a system, there could be nine or more degrees of freedom to coordinate. Such systems are generally programmed today by using teaching-pendant techniques. A planning system that could automatically synthesize the coordinated motions for such a system might be quite valuable [17,18].

Force-control simulation

In a simulated world in which objects are represented by their surfaces, it is possible to investigate the simulation of manipulator force-control strategies. This task involves the difficult problem of modeling some surface properties and expanding the dynamic simulator to deal with the constraints imposed by various contacting situations. In such an environment, it might be possible to assess various force-controlled assembly operations for feasibility [19].

Automatic scheduling

Along with the geometric problems found in robot programming, there are often difficult scheduling and communication problems. This is particularly the case if we expand the simulation beyond a single workcell to a group of workcells. Some discrete-time simulation systems offer abstract simulation of such systems [20], but few offer planning algorithms. Planning schedules for interacting processes is a difficult problem and an area of research [21,22]. An OLP system would serve as an ideal test bed for such research and would be immediately enhanced by any useful algorithms in this area.

Automatic assessment of errors and tolerances

An OLP system might be given some of the capabilities discussed in recent work in modeling positioning-error sources and the effect of data from imperfect sensors [23,24]. The world model could be made to include various error bounds and tolerancing information, and the system could assess the likelihood of success of various positioning or assembly tasks. The system might suggest the use and placement of sensors so as to correct potential problems.

Off-line programming systems are useful in present-day industrial applications and can serve as a basis for continuing robotics research and development. A large motivation in developing OLP systems is to fill the gap between the explicitly programmed systems available today and the task-level systems of tomorrow.

BIBLIOGRAPHY

[1] J. Craig, "Issues in the Design of Off-Line Programming Systems," *International Symposium of Robotics Research*, R. Bolles and B. Roth, Eds., MIT Press, Cambridge, MA, 1988.

[2] J. Craig, "Geometric Algorithms in AdeptRAPID," *Robotics: The Algorithmic Perspective: 1998 WAFR*, P. Agarwal, L. Kavraki, and M. Mason, Eds., AK Peters, Natick, MA, 1998.

[3] R. Goldman, *Design of an Interactive Manipulator Programming Environment*, UMI Research Press, Ann Arbor, MI, 1985.

[4] S. Mujtaba and R. Goldman, "AL User's Manual," 3rd edition, Stanford Department of Computer Science, Report No. STAN-CS-81-889, December 1981.

[5] T. Lozano-Perez, "Spatial Planning: A Configuration Space Approach," *IEEE Transactions on Systems, Man, and Cybernetics*, Vol. SMC-11, 1983.

[6] B. Shimano, C. Geschke, and C. Spalding, "VAL - II: A Robot Programming Language and Control System," SME Robots VIII Conference, Detroit, June 1984.

[7] R. Taylor, P. Summers, and J. Meyer, "AML: A Manufacturing Language," *International Journal of Robotics Research*, Vol. 1, No. 3, Fall 1982.

[8] Adept Technology Inc., "The Pilot User's Manual," Available from Adept Technology Inc., Livermore, CA, 2001.

[9] B. Mirtich and J. Canny, "Impulse Based Dynamic Simulation of Rigid Bodies," Symposium on Interactive 3D Graphics, ACM Press, New York, 1995.

[10] B. Mirtich, Y. Zhuang, K. Goldberg, et al., "Estimating Pose Statistics for Robotic Part Feeders," Proceedings of the IEEE Robotics and Automation Conference, Minneapolis, April, 1996.

[11] Adept Technology Inc., "AIM Manual," Available from Adept Technology Inc., San Jose, CA, 2002.

[12] B. Nelson, K. Pedersen, and M. Donath, "Locating Assembly Tasks in a Manipulator's Workspace," IEEE Conference on Robotics and Automation, Raleigh, NC, April 1987.

[13] plus 1.67pt minus 1.11pt T. Lozano-Perez, "A Simple Motion Planning Algorithm for General Robot Manipulators," *IEEE Journal of Robotics and Automation*, Vol. RA-3, No. 3, June 1987.

[14] R. Brooks, "Solving the Find-Path Problem by Good Representation of Free Space," *IEEE Transaction on Systems, Man, and Cybernetics*, SMC-13:190–197, 1983.

[15] J. Bobrow, S. Dubowsky, and J. Gibson, "On the Optimal Control of Robotic Manipulators with Actuator Constraints," *Proceedings of the American Control Conference*, June 1983.

[16] K. Shin and N. McKay, "Minimum-Time Control of Robotic Manipulators with Geometric Path Constraints," *IEEE Transactions on Automatic Control*, June 1985.

[17] J.J. Craig, "Coordinated Motion of Industrial Robots and 2-DOF Orienting Tables," Proceedings of the 17th International Symposium on Industrial Robots, Chicago, April 1987.

[18] S. Ahmad and S. Luo, "Coordinated Motion Control of Multiple Robotic Devices for Welding and Redundancy Coordination through Constrained Optimization in Cartesian Space," *Proceedings of the IEEE Conference on Robotics and Automation*, Philadelphia, 1988.

[19] M. Peshkin and A. Sanderson, "Planning Robotic Manipulation Strategies for Sliding Objects," IEEE Conference on Robotics and Automation, Raleigh, NC, April 1987.

[20] E. Russel, "Building Simulation Models with Simcript II.5," C.A.C.I., Los Angeles, 1983.

[21] A. Kusiak and A. Villa, "Architectures of Expert Systems for Scheduling Flexible Manufacturing Systems," IEEE Conference on Robotics and Automation, Raleigh, NC, April 1987.

[22] R. Akella and B. Krogh, "Hierarchical Control Structures for Multicell Flexible Assembly System Coordination," IEEE Conference on Robotics and Automation, Raleigh, NC, April 1987.

[23] R. Smith, M. Self, and P. Cheeseman, "Estimating Uncertain Spatial Relationships in Robotics," IEEE Conference on Robotics and Automation, Raleigh, NC, April 1987.

[24] H. Durrant-Whyte, "Uncertain Geometry in Robotics," IEEE Conference on Robotics and Automation, Raleigh, NC, April 1987.

EXERCISES

13.1 [10] In a sentence or two, define collision detection, collision avoidance, and collision-free path planning.

13.2 [10] In a sentence or two, define world model, path planning emulation, and dynamic emulation.

13.3 [10] In a sentence or two, define automatic robot placement, time-optimal paths, and error-propagation analysis.

13.4 [10] In a sentence or two, define RPL, TLP, and OLP.

13.5 [10] In a sentence or two, define calibration, coordinated motion, and automatic scheduling.

13.6 [20] Make a chart indicating how the graphic ability of computers has increased over the past ten years (perhaps in terms of the number of vectors drawn per second per $10,000 of hardware).

13.7 [20] Make a list of tasks that are characterized by "many operations relative to a rigid object" and so are candidates for off-line programming.

13.8 [20] Discuss the advantages and disadvantages of using a programming system that maintains a detailed world model internally.

PROGRAMMING EXERCISE (PART 13)

1. Consider the planar shape of a bar with semicircular end caps. We will call this shape a "capsule." Write a routine that, given the location of two such capsules, computes whether they are touching. Note that all surface points of a capsule are equidistant from a single line segment that might be called its "spine."

2. Introduce a capsule-shaped object near your simulated manipulator and test for collisions as you move the manipulator along a path. Use capsule-shaped links for the manipulator. Report any collisions detected.

3. If time and computer facilities permit, write routines to depict graphically the capsules that make up your manipulator and the obstacles in the workspace.

APPENDIX A

Trigonometric identities

Formulas for rotation about the principal axes by θ:

$$R_X(\theta) = \begin{bmatrix} 1 & 0 & 0 \\ 0 & \cos\theta & -\sin\theta \\ 0 & \sin\theta & \cos\theta \end{bmatrix}, \tag{A.1}$$

$$R_Y(\theta) = \begin{bmatrix} \cos\theta & 0 & \sin\theta \\ 0 & 1 & 0 \\ -\sin\theta & 0 & \cos\theta \end{bmatrix}, \tag{A.2}$$

$$R_Z(\theta) = \begin{bmatrix} \cos\theta & -\sin\theta & 0 \\ \sin\theta & \cos\theta & 0 \\ 0 & 0 & 1 \end{bmatrix}. \tag{A.3}$$

Identities having to do with the periodic nature of sine and cosine:

$$\sin\theta = -\sin(-\theta) = -\cos(\theta + 90°) = \cos(\theta - 90°),$$
$$\cos\theta = \cos(-\theta) = \sin(\theta + 90°) = -\sin(\theta - 90°). \tag{A.4}$$

The sine and cosine for the sum or difference of angles θ_1 and θ_2:

$$\cos(\theta_1 + \theta_2) = c_{12} = c_1 c_2 - s_1 s_2,$$
$$\sin(\theta_1 + \theta_2) = s_{12} = c_1 s_2 + s_1 c_2, \tag{A.5}$$
$$\cos(\theta_1 - \theta_2) = c_1 c_2 + s_1 s_2,$$
$$\sin(\theta_1 - \theta_2) = s_1 c_2 - c_1 s_2.$$

The sum of the squares of the sine and cosine of the same angle is unity:

$$c^2\theta + s^2\theta = 1. \tag{A.6}$$

If a triangle's angles are labeled a, b, and c, where angle a is opposite side A, and so on, then the "law of cosines" is

$$A^2 = B^2 + C^2 - 2BC\cos a. \tag{A.7}$$

The "tangent of the half angle" substitution:

$$u = \tan\frac{\theta}{2},$$

$$\cos\theta = \frac{1 - u^2}{1 + u^2}, \tag{A.8}$$

$$\sin\theta = \frac{2u}{1 + u^2}.$$

To rotate a vector Q about a unit vector $\hat{K}$ by θ, use **Rodriques's formula**:

$$Q' = Q\cos\theta + \sin\theta(\hat{K} \times Q) + (1 - \cos\theta)(\hat{K} \cdot \hat{Q})\hat{K}. \qquad (A.9)$$

See Appendix B for equivalent rotation matrices for the 24 angle-set conventions and Appendix C for some inverse-kinematic identities.

The 24 angle-set conventions

The 12 Euler angle sets are given by

$$R_{X'Y'Z'}(\alpha, \beta, \gamma) = \begin{bmatrix} c\beta c\gamma & -c\beta s\gamma & s\beta \\ s\alpha s\beta c\gamma + c\alpha s\gamma & -s\alpha s\beta s\gamma + c\alpha c\gamma & -s\alpha c\beta \\ -c\alpha s\beta c\gamma + s\alpha s\gamma & c\alpha s\beta s\gamma + s\alpha c\gamma & c\alpha c\beta \end{bmatrix},$$

$$R_{X'Z'Y'}(\alpha, \beta, \gamma) = \begin{bmatrix} c\beta c\gamma & -s\beta & c\beta s\gamma \\ c\alpha s\beta c\gamma + s\alpha s\gamma & c\alpha c\beta & c\alpha s\beta s\gamma - s\alpha c\gamma \\ s\alpha s\beta c\gamma - c\alpha s\gamma & s\alpha c\beta & s\alpha s\beta s\gamma + c\alpha c\gamma \end{bmatrix},$$

$$R_{Y'X'Z'}(\alpha, \beta, \gamma) = \begin{bmatrix} s\alpha s\beta s\gamma + c\alpha c\gamma & s\alpha s\beta c\gamma - c\alpha s\gamma & s\alpha c\beta \\ c\beta s\gamma & c\beta c\gamma & -s\beta \\ c\alpha s\beta s\gamma - s\alpha c\gamma & c\alpha s\beta c\gamma + s\alpha s\gamma & c\alpha c\beta \end{bmatrix},$$

$$R_{Y'Z'X'}(\alpha, \beta, \gamma) = \begin{bmatrix} c\alpha c\beta & -c\alpha s\beta c\gamma + s\alpha s\gamma & c\alpha s\beta s\gamma + s\alpha c\gamma \\ s\beta & c\beta c\gamma & -c\beta s\gamma \\ -s\alpha c\beta & s\alpha s\beta c\gamma + c\alpha s\gamma & -s\alpha s\beta s\gamma + c\alpha c\gamma \end{bmatrix},$$

$$R_{Z'X'Y'}(\alpha, \beta, \gamma) = \begin{bmatrix} -s\alpha s\beta s\gamma + c\alpha c\gamma & -s\alpha c\beta & s\alpha s\beta c\gamma + c\alpha s\gamma \\ c\alpha s\beta s\gamma + s\alpha c\gamma & c\alpha c\beta & -c\alpha s\beta c\gamma + s\alpha s\gamma \\ -c\beta s\gamma & s\beta & c\beta c\gamma \end{bmatrix},$$

$$R_{Z'Y'X'}(\alpha, \beta, \gamma) = \begin{bmatrix} c\alpha c\beta & c\alpha s\beta s\gamma - s\alpha c\gamma & c\alpha s\beta c\gamma + s\alpha s\gamma \\ s\alpha c\beta & -s\alpha s\beta s\gamma + c\alpha c\gamma & -s\alpha s\beta c\gamma - c\alpha s\gamma \\ -s\beta & c\beta s\gamma & c\beta c\gamma \end{bmatrix},$$

$$R_{X'Y'X'}(\alpha, \beta, \gamma) = \begin{bmatrix} c\beta & s\beta s\gamma & s\beta c\gamma \\ s\alpha s\beta & -s\alpha c\beta s\gamma + c\alpha c\gamma & -s\alpha c\beta c\gamma - c\alpha s\gamma \\ -c\alpha s\beta & c\alpha c\beta s\gamma + s\alpha c\gamma & c\alpha c\beta c\gamma - s\alpha s\gamma \end{bmatrix},$$

$$R_{X'Z'X'}(\alpha, \beta, \gamma) = \begin{bmatrix} c\beta & -s\beta c\gamma & s\beta s\gamma \\ c\alpha s\beta & c\alpha c\beta c\gamma - s\alpha s\gamma & -c\alpha c\beta s\gamma - s\alpha c\gamma \\ s\alpha s\beta & s\alpha c\beta c\gamma + c\alpha s\gamma & -s\alpha c\beta s\gamma + c\alpha c\gamma \end{bmatrix},$$

$$R_{Y'X'Y'}(\alpha, \beta, \gamma) = \begin{bmatrix} -s\alpha c\beta s\gamma + c\alpha c\gamma & s\alpha s\beta & s\alpha c\beta c\gamma + c\alpha s\gamma \\ s\beta s\gamma & c\beta & -s\beta c\gamma \\ -c\alpha c\beta s\gamma - s\alpha c\gamma & c\alpha s\beta & c\alpha c\beta c\gamma - s\alpha s\gamma \end{bmatrix},$$

$$R_{Y'Z'Y'}(\alpha, \beta, \gamma) = \begin{bmatrix} c\alpha c\beta c\gamma - s\alpha s\gamma & -c\alpha s\beta & c\alpha c\beta s\gamma + s\alpha c\gamma \\ s\beta c\gamma & c\beta & s\beta s\gamma \\ -s\alpha c\beta c\gamma - c\alpha s\gamma & s\alpha s\beta & -s\alpha c\beta s\gamma + c\alpha c\gamma \end{bmatrix},$$

$$R_{Z'X'Z'}(\alpha, \beta, \gamma) = \begin{bmatrix} -s\alpha c\beta s\gamma + c\alpha c\gamma & -s\alpha c\beta c\gamma - c\alpha s\gamma & s\alpha s\beta \\ c\alpha c\beta s\gamma + s\alpha c\gamma & c\alpha c\beta c\gamma - s\alpha s\gamma & -c\alpha s\beta \\ s\beta s\gamma & s\beta c\gamma & c\beta \end{bmatrix},$$

$$R_{Z'Y'Z'}(\alpha, \beta, \gamma) = \begin{bmatrix} c\alpha c\beta c\gamma - s\alpha s\gamma & -c\alpha c\beta s\gamma - s\alpha c\gamma & c\alpha s\beta \\ s\alpha c\beta c\gamma + c\alpha s\gamma & -s\alpha c\beta s\gamma + c\alpha c\gamma & s\alpha s\beta \\ -s\beta c\gamma & s\beta s\gamma & c\beta \end{bmatrix}.$$

The 12 fixed angle sets are given by

$$R_{XYZ}(\gamma, \beta, \alpha) = \begin{bmatrix} c\alpha c\beta & c\alpha s\beta s\gamma - s\alpha c\gamma & c\alpha s\beta c\gamma + s\alpha s\gamma \\ s\alpha c\beta & s\alpha s\beta s\gamma + c\alpha c\gamma & s\alpha s\beta c\gamma - c\alpha s\gamma \\ -s\beta & c\beta s\gamma & c\beta c\gamma \end{bmatrix},$$

$$R_{XZY}(\gamma, \beta, \alpha) = \begin{bmatrix} c\alpha c\beta & -c\alpha s\beta c\gamma + s\alpha s\gamma & c\alpha s\beta s\gamma + s\alpha c\gamma \\ s\beta & c\beta c\gamma & -c\beta s\gamma \\ -s\alpha c\beta & s\alpha s\beta c\gamma + c\alpha s\gamma & -s\alpha s\beta s\gamma + c\alpha c\gamma \end{bmatrix},$$

$$R_{YXZ}(\gamma, \beta, \alpha) = \begin{bmatrix} -s\alpha s\beta s\gamma + c\alpha c\gamma & -s\alpha c\beta & s\alpha s\beta c\gamma + c\alpha s\gamma \\ c\alpha s\beta s\gamma + s\alpha c\gamma & c\alpha c\beta & -c\alpha s\beta c\gamma + s\alpha s\gamma \\ -c\beta s\gamma & s\beta & c\beta c\gamma \end{bmatrix},$$

$$R_{YZX}(\gamma, \beta, \alpha) = \begin{bmatrix} c\beta c\gamma & -s\beta & c\beta s\gamma \\ c\alpha s\beta c\gamma + s\alpha s\gamma & c\alpha c\beta & c\alpha s\beta s\gamma - s\alpha c\gamma \\ s\alpha s\beta c\gamma - c\alpha s\gamma & s\alpha c\beta & s\alpha s\beta s\gamma + c\alpha c\gamma \end{bmatrix},$$

$$R_{ZXY}(\gamma, \beta, \alpha) = \begin{bmatrix} s\alpha s\beta s\gamma + c\alpha c\gamma & s\alpha s\beta c\gamma - c\alpha s\gamma & s\alpha c\beta \\ c\beta s\gamma & c\beta c\gamma & -s\beta \\ c\alpha s\beta s\gamma - s\alpha c\gamma & c\alpha s\beta c\gamma + s\alpha s\gamma & c\alpha c\beta \end{bmatrix},$$

$$R_{ZYX}(\gamma, \beta, \alpha) = \begin{bmatrix} c\beta c\gamma & -c\beta s\gamma & s\beta \\ s\alpha s\beta c\gamma + c\alpha s\gamma & -s\alpha s\beta s\gamma + c\alpha c\gamma & -s\alpha c\beta \\ -c\alpha s\beta c\gamma + s\alpha s\gamma & c\alpha s\beta s\gamma + s\alpha c\gamma & c\alpha c\beta \end{bmatrix},$$

$$R_{XYX}(\gamma, \beta, \alpha) = \begin{bmatrix} c\beta & s\beta s\gamma & s\beta c\gamma \\ s\alpha s\beta & -s\alpha c\beta s\gamma + c\alpha c\gamma & -s\alpha c\beta c\gamma - c\alpha s\gamma \\ -c\alpha s\beta & c\alpha c\beta s\gamma + s\alpha c\gamma & c\alpha c\beta c\gamma - s\alpha s\gamma \end{bmatrix},$$

$$R_{XZX}(\gamma, \beta, \alpha) = \begin{bmatrix} c\beta & -s\beta c\gamma & s\beta s\gamma \\ c\alpha s\beta & c\alpha c\beta c\gamma - s\alpha s\gamma & -c\alpha c\beta s\gamma - s\alpha c\gamma \\ s\alpha s\beta & s\alpha c\beta c\gamma + c\alpha s\gamma & -s\alpha c\beta s\gamma + c\alpha c\gamma \end{bmatrix},$$

$$R_{YXY}(\gamma, \beta, \alpha) = \begin{bmatrix} -s\alpha c\beta s\gamma + c\alpha c\gamma & s\alpha s\beta & s\alpha c\beta c\gamma + c\alpha s\gamma \\ s\beta s\gamma & c\beta & -s\beta c\gamma \\ -c\alpha c\beta s\gamma - s\alpha c\gamma & c\alpha s\beta & c\alpha c\beta c\gamma - s\alpha s\gamma \end{bmatrix},$$

$$R_{YZY}(\gamma, \beta, \alpha) = \begin{bmatrix} c\alpha c\beta c\gamma - s\alpha s\gamma & -c\alpha s\beta & c\alpha c\beta s\gamma + s\alpha c\gamma \\ s\beta c\gamma & c\beta & s\beta s\gamma \\ -s\alpha c\beta c\gamma - c\alpha s\gamma & s\alpha s\beta & -s\alpha c\beta s\gamma + c\alpha c\gamma \end{bmatrix},$$

$$R_{ZXZ}(\gamma, \beta, \alpha) = \begin{bmatrix} -s\alpha c\beta s\gamma + c\alpha c\gamma & -s\alpha c\beta c\gamma - c\alpha s\gamma & s\alpha s\beta \\ c\alpha c\beta s\gamma + s\alpha c\gamma & c\alpha c\beta c\gamma - s\alpha s\gamma & -c\alpha s\beta \\ s\beta s\gamma & s\beta c\gamma & c\beta \end{bmatrix},$$

$$R_{ZYZ}(\gamma, \beta, \alpha) = \begin{bmatrix} c\alpha c\beta c\gamma - s\alpha s\gamma & -c\alpha c\beta s\gamma - s\alpha c\gamma & c\alpha s\beta \\ s\alpha c\beta c\gamma + c\alpha s\gamma & -s\alpha c\beta s\gamma + c\alpha c\gamma & s\alpha s\beta \\ -s\beta c\gamma & s\beta s\gamma & c\beta \end{bmatrix}.$$

APPENDIX C

Some inverse-kinematic formulas

The single equation

$$\sin\theta = a \tag{C.1}$$

has two solutions, given by

$$\theta = \pm\text{Atan2}(\sqrt{1-a^2}, a). \tag{C.2}$$

Likewise, given

$$\cos\theta = b, \tag{C.3}$$

there are two solutions:

$$\theta = \text{Atan2}(b, \pm\sqrt{1-b^2}). \tag{C.4}$$

If both (C.1) and (C.3) are given, then there is a unique solution given by

$$\theta = \text{Atan2}(a, b). \tag{C.5}$$

The transcendental equation

$$a\cos\theta + b\sin\theta = 0 \tag{C.6}$$

has the two solutions

$$\theta = \text{Atan2}(a, -b) \tag{C.7}$$

and

$$\theta = \text{Atan2}(-a, b). \tag{C.8}$$

The equation

$$a\cos\theta + b\sin\theta = c, \tag{C.9}$$

which we solved in Section 4.5 with the tangent-of-the-half-angle substitutions, is also solved by

$$\theta = \text{Atan2}(b, a) \pm \text{Atan2}(\sqrt{a^2+b^2-c^2}, c). \tag{C.10}$$

The set of equations

$$a\cos\theta - b\sin\theta = c,$$
$$a\sin\theta + b\cos\theta = d, \tag{C.11}$$

which was solved in Section 4.4, also is solved by

$$\theta = \text{Atan2}(ad - bc, \ ac + bd). \tag{C.12}$$

Solutions to selected exercises

CHAPTER 2 SPATIAL DESCRIPTIONS AND TRANSFORMATIONS EXERCISES

2.1)

$$R = ROT(\hat{x}, \phi)ROT(\hat{z}, \theta)$$

$$= \begin{bmatrix} 1 & 0 & 0 \\ 0 & C\phi & -S\phi \\ 0 & S\phi & C\phi \end{bmatrix} \begin{bmatrix} C\theta & -S\theta & 0 \\ S\theta & C\theta & 0 \\ 0 & 0 & 1 \end{bmatrix}$$

$$= \begin{bmatrix} C\theta & -S\theta & 0 \\ C\phi S\theta & C\phi C\theta & -S\phi \\ S\phi S\theta & S\phi C\theta & C\phi \end{bmatrix}$$

2.12) Velocity is a "free vector" and will be affected only by rotation, not by translation:

$$^AV = {}^A_BR\,{}^BV = \begin{bmatrix} 0.866 & -0.5 & 0 \\ 0.5 & 0.866 & 0 \\ 0 & 0 & 1 \end{bmatrix} \begin{bmatrix} 10 \\ 20 \\ 30 \end{bmatrix}$$

$$^AV = \begin{bmatrix} -1.34 & 22.32 & 30.0 \end{bmatrix}^T$$

2.27)

$$^A_BT = \begin{bmatrix} -1 & 0 & 0 & 3 \\ 0 & -1 & 0 & 0 \\ 0 & 0 & 1 & 0 \\ 0 & 0 & 0 & 1 \end{bmatrix}$$

2.33)

$$^B_CT = \begin{bmatrix} -0.866 & -0.5 & 0 & 3 \\ 0 & 0 & +1 & 0 \\ -0.5 & 0.866 & 0 & 0 \\ 0 & 0 & 0 & 1 \end{bmatrix}$$

CHAPTER 3 MANIPULATOR KINEMATICS EXERCISES

3.1)

α_{i-1}	a_{i-1}	d_i
0	0	0
0	L_1	0
0	L_2	0

$$
{}^0_1T = \begin{bmatrix} C_1 & -S_1 & 0 & 0 \\ S_1 & C_1 & 0 & 0 \\ 0 & 0 & 1 & 0 \\ 0 & 0 & 0 & 1 \end{bmatrix}
$$

$$
{}^1_2T = \begin{bmatrix} C_2 & -S_2 & 0 & L_1 \\ S_2 & C_2 & 0 & 0 \\ 0 & 0 & 1 & 0 \\ 0 & 0 & 0 & 1 \end{bmatrix} \qquad {}^2_3T = \begin{bmatrix} C_3 & -S_3 & 0 & L_2 \\ S_3 & C_3 & 0 & 0 \\ 0 & 0 & 1 & 0 \\ 0 & 0 & 0 & 1 \end{bmatrix}
$$

$$
{}^0_3T = {}^0_1T\,{}^1_2T\,{}^2_3T = \begin{bmatrix} C_{123} & -S_{123} & 0 & L_1C_1 + L_2C_{12} \\ S_{123} & C_{123} & 0 & L_1S_1 + L_2S_{12} \\ 0 & 0 & 1 & 0 \\ 0 & 0 & 0 & 1 \end{bmatrix}
$$

where

$$
C_{123} = \cos(\theta_1 + \theta_2 + \theta_3)
$$
$$
S_{123} = \sin(\theta_1 + \theta_2 + \theta_3)
$$

3.8) When $\{G\} = \{T\}$, we have

$$
{}^B_WT\,{}^W_TT = {}^B_ST\,{}^S_GT
$$

So

$$
{}^W_TT = {}^B_WT^{-1}\,{}^B_ST\,{}^S_GT
$$

4.14) No. Pieper's method gives the closed-form solution for any 3-DOF manipulator. (See his thesis for all the cases.)

4.18) 2

4.22) 1

CHAPTER 5 JACOBIANS: VELOCITIES AND STATIC FORCES EXERCISES

5.1) The Jacobian in frame {0} is

$$
{}^0J(\theta) = \begin{bmatrix} -L_1S_1 - L_2S_{12} & -L_2S_{12} \\ L_1C_1 + L_2C_{12} & L_2C_{12} \end{bmatrix}
$$

$$
\begin{aligned}
\mathrm{DET}({}^0J(\theta)) &= -(L_2C_{12})(L_1S_1 + L_2S_{12}) + (L_2S_{12})(L_1C_1 + L_2C_{12}) \\
&= -L_1L_2S_1C_{12} - L_2^2S_{12}C_{12} + L_1L_2C_1S_{12} + L_2^2S_{12}C_{12} \\
&= L_1L_2C_1S_{12} - L_1L_2S_1C_{12} = L_1L_2(C_1S_{12} - S_1C_{12}) \\
&= L_1L_2S_2
\end{aligned}
$$

$\therefore$ The same result as when you start with ${}^3J(\theta)$, namely, the singular configurations are $\theta_2 = 0°$ or $180°$.

5.8) The Jacobian of this 2-link is

$$^3J(\theta) = \begin{bmatrix} L_1 S_2 & 0 \\ L_1 C_2 + L_2 & L_2 \end{bmatrix}$$

An isotropic point exists if

$$^3J = \begin{bmatrix} L_2 & 0 \\ 0 & L_2 \end{bmatrix}$$

So

$$L_1 S_2 = L_2$$

$$L_1 C_2 + L_2 = 0$$

Now, $S_2^2 + C_2^2 = 1$, so $\left(\frac{L_2}{L_1}\right)^2 + \left(\frac{-L_2}{L_1}\right)^2 = 1$.

or $L_1^2 = 2L_2^2 \rightarrow L_1 = \sqrt{2}L_2$.

Under this condition, $S_2 = \frac{1}{\sqrt{2}} = \pm.707$.

and $C_2 = -.707$.

$\therefore$ An isotropic point exists if $L_1 = \sqrt{2}L_2$, and in that case it exists when $\theta_2 = \pm 135°$.

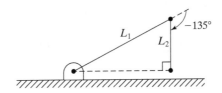

In this configuration, the manipulator looks momentarily like a Cartesian manipulator.

5.13)

$$\tau = {}^\circ J^T(\theta)\, {}^\circ F$$

$$\tau = \begin{bmatrix} -L_1 S_1 - L_2 S_{12} & L_1 C_1 + L_2 C_{12} \\ -L_2 S_{12} & L_2 C_{12} \end{bmatrix} \begin{bmatrix} 10 \\ 0 \end{bmatrix}$$

$$\tau_1 = {}^-10 S_1 L_1 - 10 L_2 S_{12}$$

$$\tau_2 = -10 L_2 S_{12}$$

CHAPTER 6 MANIPULATOR DYNAMICS EXERCISES

6.1) Use (6.17), but written in polar form, because that is easier. For example, for I_{zz},

$$I_{zz} = \int_{-H/2}^{H/2} \int_0^{2\pi} \int_0^R (x^2 + y^2) p\, r\, dr\, d\theta\, dz$$

$$x = R\cos\theta, \quad y = R\sin\theta, \quad x^2 + y^2 = R^2 (r^2)$$

$$I_{zz} = \int_{-H/2}^{H/2} \int_0^{2\pi} \int_0^R p r^3 dr\, d\theta dz$$

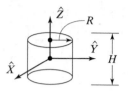

$$I_{zz} = \frac{\pi}{2} R^4 H p, \text{VOLUME} = \pi r^2 H$$

$$\therefore \quad \text{Mass} = M = p\pi r^2 H \quad \therefore \quad \boxed{I_{zz} = \frac{1}{2}MR^2}$$

Similarly (only harder) is

$$\boxed{I_{xx} = I_{yy} = \frac{1}{4}MR^2 + \frac{1}{12}MH^2}$$

From symmetry (or integration),

$$\boxed{I_{xy} = I_{xz} = I_{yz} = 0}$$

$$\therefore$$

$$c_I = \begin{bmatrix} \frac{1}{4}MR^2 + \frac{1}{12}MH^2 & 0 & 0 \\ 0 & \frac{1}{4}MR^2 + \frac{1}{12}MH^2 & 0 \\ 0 & 0 & \frac{1}{2}MR^2 \end{bmatrix}$$

6.12) $\theta_1(t) = Bt + ct^2$, so

$$\dot{\theta}_1 = B + 2ct, \ddot{\theta} = 2c$$

so

$$^1\dot{\omega}_1 = \ddot{\theta}_1 \hat{z}_1 = 2c\hat{z}_1 = \begin{bmatrix} 0 \\ 0 \\ 2c \end{bmatrix}$$

$$^1\dot{v}_{c1} = \begin{bmatrix} 0 \\ 0 \\ 2c \end{bmatrix} \otimes \begin{bmatrix} 2 \\ 0 \\ 0 \end{bmatrix} + \begin{bmatrix} 0 \\ 0 \\ \dot{\theta}_1 \end{bmatrix} \otimes \left(\begin{bmatrix} 0 \\ 0 \\ \dot{\theta}_1 \end{bmatrix} \otimes \begin{bmatrix} 2 \\ 0 \\ 0 \end{bmatrix} \right)$$

$$= \begin{bmatrix} 0 \\ 4c \\ 0 \end{bmatrix} + \begin{bmatrix} -2\dot{\theta}_1^2 \\ 0 \\ 0 \end{bmatrix}$$

$$^1\dot{v}_{c1} = \begin{bmatrix} -2(B+2ct)^2 \\ 4c \\ 0 \end{bmatrix}$$

6.18) Any reasonable $F(\theta, \dot{\theta})$ probably has the property that the friction force (or torque) on joint i depends only on the velocity of joint i, i.e.,

$$F(\theta, \dot{\theta}) = [f_1(\theta, \dot{\theta}_1) \quad F_2(\theta, \dot{\theta}_2) \quad \ldots F_N(\theta, \dot{\theta}_N)]^T$$

Also, each $f_i(\theta, \dot{\theta}_i)$ should be "passive"; i.e., the function should lie in the first & third quadrants.

** Solution written by candlelight in aftermath of 7.0 earthquake, Oct. 17, 1989!

CHAPTER 7 TRAJECTORY GENERATION EXERCISES

7.1) Three cubics are required to connect a start point, two via points, and a goal point—that is, three for each joint, for a total of 18 cubics. Each cubic has four coefficients, so 72 coefficients are stored.

7.17) By differentiation,

$$\dot{\theta}(t) = 180t - 180t^2$$

$$\ddot{\theta}(t) = 180 - 360t$$

Then, evaluating at $t = 0$ and $t = 1$, we have

$$\theta(0) = 10 \quad \dot{\theta}(0) = 0 \quad \ddot{\theta}(0) = 180$$

$$\theta(1) = 40 \quad \dot{\theta}(1) = 0 \quad \ddot{\theta}(1) = -180$$

8.3) Using (8.1), we have

$$L = \sum_{i=1}^{3}(a_{i-1} + d_i) = (0+0) + (0+0) + (0 + (U - L)) = U - L$$

$$W = \frac{4}{3}\pi U^3 - \frac{4}{3}\pi L^3 = \frac{4}{3}\pi(U^3 - L^3) \left\{ \begin{array}{l} a \\ \text{"hollow"} \\ \text{sphere} \end{array} \right.$$

$$\therefore \quad Q_L = \frac{U - L}{\sqrt[3]{\frac{4}{3}\pi(U^3 - L^3)}}$$

8.6) From (8.14),

$$\frac{1}{K_{TOTAL}} = \frac{1}{1000} + \frac{1}{300} = 4.333 \times 10^{-3}$$

$$\therefore \quad \boxed{K_{TOTAL} = 230.77 \frac{\text{NTM}}{\text{RAD}}}$$

8.16) From (8.15),

$$K = \frac{G\pi d^4}{32L} = \frac{(0.33 \times 7.5 \times 10^{10})(\pi)(0.001)^4}{(32)(0.40)} = \boxed{0.006135 \frac{\text{NTM}}{\text{RAD}}}$$

This is very flimsy, because the diameter is 1 mm!

9.2) From (9.5),

$$s_1 = -\frac{6}{2 \times 2} + \frac{\sqrt{36 - 4 \times 2 \times 4}}{2 \times 2} = -1.5 + 0.5 = -1.0$$

$$s_2 = -1.5 - 0.5 = -2.0$$

$$\therefore$$

$$quadx(t) = c_1 e^{-t} + c_2 e^{-2t} \text{ and } \dot{x}(t) = -c_1 e^{-t} - 2c_2 e^{-2t}$$

$$\text{At } t = 0 \quad x(0) = 1 = c_1 + c_2 \tag{1}$$

$$\dot{x}(0) = 0 = -c_1 - 2c_2 \tag{2}$$

Adding (1) and (2) gives

$$1 = -c_2$$

so $c_2 = -1$ and $c_1 = 2$.

$$\therefore \boxed{x(t) = 2e^{-t} - e^{-2t}}$$

9.10) Using (8.24) and assuming aluminum yields

$$K = \frac{(0.333)(2 \times 10^{11})(0.05^4 - 0.04^4)}{(4)(0.50)} = 123{,}000.0$$

Using info from Figure 9.13, the equivalent mass is $(0.23)(5) = 1.15$ kg, so

$$W_{res} = \sqrt{k/m} = \sqrt{\frac{123{,}000.0}{1.15}} \cong \boxed{327.04 \ \frac{\text{rad}}{\text{sec}}}$$

This is very high—so the designer is probably wrong in thinking that this link vibrating represents the lowest unmodeled resonance!

9.13) As in problem 9.12, the effective stiffness is $K = 32000$. Now, the effective inertia is $I = 1 + (0.1)(64) = 7.4$.

$$\therefore \quad W_{res} = \sqrt{\frac{32000}{7.4}} \cong 65.76 \frac{\text{rad}}{\text{sec}} \cong \boxed{10.47 \text{ Hz}}$$

CHAPTER 10   NONLINEAR CONTROL OF MANIPULATORS EXERCISES

10.2) Let $\tau = \alpha \tau' + \beta$

$$\alpha = 2 \quad \beta = 50\dot{\theta} - 13\dot{\theta}^3 + 5$$

and $\tau' = \ddot{\theta}_D + K_v \dot{e} + K_p e$

where $e = \theta_D - \theta$

and

$$K_p = 10$$

$$K_v = 2\sqrt{10}$$

10.10) Let $f = \alpha f' + \beta$

with $\alpha = 2$, $\beta = 5x\dot{x} - 12$

and $f' = \ddot{X}_D + k_v \dot{e} + k_p e$, $e = X_D - X$

$k_p = 20$, $k_v = 2\sqrt{20}$.

CHAPTER 11   FORCE CONTROL OF MANIPULATORS EXERCISES

11.2) The artificial constraints for the task in question would be

$$
\begin{array}{c|c}
V_z = -a_1 & F_x = 0 \\
& F_y = 0 \\
& N_x = 0 \\
& N_y = 0 \\
& N_z = 0
\end{array}
$$

where a_1 is the speed of insertion.

11.4) Use (5.105) with frames $\{A\}$ and $\{B\}$ reversed. First, find $^B_A T$, so invert $^A_B T$:

$$
^B_A T =
\begin{bmatrix}
0.866 & 0.5 & 0 & -8.66 \\
-0.5 & 0.866 & 0 & 5.0 \\
0 & 0 & 1 & -5.0 \\
0 & 0 & 0 & 1
\end{bmatrix}
$$

Now,

$$
^B F = ^B_A R \, ^A F = \begin{bmatrix} 1 & 1.73 & -3 \end{bmatrix}^T
$$

$$
^B N = ^B P_{AORG} \otimes ^B F + ^B_A R \, ^A N = \begin{bmatrix} -6.3 & -30.9 & -15.8 \end{bmatrix}^T
$$

$$
\therefore \quad ^B F = \begin{bmatrix} 1.0 & 1.73 & -3 & -6.3 & -30.9 & -15.8 \end{bmatrix}^T
$$

Index